Quantum Science and Technology

The book series Quantum Science and Technology is dedicated to one of today's most active and rapidly expanding fields of research and development. In particular, the series will be a showcase for the growing number of experimental implementations and practical applications of quantum systems. These will include, but are not restricted to: quantum information processing, quantum computing, and quantum simulation; quantum communication and quantum cryptography; entanglement and other quantum resources; quantum interfaces and hybrid quantum systems; quantum memories and quantum repeaters; measurement-based quantum control and quantum feedback; quantum nanomechanics, quantum optomechanics and quantum transducers; quantum sensing and quantum metrology; as well as quantum effects in biology. Last but not least, the series will include books on the theoretical and mathematical questions relevant to designing and understanding these systems and devices, as well as foundational issues concerning the quantum phenomena themselves. Written and edited by leading experts, the treatments will be designed for graduate students and other researchers already working in, or intending to enter the field of quantum science and technology.

Claudio Conti

Quantum Machine Learning

Thinking and Exploration in Neural Network Models for Quantum Science and Quantum Computing

Springer

Claudio Conti
Physics Department
Sapienza University of Rome
Rome, Italy

ISSN 2364-9054 ISSN 2364-9062 (electronic)
Quantum Science and Technology
ISBN 978-3-031-44225-4 ISBN 978-3-031-44226-1 (eBook)
https://doi.org/10.1007/978-3-031-44226-1

This Springer imprint is published by the registered company Springer Nature Switzerland AG
The registered company address is: Gewerbestrasse 11, 6330 Cham, Switzerland

Paper in this product is recyclable.

To Stefania, Candy, and Bella. They are my feature space.

Preface

These are exciting years for physicists. Two significant revolutions are changing our tools and perspectives. I have in mind quantum technologies and artificial intelligence. While the former seems still far-fetched but involves a growing number of scientists, it is undoubtful that the latter is rapidly changing the world. So what is the link between these apparently unrelated directions?

Big companies like Google^{TM} and IBM^{TM} are highly active in quantum technology and artificial intelligence. Quantum machine learning is probably the synthesis of both; however, it is a niche in the huge background ranging from quantum cryptography and quantum sensing on the one hand and generative artificial intelligence and robots on the other hand.

However, quantum machine learning is potentially vibrant, although only the surface has been scratched, and investigations have been limited to enthusiasts. The modern tools of deep learning enable a new way to tackle old problems, such as finding the eigenstates of a many-body system, with applications from chemistry to combinatorial optimization.

Also, there are fundamental open issues, such as the role of the quantum in the human brain or quantum physics on graphs, which can be a route to quantum gravity.

But it is not only artificial intelligence affecting quantum technologies (and practically all the subfields in modern physics). On the contrary, the mathematical techniques and the way of thinking in quantum mechanics have strongly influenced the models in deep learning, from the perceptron to large language models.

If one looks at the mathematical background, one can quickly identify some elementary quantum mechanics in things like kernel machines and the encoder/decoder in a transformer. It would be an exaggeration to affirm who comes first or only put quantum mechanics into the game, forgetting linear algebra or functional analysis. But the idea here is that these two directions, artificial intelligence, and quantum technology, will enormously benefit for many years to come from their interaction.

There is no better way to deepen these connections than making some practice and examples, filling our gaps in deep learning and quantum mechanics, and applying methods from the first to the second and vice-versa. Moreover, the topic is so new and exciting that we quickly find ourselves doing new things, such as new algorithms. Also, if we like coding, we can play with some of the most advanced tools humans have invented to process unprecedented amounts of information, such as large datasets with images or complex protein structures.

The book originates from lectures I gave in 2018 and the following years. The book also emerges from my interest in machine learning models and the link between their mathematical structure and the fundamentals of quantum mechanics.

Machine learning and quantum mathematics are strongly related. Beyond this, it is indisputable that the 2020s are characterized by the explosions of deep learning applications with algorithms reaching the scale of billions of weights and quantum computing. In the latter case, the current scale is 100 qubits, but many companies and researchers promise 100,000 qubits in 10 years.

The book is an introductory text containing various directions and ideas. It reports the fundaments of quantum mechanics but beyond the typical level in quantum computing books. For example, it does not focus on gates but on the representation of states in a way compatible with their transposition to a `TensorFlow`™ code. Innovations are present in the model of qubits and in the use of continuous variables. The reader can find the basic notions and some examples. And start from them for new applications and innovative research work.

A basic knowledge of quantum physics and `Python` programming is needed. Any chapter is self-sustaining, and various coding examples are given. Further readings are suggested at the end of each chapter to guide you through the gigantic forest of published papers and books.

Theoretical concepts are interleaved with coding examples. The material is presented at an introductory level. The interested scholar will eventually have the notions and the tools to develop their interests and applications. The book is intended for experimentalists who want to start playing with coding that can be interfaced with laboratory equipment. The book is also designed for theoreticians and computer scientists to get an introduction to the subject for developing their projects.

Chapter 1 introduces basic ideas, linking quantum mechanics with kernel methods. Kernel methods are a robust mathematical framework that enables one to settle some aspects rigorously and directly show the connection between quantum mechanics and learning machines. Kernel methods with quantum states are the subject of Chapter 2, which introduces concepts related to Gaussian states, some basic gates, and boson sampling.

Chapter 3 focuses on qubits concerning some pioneering experimental work on quantum feature mapping and supervised learning. Here we deepen the connection between tensors and qubits, which is the starting point for Chap. 4, where we introduce optimization and the Ising model. Chapter 4 contains the first working examples of variational neural network states.

When considering optimization in a two-qubit Hilbert space, we readily find the occasion of considering entanglement and its rule in feature mapping. In Chap. 5, we introduce the entropy of entanglement and its computation by tensors. Chapter 6 focuses on neural network states with examples of minimizing a many-body Hamiltonian with qubits, precisely the transverse field Ising model. Different variational quantum circuits are presented, ranging from experimentally accessible unitary maps to idealized neural-network quantum states.

Chapter 7 starts a new book part based on phase space representations. Characteristic functions readily map to neural networks, and one can use this aspect to deepen the case of continuous variables. Chapter 7 is mainly oriented to the fundamental theoretical concepts. Chapter 8 deals with the connections between many-body states in the phase space and the correspondence between linear transformations and deep neural networks. Chapter 9 discusses a simple form of reservoir computing as an example for developing the training of neural network models with Gaussian states and continuous variables. Chapter 10 introduces gates representing squeezed states, beam splitters, and dedicated neural layers to mimic detection in quantum machine learning. Chapter 11 gives the tools for investigating advanced aspects, such as measuring uncertainties and entanglement with Gaussian and more general states.

Finally, Chaps. 12 and 13 report two advanced applications focusing on Gaussian states, as these enable feasible computing in simple hardware. Generalizations to non-Gaussian states can be tackled by the reader at the expense of a more significant computational effort.

Chapter 12 is about Gaussian boson sampling, the most critical direction in the context of the so-called quantum advantage, and its considerable link with machine learning. The link is developed in several experiments at a growing number of qubits.

Chapter 13 is about minimizing many-body Hamiltonians with continuous variables, a direction linked (as Gaussian boson sampling) with quantum chemistry and fundamental aspects, but here developed with a simple application in studying nonlinear localizations, such as quantum solitons.

Sections with an asterisk can be omitted at a first reading.

Despite many powerful packages on quantum computing being available throughout the book, I prefer to develop the codes from scratch, starting from the application programming interface (as TensorFlow™ and Python, with some theoretical exercises in MATLAB™ and Mathematica™). This is fun if you love programming and numerical science. Even if a mantra in computer science is "Do not reinvent the wheel," if you do not know how the wheel works or how to build the wheel, you cannot invent more clever wheels for specialized things, such as a rover on the moon or Mars.

In conclusion, I am not a zealot of quantum computing, and I have many doubts about the future of quantum processors. However, to be optimistic, one could outline that recent literature has shown a continuous increase in the number of qubits and the scale of quantum processors, as it was in the 1980s for the size of conventional processors. However, a key difference is that while in the 1980s, it was evident that semiconductors were the best technology, nowadays, we are still struggling with the most scalable approach for quantum systems. Moreover, many scholars think that all these efforts risk being pointless after the initial hype. In the long run, we will determine if any quantum computer is equivalent to a classical one with a proper feature space. But I agree that scientific and technological revolutions have been triggered by apparently useless investigations, which can develop branches in many directions to continuously appear as a novelty and attract young minds. This is undoubtedly the case with quantum machine learning, a trend that inspires artificial intelligence and fundamental physics. Also, quantum machine learning routinely pops out when finding applications for quantum experiments and new quantum devices.

This book aims to be an original and slim introduction that can inspire the interested reader.

Rome, Italy
June 2023

Claudio Conti

Supplementary Code

Code developed by the author and referenced in the book is available at

https://github.com/nonlinearxwaves/thqml-1.0

under the 3-Clause BSD license.

Files from the TensorFlow library referenced in this book are subject to the following license:

Files from the scikit-learn library referenced in this book are covered by the 3-Clause BSD and available at https://scikit-learn.org/stable/

Files from the qutip library referenced in this book are covered by the 3-Clause BSD and available at https://qutip.org/

Contents

Abbreviations and Acronyms

w.r.t.	with respect to
r.h.s.	right-hand side
l.h.s.	left-hand side
QML	Quantum Machine Learning
NN	Neural Network
QNN	Quantum Neural Network
HO	Harmonic Oscillator
API	Application Programming interface
RBF	Radial Basis Functions
SVM	Support Vector Machine
SVC	Support Vector Classifier
GBS	Gaussian Boson Sampling
SVD	Singular Value Decomposition
`tf`	`tensorflow`
QUBO	Quantum Unconstrained Binary Optimization
TIM	Transverse-Field Ising Model
QAOA	Quantum Approximate Optimization Algorithm
NNQS	Neural Network Quantum State
API	Application Programming Interface
BSVA	Boson Sampling Variational Ansatz
QDNLS	Quantum Discrete Nonlinear Schrödinger Equation

Definitions and Symbols

$\mathcal{E}$	Mean information
D	Number of qubits in a quantum register
N_X	Number of elements in the dataset
$\hat{\mathrm{X}}, \hat{\mathrm{Y}}, \hat{\mathrm{Z}}, \hat{\mathrm{I}}$	One-qubit Pauli matrices
$\hat{\mathrm{H}}$	One-qubit Hadamard matrix
$\mathbf{X}^{(j)}$	An observation in the dataset
$y^{(j)}$	A label in the dataset
$\Re$	Real part symbol, example $\Re\alpha$
$\Im$	Real part symbol, example $\Im\alpha$
χ	Characteristic function, scalar, complex
χ_R	Real part of the characteristic function, equivalent to $\Re\chi$
χ_I	Imaginary part of the characteristic function, equivalent to $\Re\chi$
$\boldsymbol{d}$	Weights, displacement vector, $N \times 1$, real
$\boldsymbol{g}$	Weights, covariance matrix, $N \times N$, real
$\mathbf{J}$	Constant, symplectic matrix, $N \times N$, real
$\mathbf{J}_1$	Constant, single mode symplectic matrix, 2×2, real
$\boldsymbol{x}$	Variable vector, $1 \times N$, real
q_k	Real-valued quadrature (position) for mode k
p_k	Real-valued quadrature (momentum) for mode k
$\delta_{pq}, \delta^i{}_j, \delta_i{}^j$	Kronecker delta
$\mathbf{0}_n$	$n \times n$ matrix with all elements equal to 0
$\mathbf{1}_n$	$n \times n$ identity matrix

Chapter 1
Quantum Mechanics and Data-Driven Physics

Don't ask what quantum computers can do for you, but what you can do for quantum computers.

Abstract Why quantum machine learning? Quantum mechanics and data-driven physics are similar. We describe their connections by focusing on their mathematical paradigms. We introduce quantum feature maps, quantum kernels, and examples of quantum classifiers. We discuss the basics of kernel methods with coding examples.

1.1 Introduction

Nearly one century ago, Paul Adrien Maurice Dirac [1] outlined that the importance of quantum mechanics is in its quantitative agreement with experimental data. We may eventually miss a physical intuition, or a simple understanding, of our observations. What is critical is that we have a theory that numerically predicts our experiments.

Following Richard Feynman [2], quantum mechanics is a list of rules that enable one to fit experimental data with unprecedented precision. In the early days of quantum theory, the Bohr model gave the spectral lines of the hydrogen atom. Later, quantum electrodynamics predicted the electron magnetic moment with 12 decimal digits [3, 4]. We could mention many other examples of successful applications of quantum rules. Reproducing experimental data with unprecedented precision is the main achievement of quantum mechanics, and we have to accept that quantities exist (as the wave function) that cannot be accessed, but we can use them for predictions. Feynman [2] also stated that it is impossible to understand quantum mechanics. Even without such an "understanding," quantum mechanics is the most successful theory in explaining experimental data.

Nowadays, we use identical arguments for the so-called data-driven physics [5]. Without resorting to general principles, we apply models and algorithms to fit and predict the behavior of complex systems, as in biophysics or social sciences. In some cases, we also infer laws relating the observable quantities without using

C. Conti, *Quantum Machine Learning*, Quantum Science and Technology,
https://doi.org/10.1007/978-3-031-44226-1_1

deductions from principles, at variance with the common practice at the beginning of the twentieth century. Data-driven physics uses machine learning and artificial intelligence to infer physical laws and the leading elements to describe complex phenomena. Machine learning is releasing huge mathematical models with billions of parameters to fit a large amount of data.

There is no evident magic or intelligence in machine learning. It is just a colossal fit with some human interpretation. However, this fitting is so clever that we can learn something about the data and make predictions. Even if we may question data-driven physics as an approach to understanding nature, there are no doubts that methodological similarities exist with quantum mechanics. Such similarities are also present in the mathematical models. Unveiling this connection is one of the goals of this book.

To convince you about the connection between centenarian quantum mechanics and the modern language of machine learning, let us consider their mathematical formulations. For example, we can formulate quantum mechanics in a system with a finite number of levels as an algebra of matrices in which components of state vectors are scalar products. Seemingly, in *transformers* [6], a leading algorithm for language processing and other large-scale computational tasks, the input data are matrices, which we apply to reference vectors to extract patterns. This similitude suggests that we can use quantum systems as accelerators for physically implementing modern computational models. Reciprocally, we can use quantum mechanics as an inspiration to improve our models of artificial intelligence.

Even if we have the best quantum computer, the important fact to keep in mind is that we need to perform computations with data that we can manage. As we live in a classical world, inputs and outputs that we directly manipulate are classical data. For example, an image on the screen is a sequence of classical bits in a random-access memory. To use a quantum computer to process the image, we need to encode the classical input into a quantum memory (i.e., a quantum state) and then retrieve the output after the quantum processing.

Describing the computational models that elaborate classical data by quantum accelerators to perform machine learning tasks is the other goal of this book.

1.2 Quantum Feature Mapping and the Computational Model

We map our classical data $\mathbf{X}$ in a state $|\phi\rangle$ belonging to a Hilbert space (see Fig. 1.1), an operation also denoted as `quantum encoding`. This implies that for a given input $\mathbf{X}$, the quantum state $|\phi\rangle$ of the system will have as parameters the input datum, namely,

$$|\phi\rangle = |\phi(\mathbf{X})\rangle \ . \tag{1.1}$$

$$\mathbf{X} \rightarrow |\phi(\mathbf{X})\rangle$$

$$\mathbf{X} \rightarrow |\alpha(\mathbf{X})\rangle$$

$$\mathbf{X} \rightarrow |c(\mathbf{X}), \varphi(\mathbf{X})\rangle$$

Fig. 1.1 Quantum feature maps from the input dataset: a vector of real variables $\mathbf{X}$ to a vector of a quantum Hilbert space $|\phi\rangle$. Examples show feature mapping to coherent states $|\alpha\rangle$ and squeezed vacuum $|c, \varphi\rangle$ [7]

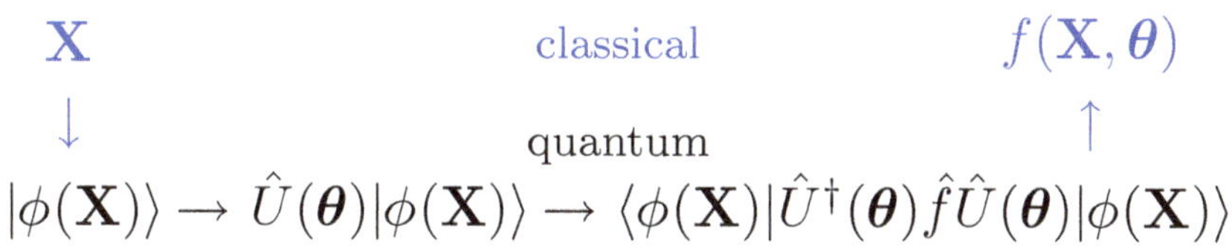

Fig. 1.2 An overview of the quantum machine learning model for approximating arbitrary classical functions. For each different input $\mathbf{X}$, we have a quantum state $|\phi(\mathbf{X})\rangle$. Evolution occurs by the unitary operator $\hat{U}(\boldsymbol{\theta})$, realized by a quantum circuit with parameters $\boldsymbol{\theta}$. The mean value of the observable $\hat{f}$ is the output and depends on both the input $\mathbf{X}$ and the parameters $\boldsymbol{\theta}$. In supervised quantum learning, one has the goal to find $\boldsymbol{\theta}^*$ to approximate a target function $f_T(\boldsymbol{x}) = f(\boldsymbol{x}, \boldsymbol{\theta}^*)$

The quantum computer elaborates the input datum by a controlled unitary operator. At the end of the evolution, we access the data by a quantum measurement. Then, further classical processing is required to extract the classical output, which is the `quantum decoding`.

The quantum evolution after the encoding is a unitary operator $\hat{U}$, which is dependent on some parameter vector $\boldsymbol{\theta}$. The components of $\boldsymbol{\theta}$ are the parameters in the quantum device, which we control to optimize the operation. The evolved vector is

$$\hat{U}(\boldsymbol{\theta})|\phi(\mathbf{X})\rangle. \tag{1.2}$$

By measuring an observable $\hat{f}$, we extract information from the evolved state vector; for example, we measure the mean value

$$\langle\hat{f}\rangle = \langle\phi(\mathbf{X})|\hat{U}^\dagger(\boldsymbol{\theta})\hat{f}\hat{U}(\boldsymbol{\theta})|\phi(\mathbf{X})\rangle = f(\mathbf{X}, \boldsymbol{\theta}) . \tag{1.3}$$

A quantum neural network (QNN) is a device such that specific value $\boldsymbol{\theta}^*$ exists for the vector of parameters $\boldsymbol{\theta}$ satisfying

$$f(\mathbf{X}, \boldsymbol{\theta}^*) = f_T(\mathbf{X}) , \tag{1.4}$$

with $f_T(\mathbf{X})$ a target function (Fig. 1.2).

We will see that quantum systems are equivalent to classical models for neural networks. The *universal approximation theorems* indicate that QNNs are universal approximators if properly designed [8]. QNNs approximate arbitrary functions and realize different devices and protocols as classifiers and supervised learning.

Classical NNs perform an impressive amount of tasks. NNs are *universal approximators*; in a sense, they are fitting machines such that they can fit the given dataset, i.e., a list of inputs and outputs, and extrapolate to unknown inputs.

We can use quantum systems to realize NNs and machine learning. *But why should we exert quantum physics for universal approximation and artificial intelligence?* This is what we discuss in the next sections. We start considering some examples.

1.3 Example 1: The First Supervised Learning by Superconducting Qubits

Superconducting qubits are the first successful example of qubits in the real world. In a superconducting qubit, energy is stored in an excited level of a harmonic oscillator "in disguise," i.e., with non-equispaced energy levels [9]. The harmonic oscillator is realized by a low-temperature electrical circuit. The qubits are couples of electrons at the edge of material junctions.

A conventional harmonic oscillator is not a qubit as it has infinite energy levels. On the contrary, a qubit has only two quantum levels $|0\rangle$ and $|1\rangle$. The weirdness in the superconducting harmonic oscillators is in that they are not at all *harmonic*, i.e., the energy levels are not equispaced. The energy gap between the level $|1\rangle$ and the level $|2\rangle$ is so large that only the first excited state comes into play, and the system operates as a qubit.

This works in practice, and microwave signals excite and probe the state of the superconducting qubit. Figure 1.3 shows a sketch of a superconducting circuit, with wiggled lines representing the waveguides to drive the signals to the junctions represented by the squares. Each square is a qubit. Two qubits `Q1` and `Q2` are indicated.

In the early years, superconducting qubits demonstrated the first simple gates, as rotations or controlled NOT gates [9]. However, this technology also enabled quantum supervised learning [10]. This early demonstration is a proof of concept with only two qubits; it may be retained as a milestone toward the employment of quantum circuits for machine learning.

1.4 Example 2: Photonic Quantum Processors, Boson Sampling, and Quantum Advantage

Superconducting qubits need low temperatures and engineered microwave signals; quantum lies in the electron pairs that accumulate at the boundaries of a junction.

Fig. 1.3 A schematic showing the quantum circuit [10] for the first demonstration of a quantum supervised learning with an entangled feature map. The squares represent the superconducting junctions encoding the qubits `Q1` and `Q2`. The wiggled lines are the waveguides that drive and couple the qubits

In this framework, the microwave signals are just perturbations treated classically. In a completely different quantum technology, one leverages the fact that also electromagnetic waves are quantum systems. This is the realm of quantum optics [9], which relies on the fact that any mode in light is a harmonic oscillator.

The good news is that we can control to a certain degree the quantum state of a laser beam. We can also build large-scale states with entanglement and other nonclassical features. However, we are in trouble as the photon energy levels cannot be put in disguise in a way such that only two states $|0\rangle$ and $|1\rangle$ dominate.

In quantum optics, we always have to consider systems with many levels. Hence the basic notions of quantum information must often be enriched by other concepts. For example, one has to consider Gaussian states [7], i.e., infinite combinations of number states, to describe the situation of photons in modes propagating in a system. Nevertheless, intriguing and unexpected outcomes may arise.

An example is the Gaussian boson sampling, a task running in specific quantum devices order of magnitudes faster than in a classical computer [11, 12]. But the real quantum advantage of boson sampling is still debated. Nevertheless, many scholars believe that quantum optical experiments on boson sampling are the first historical evidence of quantum advantage.[1]

The goal of boson sampling is to predict the occurrence of certain particle combinations at the output channels of multimodal circuits, which include squeezers and interferometers. Determining the output probability of specific events requires computing special determinants with numerous combinatorial terms. This computation is believed to be hard classically, even though new algorithms are emerging that challenge this statement [13, 14].

But if we build a device, a quantum optical device, made of many optical channels (e.g., waveguides), and nonlinear systems able to squeeze light, as resonators, we can launch photons in it and count the frequency of their occurrence at the output. We use this specific quantum optical computer to solve a computational

[1] We will not elaborate on the details of boson sampling, as it is considered in later chapters.

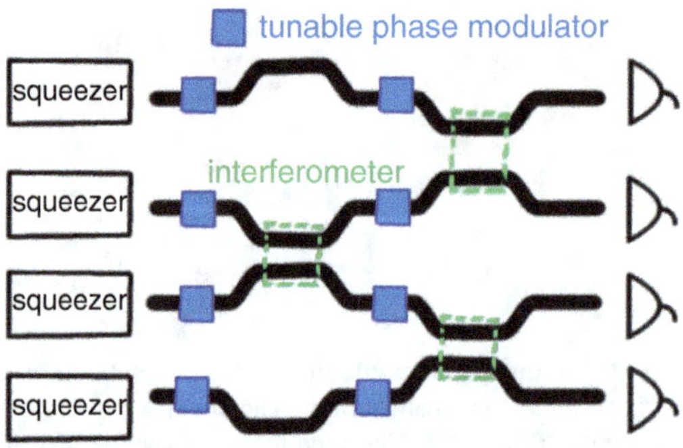

Fig. 1.4 A schematic showing the optical integrated circuit for the quantum advantage with boson sampling, which is a backbone for quantum machine learning. One can spot multiple waveguides representing the channels of the optical signals, tunable interferometers and detectors. This device is a promising tool for feature mapping in quantum machine learning

task (computing the probability of boson patterns) in a time much smaller than a conventional computer.

Xanadu, among others, has developed a class of quantum optical processors specialized for boson sampling. Figure 1.4 shows a schematic of the processor.

One can also perform similar sampling experiments [15] with superconducting qubits [14]. The challenge for the best technology in these specific tasks is still open.

The strongest criticism of boson sampling is that it does not appear a useful or programmable task. But new applications are emerging as, e.g., computing specific energy levels in quantum chemistry [16], and programmable boson sampling circuits were also demonstrated [17, 18].

A significant aspect of programmable boson sampling is that one can use the device to make a feature mapping for quantum machine learning [19] – in particular, the kind of quantum machine learning that is intrinsically hard to implement classically. For this reason, a promising application of Xanadu and similar quantum optical processors is quantum machine learning.

1.5 Quantum State Representations

We have taken for granted that our quantum system is described by a state denoted by a "ket" $|\psi\rangle$ in the Hilbert space, following the original notation of Dirac [1]. We know other representations exist, such as the density matrix ρ.

ρ is, in a sense, more precise than the ket $|\psi\rangle$. Indeed, the latter identifies a state within an arbitrary phase, the *absolute phase* ϕ, such that $|\psi\rangle$ and $e^{i\phi}|\psi\rangle$ are the same quantum state for any value of ϕ. On the contrary, for a pure state, the density matrix is

$$\rho = |\psi\rangle\langle\psi| \tag{1.5}$$

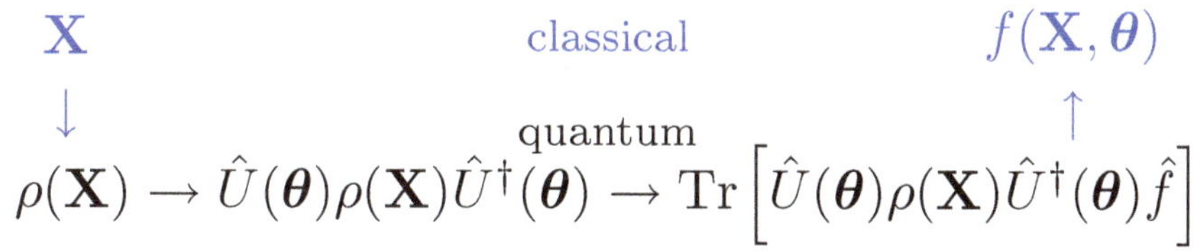

Fig. 1.5 Quantum feature mapping, as in Fig. 1.2, by using the density matrix ρ to represent the quantum state. Classical data $\mathbf{X}$ map to $\rho(\mathbf{X})$, which evolves with the unitary operator $\hat{U}(\boldsymbol{\theta})$. A measurement $\mathrm{Tr}(\rho\hat{f})$ of the mean value of the observable $\hat{f}$ returns the classical value $f(\mathbf{X}, \boldsymbol{\theta})$

and is independent of the absolute phase ϕ. But this is not the only significant aspect of the density matrix.

Density matrix operators describe a wider class of quantum systems as mixed states. Steven Weinberg [20] outlines the significance of ρ and the possibility of developing quantum theory without using the concept of state but just using the density matrix.

Notably, the computational model in Fig. 1.2 also applies with reference to ρ. Figure 1.2 can be redrawn as in Fig. 1.5. For the feature space, we use the density matrix space $\rho(\boldsymbol{x})$. The unitary evolution $\hat{U}(\boldsymbol{\theta})$ transforms the density matrix as

$$\tilde{\rho}(\mathbf{X}, \boldsymbol{\theta}) = \hat{U}(\boldsymbol{\theta})\rho(\mathbf{X})\hat{U}^\dagger(\boldsymbol{\theta}). \tag{1.6}$$

For the measurement of the expected value of an operator $\hat{f}$, we use the trace as

$$f(\mathbf{X}, \boldsymbol{\theta}) = \langle \hat{f} \rangle = \mathrm{Tr}[\tilde{\rho}\hat{f}] = \mathrm{Tr}[\hat{U}\rho\hat{U}^\dagger \hat{f}]\,. \tag{1.7}$$

The use of the density matrix will become handy in a later section, as it plays a significant role in QML.

1.6 Pros and Cons of Quantum Neural Networks

Quantum neural networks can be kept in mind as an implementation of classical computing paradigms as the so-called kernel machines, or the exponential machines.

In these machines, one maps the input datum to an abstract feature space and then retrieves the output by projections to the smaller space of observables. The projection includes a large number of parameters, which enable good fitting of target functions.

To have a large feature quantum space, we use a quantum register $|\phi\rangle$, for example, D qubits, such that the Hilbert space has dimension $N = 2^D$. For $D = 100$, we have $N \simeq 10^{30}$, orders of magnitudes larger than the number of stars in the universe. It is impressive that we have access to quantum computers nowadays at such a scale.[2] However, it is difficult to exploit all the states. Their majority is entangled and large-scale entanglement is hard to support when interacting with the classical world. Within the many inaccessible entangled states, the much more friendly product states span a smaller space with dimension $2D$.

If we consider $D = 3$ qubits, $|j\rangle$, with $j = 0, 1, 2$, the product states are

$$|\psi\rangle = |\psi_0\rangle|\psi_1\rangle|\psi_2\rangle \; . \tag{1.8}$$

Each state $|\psi_j\rangle$ is determined by two real parameters (the two angles in the Bloch sphere). We need $2D = 6$ parameters to identify the product state $|\psi\rangle$. $2D$ is the dimension of the subspace spanned by the product states at a first approximation ($2D = 6$ in Eq. (1.8)). If we also account for the normalization condition and the arbitrary phase, the actual dimension of the product space is $2D - 2 = 4$.

We cannot write the majority of the states as in Eq. (1.8), because they are entangled. With $D = 3$ qubits the basis of the Hilbert space is given by the following vectors:

$$|000\rangle, |001\rangle, |010\rangle, |011\rangle, |100\rangle, |101\rangle, |110\rangle, |111\rangle \; . \tag{1.9}$$

Each state is a linear combination of the basis vectors in (1.9) and determined by $2^3 = 8$ complex numbers, or 16 real numbers. When including the constraint of normalization and the absolute phase of the state, the number of real free parameters is $2^{D+1} - 2 = 16 - 2 = 14$. For D qubits, we need $2^D - 1$ complex parameters for identifying a quantum state.

Hence, on the one hand, quantum systems have the advantage of populating a space spanned by a large number of parameters; on the other hand, we need to support entanglement against decoherence to exploit such a large scale.

Without entanglement, the quantum system is equivalent to a classical one with the same number of real parameters as the product-state subspace, i.e., $2D - 2$. Entanglement requires isolation from the environment. Any interaction with the environment destroys entanglement; this circumstance makes quantum computers expensive and hard to build.

A further issue is that quantum measurements are probabilistic. We need repeated trials to get a certain confidence in the output, e.g., to reach a given accuracy in estimating statistical quantities such as the mean value in Eq. (1.3). For this reason, the details of a specific implementation determine if a large-scale quantum computer is advantageous with respect to the classical counterpart.

The common wisdom is that if you have a task that is solved efficiently by a classical computer (i.e., it scales polynomially), you do not need a quantum computer, which – at best – will also scale polynomially [21]. Quantum computers are needed for those tasks that we cannot solve efficiently by a classical computer.

[2] https://research.ibm.com/blog/127-qubit-quantum-processor-eagle.

Accuracy in a quantum computer is also limited by uncertainties in the parameters, which are due to fabrication imperfections and tolerances. This aspect becomes progressively more critical as the number of qubits increases because of the resulting circuit complexity.

Finally, there are intrinsic limitations [22]. We cannot copy a quantum register (as stated by the *Quantum no-cloning* theorem [23, 24]), or we cannot access the precise content of a quantum register without collapsing its state. These limitations hamper our quantum algorithms but also stimulate new kinds of programming protocols.

The previous arguments make the challenge risky, but there are experimental demonstrations of quantum neural networks. These demonstrations empower our motivation to study and develop the subject.

1.7 Quantum Mechanics and Kernel Methods

Why kernel methods? We are often exposed to deep learning in any introduction to artificial intelligence. But deep learning is an heuristic approach. The trial-and-error nature of deep learning is evident in the design of models that grow in complexity, as it happens in *transformers* [6] or more sophisticated models for large-scale tasks, such as natural language processing.

A heuristic approach is not helpful in unveiling the fundamental links between quantum mechanics and machine learning. Recent investigations tag kernel methods as a good starting point [8].

In kernel methods, the goal is approximating an unknown function $f(\mathbf{X})$ by using a dataset $\left[\mathbf{X}^{(i)}, y_i\right]$ determined by observations $\mathbf{X}^{(i)}$ (i.e., the input data) and labels y_i, with $i = 0, 1, \ldots, N_X - 1$. General theorems [25] guarantee that the solution is

$$f(\mathbf{X}) = \sum_{i=0}^{N_X-1} \alpha_i y_i K\left[\mathbf{X}, \mathbf{X}^{(i)}\right] , \tag{1.10}$$

given a function of two observations $K(\mathbf{X}, \mathbf{Y})$, named "kernel" and weights α_i determined by fitting the labels y_i.

The kernel function satisfies a series of properties [26]: It is symmetric $K(\mathbf{X}, \mathbf{Y}) = K(\mathbf{Y}, \mathbf{X})$, and, for any set of real numbers $c_0, c_1, \ldots, c_{N_X-1}$, we have

$$\sum_{i,j} c_i c_j K\left[\mathbf{X}^{(i)}, \mathbf{X}^{(j)}\right] \geq 0 . \tag{1.11}$$

Equation (1.11) implies that the "kernel matrix" (also named the "Gram matrix"),

$$K_{ij} = K\left[\mathbf{X}^{(i)}, \mathbf{X}^{(j)}\right] , \tag{1.12}$$

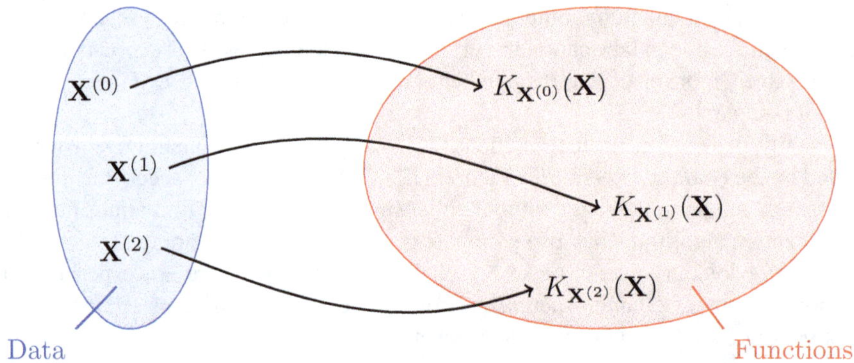

Fig. 1.6 At the inner core of kernel methods, there is a mapping from the observations $\mathbf{X}^{(j)}$ to functions $K_{\mathbf{X}^{(j)}}(\mathbf{X})$ tagged by the observation. This mapping links the space of observations and the linear space of tagged functions. Any target function belongs to the space of the linear combinations of the tagged functions

is positive definite. The fact that Eq. (1.10) is the best solution to the considered supervised learning process is guaranteed by the "representer theorem," a milestone in kernel methods [25].

We describe Eq. (1.10) as follows: we have a set of functions, each tagged by one observation $\mathbf{X}^{(i)}$, $K_{\mathbf{X}^{(i)}}(\mathbf{X}) = K\left[\mathbf{X}, \mathbf{X}^{(i)}\right]$, and our solution is a linear combination of these functions. Figure 1.6 shows the mapping from the space of the observations to the space of the corresponding functions. The best approximation to an unknown $f(\mathbf{X})$ is just a linear combination of these functions.

We will describe below how such a situation emerges in the context of "support vector machines," but let us focus here on the similarities with quantum mechanics.

Equation (1.10) is fascinating if compared with elementary quantum mechanics. In quantum mechanics, we know that the state is a linear combination of special functions, such as the energy eigenstates.

Consider the simplest case of the harmonic oscillator (HO). We have a set of observations that are the HO energy levels E_n with $n = 0, 1, 2, \ldots$. We also have a set of real-valued normalized functions $\phi_{E_n}(x)$, where x is the one-dimensional spatial coordinates. In the position representation, any quantum state is a complex function $\phi(x)$ that is a linear combination of HO eigenstates, i.e.,

$$\phi(x) = \sum_{n=0}^{\infty} c_n \phi_{E_n}(x) . \tag{1.13}$$

Equation (1.13) is resembling Eq. (1.10) in a way such that we have a set of real-valued observations, the energy levels E_n and associated real function $\phi_{E_n}(x) = \phi_n(x)$.

We know that the quantum system is completely determined by a combination of these functions, in the sense that we can use them to predict the mean values of observable quantities and other physical properties.

However, at variance with (1.10), in quantum mechanics we have an infinite number of terms in (1.13). The infinite number arises from the mathematical structure of the theory. But, in practice, the energy involved is always upper limited (we never observe infinite energy in the real world), and a finite number of terms is only significant.

Also, the wave function $\phi(x)$ is complex valued, and the coefficients c_n are complex valued. The c_n are determined by the scalar products in the Hilbert space spanned by ϕ_n, namely,

$$c_n = \langle n|\psi\rangle = \langle\phi_n(x), \phi(x)\rangle = \int \phi(x)\phi_n(x)\mathrm{d}x \ . \tag{1.14}$$

These differences relate to the fact the probability of observing a quantum state at a specific energy level $\Pr(E_n)$ is given by the "Born rule"

$$\Pr(E_n) = |c_n|^2 = |\langle\phi_n(x), \phi(x)\rangle|^2. \tag{1.15}$$

Let us consider an observable quantity, for example, the energy, whose quantum operator is the Hamiltonian $\hat{H}$. The mean value of the energy is

$$\langle\hat{H}\rangle = \langle\psi|\hat{H}|\psi\rangle = \sum_n E_n|\langle n|\psi\rangle|^2 \ , \tag{1.16}$$

with the sum involving all the energy basis.

By comparing (1.16) and (1.10), we could think to use the quantum system as kernel machine by tailoring the energy levels E_n as the labels y_i and exploiting the $\Pr(E_n) = |\langle\phi_n(x), \phi(x)\rangle|^2$ as some sort of kernel function. But we are still missing the weights of the fit α_n.

To introduce the weights, we consider a further observable in the energy basis

$$\hat{O} = \sum_n \alpha_n E_n |n\rangle\langle n| \ . \tag{1.17}$$

$\hat{O}$ is an observable that commutes with $\hat{H}$ (they have the same eigenstates $|n\rangle$), with unknown coefficients α_n. $\hat{O}$ can also be represented as a function of the energy $\hat{O} = \theta(\hat{H})$, such that $\theta_n = \theta(E_n) = \alpha_n E_n$ are its eigenvalues. We have

$$\langle\psi|\hat{O}|\psi\rangle = \sum \alpha_n E_n|\langle n|\psi\rangle|^2 = \sum \theta_n|\langle n|\psi\rangle|^2 \ , \tag{1.18}$$

which reminds Eq. (1.10) as we identify the kernel function with

$$K_n(\mathbf{X}) = |\langle n|\psi\rangle|^2. \tag{1.19}$$

To fix the formal identity between the kernel methods (Eq. 1.10) and (1.18), we have to set a correspondence between each datum $\mathbf{X}^{(n)}$ and each energy level E_n, and we need a state $|\psi\rangle$ dependent on the input $\mathbf{X}$.

Fitting the dataset fixes the α_n and corresponds to finding an operator $\hat{O}$ such that its mean value gives the target function, namely,

$$\langle\psi(\mathbf{X})|\hat{O}|\psi(\mathbf{X})\rangle = f_T(\mathbf{X}) \ . \tag{1.20}$$

The former arguments show that a quantum mechanical system is equivalent to a kernel method and the coefficients α_n fix the target operator $\hat{O}$.

However, it is not straightforward to realize a quantum mechanical system with given energy levels $E_n = y_n$, neither measuring a target operator $\hat{O}$. In addition, we need a number of energy levels equal to the elements in the dataset (see (1.21)). But we have the insight that we can use quantum mechanical systems to compute kernel functions. In other words, we have the first evidence that we can exploit quantum systems to accelerate learning algorithms as – specifically – kernel methods.

1.8 More on Kernel Methods

The simple message of kernel methods is: if you have a supervised classification problem, the best classifier is

$$f(\mathbf{X}) = \sum_{j=0}^{N_X-1} \alpha_j y_j K\left[\mathbf{X}, \mathbf{X}^{(j)}\right] \tag{1.21}$$

where we assume to be dealing with a dataset of N_X observations $\mathbf{X}^{(j)}$ and the corresponding labels y_j. The data are D dimensional vectors $\mathbf{X}$ with components $X^0, X^1, \ldots, X^{D-1}$.

In Eq. (1.21), $K(\mathbf{X}, \mathbf{Y})$ is the kernel function chosen, e.g., a Gaussian function $K(\mathbf{X}, \mathbf{Y}) = \exp(-|\mathbf{X} - \mathbf{Y}|^2)$.

We have outlined the similarity between (1.21) and (1.18), and we can identify in the kernel matrix the modulus square of the quantum mechanical scalar products.

In Eq. (1.21) the α_j are a set of weights determined by conventional regression algorithms by the labels y_j.

For example, as we must have $f\left[\mathbf{X}^{(i)}\right] = y_i$, we have from (1.21)

$$y_i = \sum_{j=0}^{N_X-1} \alpha_j y_i K\left[\mathbf{X}^{(i)}, \mathbf{X}^{(j)}\right] = \sum K_{ij}\theta_j \tag{1.22}$$

letting $\theta_j = \alpha_j y_j$ with

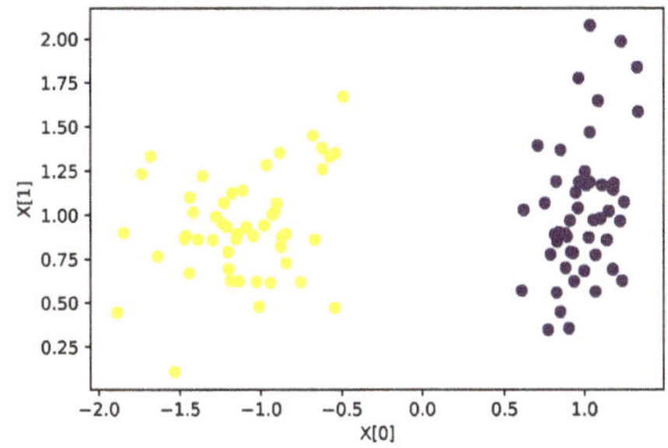

Fig. 1.7 Example dataset created by `scikit-learn`. Two clusters of points are generated. One has to find an algorithm able to distinguish the two groups. In this example, we have a linearly separable dataset

$$K_{ij} = K\left[\mathbf{X}^{(i)}, \mathbf{X}^{(j)}\right] , \tag{1.23}$$

the elements of the Gram matrix. We find the θs and the αs by inverting the K_{ij} matrix.[3]

One way to understand Eq. (1.21) is to consider a classification problem as, e.g., in Fig. 1.7 with $D = 2$. We have a set of points corresponding to labels $y = 1$ and to labels $y = -1$. All those points with label $y = 1$ lie above a line in the data space; the data are *linearly separable*.

If we need to assign a label to a new observation $\mathbf{X}$, we need a *decision rule*. One strategy is looping the points to find the nearest neighbor and assigning the label of the nearest neighbor. However, this approach is demanding in terms of computational time. Also, if we add a new point in the dataset, we need to update the information about the neighbors.

One hence introduces a similarity measure, the kernel function $K(\mathbf{X}, \mathbf{Y})$, and computes the labels as linear combination of similarities. For example, $K(\mathbf{X}, \mathbf{Y})$ can be determined by the Euclidean distance $d(\mathbf{X}, \mathbf{Y}) = |\mathbf{X} - \mathbf{Y}|$, such that the label of the nearest neighbor has a larger weight, as in

$$f(\mathbf{X}) = \frac{1}{N_X} \sum_j y_j e^{-d(\mathbf{X},\mathbf{Y})^2} . \tag{1.24}$$

In this case, looking at Eq. (1.21), we have $\alpha_j = 1/N_X$.

We can generalize this expression and retain the α_j as coefficients to be optimized to improve the fitting. However, the argument above does not explain why we should limit our fitting function to a summation over the data. We could consider other averages as geometric averages or generalize the similarity function to more points as, say, $K\left[\mathbf{X}, \mathbf{X}^{(i)}, \mathbf{X}^{(i+1)}\right]$.

Expressions like (1.21) are advantageous: we only need to use input data, the labels, and a linear combination, easily updated with new entries in the dataset.

[3] The inversion of large matrices is often ill-conditioned, and linear regression is adopted [25].

The fact that (1.21) is the optimal solution to the supervised learning problem is demonstrated when considering the *support vector machines* in Sec. 1.10.

1.9 Coding Example of Kernel Classification

Packages such as `scikit-learn` [27, 28] furnish powerful tools to play with machine learning and kernel methods. The following is adapted from an available example[4] in `scikit-learn` covered by BSD 3-Clause licens. We start considering a linearly separable dataset as in Fig. 1.7, which we generate as follows.[5]

```
#create a dataset with 2 sets of data centered around 2 regions
# details in
# https://scikit-learn.org/stable
# /modules/generated/sklearn.datasets.make_classification.html
# the shape for X is (100, 2)
# the for y is  (100, )
X, y = make_classification(
    n_features=2, n_redundant=0,
    n_informative=2, random_state=1,
    n_clusters_per_class=1
)
# plot dataset
plt.scatter(X[:, 0], X[:, 1], c=y)
plt.xlabel("X[0]")
plt.ylabel("X[1]")
```

Figure 1.7 shows the generated dataset.

We create and train a linear kernel machine as follows.

```
# SVC is a python class for support vector classification
# kernel =``linear''
# is a linear classifies, and C is a regularization parameter
clf=SVC(kernel="linear", C=0.05)
# fit the model to the generated train points
clf.fit(X_train, y_train)
  # compute the score with the test points
score = clf.score(X_test, y_test)
# returns score = 1.0
```

[4] https://scikit-learn.org/stable/auto_examples/classification/plot_classifier_comparison.html.

[5] The code is in `jupyternotebooks/quantumfeaturemap/kernelexample.ipynb`.

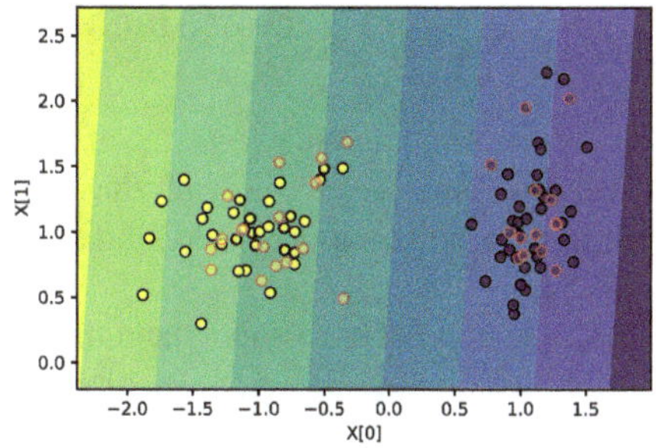

Fig. 1.8 Linear classification of the linearly separable dataset in Fig. 1.7. Training and testing (red boundaries) portions of the dataset are evidenced

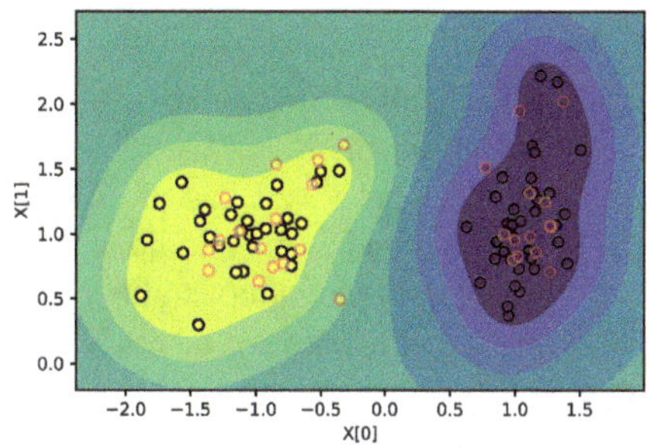

Fig. 1.9 Classification of the linearly separable dataset in Fig. 1.7 by a Gaussian kernel. Training and testing (red boundaries) portions of the dataset are evidenced

As the data are linearly separable, a simple linear classifier (detailed in the next session) returns optimal classification performance. Figure 1.8 shows the classified dataset, with the data separated by a linear boundary.

As a second example, we consider a Gaussian kernel, a nonlinear function, as follows.

```
# Define a classifier with a Gaussian kernel
# here gamma is the coefficient in the kernel
# exp(-gamma x^2)
clf=SVC(gamma=4, C=1)
# fit the model to the generated train points
clf.fit(X_train, y_train)
  # compute the score with the test points
score = clf.score(X_test, y_test)
# also returns score = 1.0
```

As for the linear classifier, we obtain the optimal classification. Classified data are plotted in Fig. 1.9; we also report the nonlinear classification boundary to be compared with the straight lines in Fig. 1.8.

In later sections, we see how things change when data are not linearly separable. But first, in the next section, we detail the theoretical basis of linear classifiers, which generalizes to nonlinearly separable datasets by feature mapping.

1.10 Support Vector Machine: The Widest Street Approach*

We detail mathematical aspects of the *support vector machine*, an algorithm to classify linearly separable data that is a cornerstone of modern machine learning. We report on the simplest derivation and leave extensions to specialized textbooks, as [25].

The main goal is to understand the origin of Eq. (1.21). We start considering the case in which the similarity measure is just the scalar product, namely,

$$K_{ij} = \mathbf{X}^{(i)} \cdot \mathbf{X}^{(j)} . \tag{1.25}$$

We consider a two-dimensional space (as in Fig. 1.10), such that each observation has two components $\mathbf{X} = (X^0, X^1)$. We assume a line exists separating two classes, the data with label $y = 1$ and those with $y = -1$. Such a line (dashed in Fig. 1.10) can be kept in mind as the middle of the *widest street* separating the two classes. The street boundaries (continuous blue lines in Fig. 1.10) are determined by the nearest observations with different labels $\mathbf{X}_+$ and $\mathbf{X}_-$ in Fig. 1.11. This is *the widest street approach.*[6]

Letting $\mathbf{w}$ the vector orthogonal to the line (see Fig. 1.11), a decision rule is

$$\mathbf{X} \cdot \mathbf{w} \geq C . \tag{1.26}$$

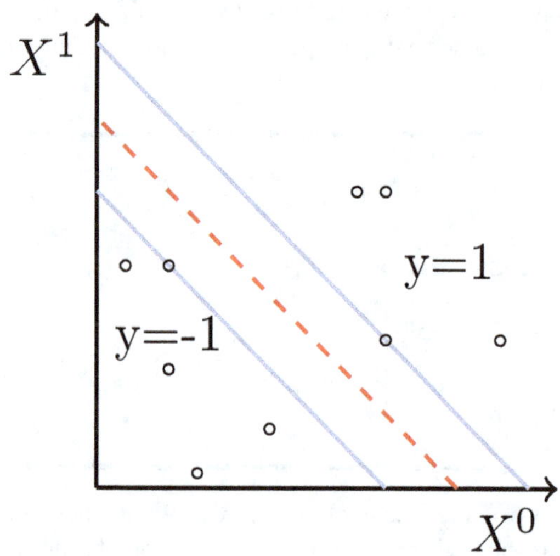

Fig. 1.10 The widest street approach in the classification of a linearly separable dataset by support vector machines. The dataset (black circles) is linearly separable in two subsets by the middle line (dashed). One looks for the biggest margin separating the two subsets by maximizing the distance among all the possible choices of the middle line. The points lying on the margins (blue lines) are the support vectors whose coordinates determine the widest street

[6] This derivation of the support vector machine follows Patrick H. Winston in https://www.youtube.com/watch?v=_PwhiWxHK8o.

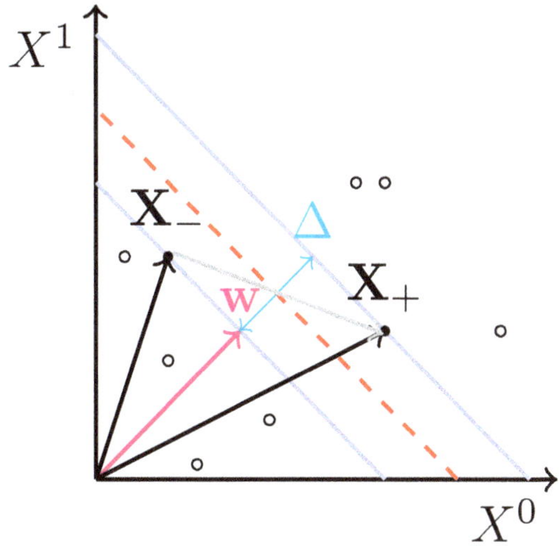

Fig. 1.11 In the widest street approach (Fig. 1.10), one looks for the biggest margin separating the two subsets by maximizing the distance Δ among all the possible choices of the middle line. The points $\mathbf{X}_+$ and $\mathbf{X}_-$ lying on the margins (filled circles) are the support vectors whose coordinates determine the widest street. The vector $\mathbf{w}$ is orthogonal to the middle and the margin lines. Δ is the projection of $\mathbf{X}_+ - \mathbf{X}_-$ (gray vector) in the direction of $\mathbf{w}$

Equation (1.26) states that the projection of the observation vector $\mathbf{X}$ in the $\mathbf{w}$ direction must be greater than a threshold when the label is $y = 1$, or otherwise $y = -1$.

Letting $b = -C$, Eq. (1.26) is equivalently written as

$$\mathbf{X} \cdot \mathbf{w} + b \geq 0\,. \tag{1.27}$$

Given the dataset $\left[\mathbf{X}^{(j)}, y_j\right]$, we need to find $\mathbf{w}$ and b to classify any point $\mathbf{X}$. We use the known dataset to find out $\mathbf{w}$ and b that minimize a loss function and fix the decision rule in (1.27).

If we consider an observation $\mathbf{X}_+$ with positive label $y = 1$, we have

$$\mathbf{X}_+ \cdot \mathbf{w} + b \geq 0\,. \tag{1.28}$$

The data are typically noisy, and the points near the threshold may be subject to errors in classification. This situation happens because precision in the data is always limited (e.g., number of available digits) or because of uncertainty in their measurement.

Thus, it is convenient to introduce a certain classification margin and to require that data with labels $y = 1$ are such that

$$\mathbf{X}_+ \cdot \mathbf{w} + b \geq 1\,. \tag{1.29}$$

In Eq. 1.29, we use "1" in r.h.s. as a working choice; we could use any other positive number. One can realize that a scaling of $\mathbf{w}$ and b can fix that r.h.s. number, and Eq. (1.29) fixes the scale for $\mathbf{w}$ and b, which is undetermined in Eq. (1.28).

Seemingly, we require that data $\mathbf{X}_-$ with negative label $y = -1$ satisfy

$$\mathbf{X}_- \cdot \mathbf{w} + b \leq -1 \ . \tag{1.30}$$

We write Eqs. (1.29) and (1.30) as a single equation. By multiplying (1.29) and (1.30) by the corresponding labels, we have

$$y_i \left[\mathbf{X}^{(i)} \cdot \mathbf{w} + b \right] \geq 1 \ , \tag{1.31}$$

for any observation $\mathbf{X}^{(i)}$ with label y_i. Equation (1.31) fixes the classification constraint for all the points in the dataset.

In addition, we want to maximize the margin for the classification Δ, which is the distance between the nearest observations $\mathbf{X}_+$ and $\mathbf{X}_-$ to the middle line (i.e., the width of the street in Fig. 1.11). Δ is the length of the projection of the vector $\mathbf{X}_+ - \mathbf{X}_-$ in the direction of $\mathbf{w}$. The direction is given by the unit vector $\mathbf{w}/|\mathbf{w}|$, with $|\mathbf{w}| = (\mathbf{w} \cdot \mathbf{w})^{1/2}$.

By using (1.31), we have

$$\Delta = (\mathbf{X}_+ - \mathbf{X}_-) \cdot \frac{\mathbf{w}}{|\mathbf{w}|} = \frac{2}{|\mathbf{w}|} \ . \tag{1.32}$$

Equation (1.32) implies that to maximize the classification margin, we need to minimize the length of the vector $|\mathbf{w}|$. For mathematical convenience, we choose equivalently to minimize the quantity $|\mathbf{w}|^2/2$.

As $\mathbf{w}$ and b are also subject to the constraints (1.31), the problem of finding $\mathbf{w}$ and b given the dataset $\left[\mathbf{X}^{(i)}, y_i\right]$ corresponds to a constrained minimization problem for $|\mathbf{w}|^2/2$. Such a constrained minimization problem is tackled by the method of Lagrange multipliers [25] .

We introduce the *dual* positive coefficients $\alpha_j \geq 0$ and consider the *Lagrangian* $\mathcal{L}$ including the constraints in Eq. (1.31).

$$\mathcal{L} = \frac{|\mathbf{w}|^2}{2} - \sum_{j=0}^{N_X-1} \alpha_j \left\{ y_j \left[\mathbf{X}^{(i)} \cdot \mathbf{w} + b \right] - 1 \right\} \tag{1.33}$$

$$= \sum_{p=0}^{D-1} \frac{(w^p)^2}{2} - \sum_{j=0}^{N_X-1} \alpha_j \left\{ y_j \left[\mathbf{X}^{(i)} \cdot \mathbf{w} + b \right] - 1 \right\} \ , \tag{1.34}$$

where the components w^p of $\mathbf{w}$ and b are the *primal* parameters.

We find the unknown parameters $\mathbf{w}$ and b by minimizing $\mathcal{L}$ with respect to the primal parameters and then maximizing with respect to the dual parameters. The rationale beyond this approach is that if some constraint is not satisfied, the value of $\mathcal{L}$ is higher. The larger the α_j, the more significant is the weight of the unsatisfied constraints. Correspondingly, one varies w^p and b as far as all the constraints are satisfied.

When all the constraints are fixed, we realize that the minimum of $\mathcal{L}$ is when $\alpha_j = 0$ for all the observations such that $y_j \left[\mathbf{X}^{(j)} \cdot \mathbf{w} + b\right] > 1$. Thus, only the terms corresponding to the margins (the boundaries of the street), such that $y_j \left[\mathbf{X}^{(i)} \cdot \mathbf{w} + b\right] = 1$, are present in $\mathcal{L}$ with $\alpha_j \neq 0$. These terms correspond to the *support vectors* and they only concur in determining $\mathbf{w}$ and b.

The fact that only the support vectors are significant in the optimization is understood by observing that the separation plane does not change if we add datapoints outside the margins, i.e., out of the street. On the contrary, if we add points in the margins, we need to reoptimize both $\mathbf{w}$ and b.

Given the strategy, we compute the derivatives of $\mathcal{L}$, with respect to w^p, b, and α_j. We write Eq. (1.34) as

$$\mathcal{L} = \frac{1}{2} \sum_{p=0}^{D-1} (w^p)^2 - \sum_{j=0}^{N_X-1} \alpha_j \left\{ y_j \left[\sum_{q=0}^{D-1} X_q^{(j)} w^q + b \right] - 1 \right\} . \tag{1.35}$$

Setting the derivatives to zero, we obtain

$$\frac{\partial \mathcal{L}}{\partial \alpha_j} = y_j \left[\mathbf{X}^{(i)} \cdot \mathbf{w} + b\right] - 1 = 0 \tag{1.36}$$

$$\frac{\partial \mathcal{L}}{\partial w^p} = w^p - \sum_{j=0}^{N_X-1} \alpha_j y_j X_p^{(j)} = 0 \tag{1.37}$$

$$\frac{\partial \mathcal{L}}{\partial b} = - \sum_{j=0}^{N_X-1} \alpha_j y_j = 0 . \tag{1.38}$$

Equation (1.36) corresponds to the constraints and holds only if $\alpha_j \neq 0$; otherwise the term in the Lagrangian is not present, and the derivative is not present. As anticipated, this implies that $\alpha_j \neq 0$ only for the support vectors.

Equation (1.37) implies for $p = 0, 1, \ldots, D-1$

$$w^p = \sum_{j=0}^{N_X-1} \alpha_j y_j X_p^{(j)} , \tag{1.39}$$

which we write as

$$\mathbf{w} = \sum_{j=0}^{N_X-1} \alpha_j y_j \mathbf{X}^{(j)} . \tag{1.40}$$

Hence $\mathbf{w}$ is a linear combination of the datapoints. More precisely, as $\alpha_j \neq 0$ only for the support vectors, $\mathbf{w}$ is a linear combination of the support vectors. Once we have $\mathbf{w}$, b is determined by any of the (1.36).

Finally, Eq. (1.38) corresponds to

$$\sum_{j=0}^{N_x-1} \alpha_j y_j = 0 \,, \tag{1.41}$$

which holds for the support vectors with $\alpha_j \neq 0$.

Equation (1.41) has an interesting mechanical analogy: if the α_j are interpreted as forces acting on an imaginary rigid body corresponding to the separation plane, the resultant of the forces must be zero, i.e., it is an equilibrium condition (from which the name "support vectors" originated). Seemingly, Eq. (1.37) can be retained as the equilibrium condition for the momenta.

By Eq. (1.40), the decision rule (1.27) reads as

$$\sum_{j=0}^{N_X-1} \alpha_j \mathbf{X} \cdot \mathbf{X}^{(j)} + b \geq 0 \,, \tag{1.42}$$

which shows that only the scalar products $\mathbf{X} \cdot \mathbf{X}^{(j)}$ are significant and, more precisely, only the scalar product with the support vectors.

If we assign the label according to

$$y = \text{sign}(\mathbf{X} \cdot \mathbf{w} + b) = \text{sign}\left[\sum_{j=0}^{N_x-1} \alpha_j y_j \mathbf{X} \cdot \mathbf{X}^{(j)} + b\right] \,, \tag{1.43}$$

we realize that the classifying function is completely determined by the values of the kernel function

$$K[\mathbf{X}, \mathbf{X}^{(j)}] = \mathbf{X} \cdot \mathbf{X}^{(j)} \,, \tag{1.44}$$

which is the linear scalar product.

What we are still missing are the values of the dual parameters α_j, obtained by Eqs. (1.37) and (1.38) in the $\mathcal{L}$. This provides the dual Lagrangian $\mathcal{W}$

$$\mathcal{W} = \sum_{j=0}^{N_X-1} \alpha_j - \frac{1}{2} \sum_{j,k=0}^{N_X-1} \alpha_j \alpha_k y_j y_k \mathbf{X}^{(j)} \cdot \mathbf{X}^{(k)} \,, \tag{1.45}$$

which needs to be maximized with respect to the αs with the constraint $\sum_{j=0}^{N_X-1} \alpha_j y_j = 0$, after Eq. (1.38). It is possible to show that this maximization problem is convex and has a single minimum, which is computed by conventional linear regression algorithms [25]. This furnishes a proof that the kernel machine is the best solution to the supervised learning problem.

1.11 The Link of Kernel Machines with the Perceptron

Kernel machines, intended as the learning models for optimization based on the ansatz

$$f(\mathbf{X}) = \sum_{j=0}^{N_X-1} \alpha_j y_j K\left[\mathbf{X}, \mathbf{X}^{(j)}\right] , \tag{1.46}$$

are computational paradigms sustained by solid mathematical arguments [25]. This circumstance is to be compared with the successful approach based on deep learning and neural networks, for which many mathematical aspects are still open.

Generally speaking, big neural networks for demanding tasks, such as natural language processing, are sustained by heuristic arguments and empirical evidence. However, kernel methods and neural networks are not independent. In particular, there is a link between the support vector machine and the perceptron.

The perceptron is the simplest one-layer neural network. Given the input vector $\mathbf{X}$, the perceptron output is a nonlinear function of linear combinations of its components X^q:

$$f(\mathbf{X}) = g\left(\mathbf{w} \cdot \mathbf{X}\right) = g\left(\sum_{q=0}^{D-1} w^q X^q\right) , \tag{1.47}$$

g being the *activation function* and $\mathbf{w}$ a weight vector whose components w^q are obtained by optimization. One possibility for g is $g(x) = 1$ for $x > 0$ and $g(x) = 0$ otherwise (a step function). In some formulation, the input to g is written as $\mathbf{w}\cdot\mathbf{X}+b$, where the b is the bias, which can be included in the inputs in (1.47) by a new component of the input vector $X^D = 1$ and $w^D = b$.

This argument shows that the support vector machine is a perceptron such that $\mathbf{w}$ is a linear combination of the input vectors $\mathbf{X}^{(j)}$.

One can also outline the links with other modern paradigms as the *transformers*, large-scale multilayer models that adopt similarity measures expressed as scalar products with vectors determined by regression from the input dataset [6].

1.12 Kernel Classification of Nonlinearly Separable Data

In the examples of Figs. 1.2 and 1.9, the dataset is linearly separable and can be classified by a linear or a nonlinear kernel. Let us consider a dataset that is not linearly separable, as the one generated by the following code.[7]

[7] The code is in `jupyternotebooks/quantumfeaturemap/kernelexample.ipynb`.

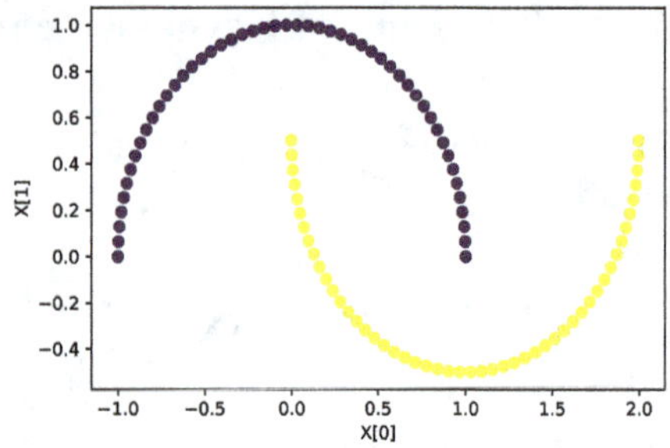

Fig. 1.12 Example of dataset that is not linearly separable created by `scikit-learn`

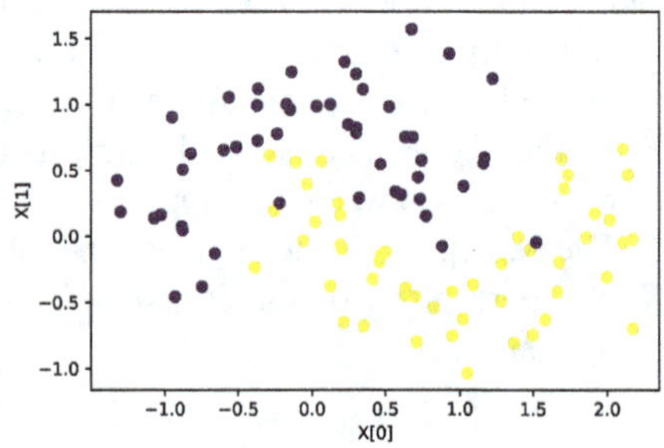

Fig. 1.13 Dataset in Fig. 1.12 with some noise

```
# create a dataset with nonlinearly separable states
dm=make_moons(noise=0.0, random_state=0)
X, y = dm
```

Figure 1.12 shows the generated points with binary labels. We obtain a randomized version of the dataset by (see Fig. 1.13)

```
# create a dataset with nonlinearly separable states
# we add some noise
dm=make_moons(noise=0.3, random_state=0)
X, y = dm
# random_state is a seed to generate the same noise
# and make the result reproducible
# if we change random_state we have different noise
```

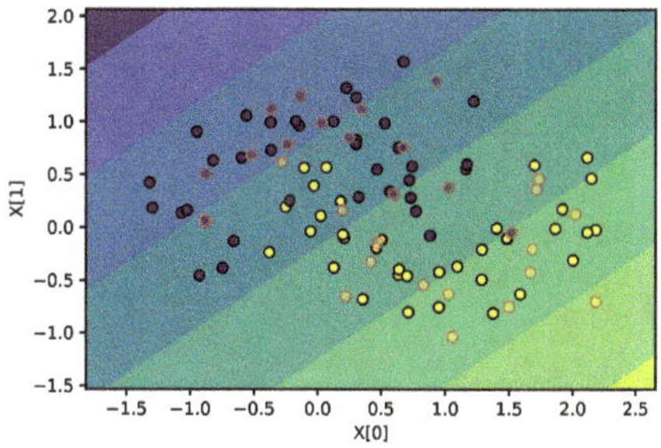

Fig. 1.14 Trying to classify a nonlinearly separable dataset with a linear classifier. The clusters of points are not well discriminated by straight lines

First, we try to classify the data by a linear classifier with the code

```
# split data into train and test
X_train, X_test, y_train, y_test = train_test_split(
        X, y, test_size=0.3, random_state=42
    )
# create a linear classifier
clf=SVC(kernel="linear", C=0.05)
# try to fit the new dataset
clf.fit(X_train, y_train)
score = clf.score(X_test, y_test)
# return the score 0.87
```

The score is 0.87, and we check the classified data in Fig. 1.14. The score is the *coefficient of determination* $\mathrm{R}^2 \in [0, 1]$, which is a statistical measure of how well the model replicates the label. $\mathrm{R}^2 = 1$ means that the estimated labels are indistinguishable from the input labels y_j. R^2 is defined by the following equations:

$$\langle y_j \rangle = \frac{1}{N_X} \sum_{j=0}^{N_X-1} y_j \,, \tag{1.48}$$

$$u = \sum_{n=0}^{N_X-1} \left\{ y_j - f\left[\mathbf{X}^{(j)}\right] \right\}^2 \,, \tag{1.49}$$

$$v = \sum_{n=0}^{N_X-1} \left(y_j - \langle y_j \rangle \right)^2 \,, \tag{1.50}$$

$$\mathrm{R}^2 = 1 - \frac{u}{v} \,, \tag{1.51}$$

where $f\left[\mathbf{X}^{(j)}\right]$ is the label estimated by the model and $\langle y_j \rangle$ is the mean value of the labels. Notice that the straight lines cannot separate the two clusters of points.

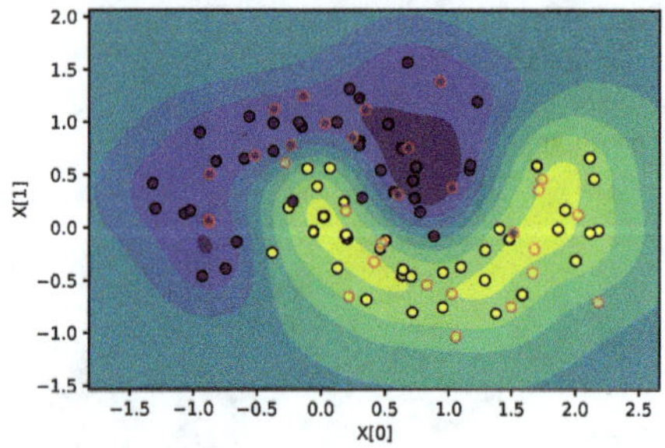

Fig. 1.15 A kernel model with a Gaussian kernel returns a much better performance in classifying the noisy nonlinearly separable dataset. This is the first example of feature mapping

Next, we use a nonlinear classifier, a Gaussian kernel, as in the following code.

```
# create a classifier with a Gaussian kernel
clf=SVC(gamma=4, C=1)
# train the model
clf.fit(X_train, y_train)
# check the score
score = clf.score(X_test, y_test)
# returns 0.93
```

The resulting score is 0.93. Figure 1.15 shows the dataset and the decision boundaries fitting the two clusters of points. The fact that a nonlinear kernel works better is understood by the concept of feature mapping, described in the following.

1.13 Feature Mapping

The representation potential of linear support vector machines is limited to datasets that are linearly classifiable. For data that are not linearly separable as in Fig. 1.15, one strategy is using a mapping from the original data space to another where they are linearly separable.

We sketch the idea in Fig. 1.16, which shows that data nonlinearly separable in a two-dimensional space may be separated by a smart transformation Φ' that keeps the number of dimensions unchanged, or by Φ'' that maps the points into a three-dimensional space. In the former case, Φ' may be demanding to design. Increasing the number of dimensions is the typical approach and can employ linear (as a rotation), or nonlinear, maps.

For example, for two labels $y = -1$ and $y = 1$, and 2D observations with coordinates X^0 and X^1, the map may leave unchanged the coordinates and add a dimension $X^3 = y$, i.e., assign the label as a further coordinate. As a result, the dataset is separated by the plane $X^3 = 0$.

The feature mapping is represented by

$$\mathbf{X}' = \Phi(\mathbf{X}) \,, \tag{1.52}$$

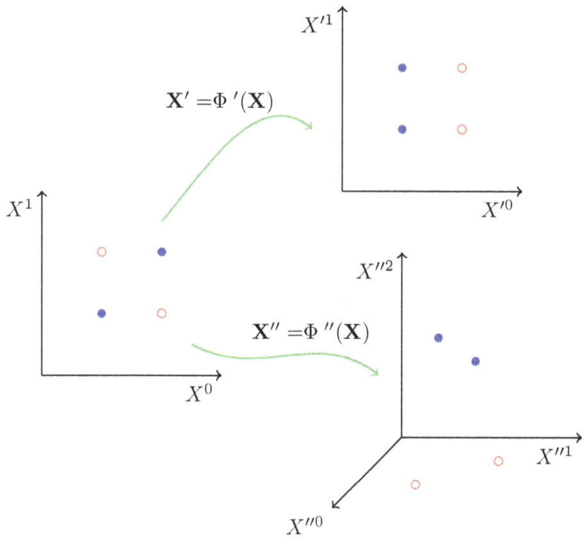

Fig. 1.16 The idea of feature mapping is the existence of a space in which nonlinearly separable data (left panel) map to linearly separable ones (right panels). One can image mappings from two-dimensional vectors $\mathbf{X}$ to two-dimensional vectors $\Phi'(\mathbf{X}) = \left[\Phi'^0(\mathbf{X}), \Phi'^1(\mathbf{X})\right]$ as in the top panel, or more complicated mappings that change the data space. A typical approach is increasing the number of dimensions as in $\Phi''(\mathbf{X}) = \left[\Phi''^0(\mathbf{X}), \Phi''^1(\mathbf{X}), \Phi''^2(\mathbf{X})\right]$ (bottom panel). One can also consider mapping to infinite-dimensional spaces (not represented in the figure). The existence of the map is guaranteed by the universality of kernel models. One has only to specify the kernel function without knowledge of the mapping (this is the so-called kernel trick)

with $\mathbf{X}'$ the image of $\mathbf{X}$ in the new space. In the feature space (see Fig. 1.16), we repeat the arguments of the linear support vector machine. The decision rule reads

$$\mathbf{w} \cdot \Phi(\mathbf{X}) + b \geq 0\,, \tag{1.53}$$

where we assume that $\Phi(\mathbf{X})$ is in a Euclidean space with a conventional scalar product. We may also consider more sophisticated maps where $\Phi(\mathbf{X})$ belongs to an infinite-dimensional Hilbert space equipped with a scalar product. In this case, the similarity measure reads

$$K\left[\mathbf{X}^{(i)}, \mathbf{X}^{(j)}\right] = \langle \Phi(\mathbf{X}^{(i)}), \Phi(\mathbf{X}^{(j)}) \rangle\,, \tag{1.54}$$

where the brackets with the comma $\langle\,,\,\rangle$ indicate a scalar product between the two images in the feature space $\Phi\left[\mathbf{X}^{(i)}\right]$ and $\Phi\left[\mathbf{X}^{(j)}\right]$.

It turns out that this approach is very effective in classifying complex datasets, and it can be adopted in quantum systems.

1.14 The Drawbacks in Kernel Methods

Kernel methods can be described as those methods that use a linear superposition as

$$f(\mathbf{X}) = \sum_{j=0}^{N_X-1} \alpha_j K\left[\mathbf{X}, \mathbf{X}^{(j)}\right] \tag{1.55}$$

as an ansatz for the solution of a supervised learning problem.

In general, the αs depend on the labels. The other terms of the expansion in (1.55) only depend on the observation $\mathbf{X}_j$. Once we have fixed the kernel function, we can compute the weights α_j by inverting the Gram matrix or by a pseudo-inverse (the approach typically used). We need a number of evaluations of the kernel of the order of N_X^2 for the Gram matrix. Also, the computational effort to invert the Gram matrix is of the order of N_X^3. This effort is needed to compute the unknown N_X weights α_j.

On the contrary, deep learning requires an effort to determine the weights of the order of N_X^2, but the solution is typically suboptimal [29].

Hence, kernel methods have a solid mathematical background that guarantees optimal convergence, but deep learning enables much more convenient heuristics, which is significant when the number of weights is large. Indeed, nowadays, we work with networks with billions of tokens.

Loosely speaking, kernel methods are not the most efficient approach in modern learning techniques. However, they are well suited to match the math of quantum mechanics. Also, one can argue if, by using quantum systems, we can accelerate computing the kernel function or its inverse needed for the weights. Are quantum systems advantageous to implement learning with kernels?

The potential existence of an advantage was established in 2014 [30]. The gain arises from the acceleration of linear algebra, which – however – is unfeasible because of the limited number of qubits today available in quantum processors.

Another source of potential advantage is using physical systems for computing kernel functions that are complex and demanding for a conventional computer. This is the case of feature mapping using squeezed states or entangled qubit maps that we will consider in Chap. 2. Also, one can use the parallelism of a quantum system to speed up the computation of the kernel matrix terms.

Lastly, the Hilbert space of quantum mechanical systems is gigantic. If one exploits entanglement at a large scale to cover the entire Hilbert space, one can use a quantum system as a physical accelerator to implement feature mapping at high dimensions. We will explore quantum feature mapping in the following chapters.

1.15 Further Reading

- An advanced book on kernel machines is [25].
- A pioneering book on quantum supervised learning is by Schuld and Petruccione [8].

- A book on the theoretical fundaments of machine learning and related fields is [31].
- A study of the scaling of large deep learning models is [29].

References

1. P.A.M. Dirac, *The Principles of Quantum Mechanics*, 4th edn. (Clarendon Press, Oxford, 1982)
2. R. Feynman, *QED* (Princeton University Press, Princeton, 1988)
3. D. Hanneke, S. Fogwell, G. Gabrielse, Phys Rev Lett. **100**(12) (2008). https://doi.org/10.1103/physrevlett.100.120801
4. T. Aoyama, T. Kinoshita, M. Nio, Phys. Rev. D **97**(3) (2018). https://doi.org/10.1103/physrevd.97.036001.
5. S.L. Brunton, J.N. Kutz, *Data-Driven Science and Engineering: Machine Learning, Dynamical Systems, and Control*, 2nd edn. (Cambridge University Press, Cambridge, 2022). https://doi.org/10.1017/9781009089517
6. A. Vaswani, N. Shazeer, N. Parmar, J. Uszkoreit, L. Jones, A.N. Gomez, L. Kaiser, I. Polosukhin, Attention is all you need (2017). https://doi.org/10.48550/ARXIV.1706.03762
7. R. Loudon, *The Quantum Theory of Light*, 2nd edn. (Clarendon Press, Oxford, 1983)
8. M. Schuld, F. Petruccione, *Supervised Learning with Quantum Computers* (Springer, Berlin, 2018)
9. C. Gardiner, P. Zoller, *The Quantum World of Ultra-Cold Atoms and Light Book II: The Physics of Quantum-Optical Devices* (Imperial College Press, London, 2015). https://doi.org/10.1142/p983
10. V. Havlíček, A.D. Córcoles, K. Temme, A.W. Harrow, A. Kandala, J.M. Chow, J.M. Gambetta, **567**, 209 (2019). https://doi.org/10.1038/s41586-019-0980-2
11. C.S. Hamilton, R. Kruse, L. Sansoni, S. Barkhofen, C. Silberhorn, I. Jex, Phys. Rev. Lett. **119**, 170501 (2017). https://doi.org/10.1103/PhysRevLett.119.170501
12. R. Kruse, C.S. Hamilton, L. Sansoni, S. Barkhofen, C. Silberhorn, I. Jex, Phys. Rev. A **100**, 032326 (2019)
13. Y. Li, M. Chen, Y. Chen, H. Lu, L. Gan, C. Lu, J. Pan, H. Fu, G. Yang (2020)
14. Y. Wu, W.S. Bao, S. Cao, F. Chen, M.C. Chen, X. Chen, T.H. Chung, H. Deng, Y. Du, D. Fan, M. Gong, C. Guo, C. Guo, S. Guo, L. Han, L. Hong, H.L. Huang, Y.H. Huo, L. Li, N. Li, S. Li, Y. Li, F. Liang, C. Lin, J. Lin, H. Qian, D. Qiao, H. Rong, H. Su, L. Sun, L. Wang, S. Wang, D. Wu, Y. Xu, K. Yan, W. Yang, Y. Yang, Y. Ye, J. Yin, C. Ying, J. Yu, C. Zha, C. Zhang, H. Zhang, K. Zhang, Y. Zhang, H. Zhao, Y. Zhao, L. Zhou, Q. Zhu, C.Y. Lu, C.Z. Peng, X. Zhu, J.W. Pan, Phys. Rev. Lett. **127**(18), 180501 (2021). https://doi.org/10.1103/physrevlett.127.180501
15. S. Boixo, S.V. Isakov, V.N. Smelyanskiy, R. Babbush, N. Ding, Z. Jiang, M.J. Bremner, J.M. Martinis, H. Neven, Nat. Phys. **14**(6), 595 (2018). https://doi.org/10.1038/s41567-018-0124-x
16. J.M. Arrazola, V. Bergholm, K. Brádler, T.R. Bromley, M.J. Collins, I. Dhand, A. Fumagalli, T. Gerrits, A. Goussev, L.G. Helt, J. Hundal, T. Isacsson, R.B. Israel, J. Izaac, S. Jahangiri, R. Janik, N. Killoran, S.P. Kumar, J. Lavoie, A.E. Lita, D.H. Mahler, M. Menotti, B. Morrison, S.W. Nam, L. Neuhaus, H.Y. Qi, N. Quesada, A. Repingon, K.K. Sabapathy, M. Schuld, D. Su, J. Swinarton, A. Száva, K. Tan, P. Tan, V.D. Vaidya, Z. Vernon, Z. Zabaneh, Y. Zhang, Nature **591**(7848), 54 (2021). https://doi.org/10.1038/s41586-021-03202-1
17. F. Hoch, S. Piacentini, T. Giordani, Z.N. Tian, M. Iuliano, C. Esposito, A. Camillini, G. Carvacho, F. Ceccarelli, N. Spagnolo, A. Crespi, F. Sciarrino, R. Osellame (2021). arXiv:2106.08260
18. J. Jašek, K. Jiráková, K. Bartkiewicz, A. Černoch, T. Fürst, K. Lemr, Opt. Express **27**(22), 32454 (2019)
19. C. Conti, Quantum Mach. Intell. **3**(2), 26 (2021). https://doi.org/10.1007/s42484-021-00052-y

20. S. Weinberg, Phys. Rev. A **90**, 042102 (2014). https://doi.org/10.1103/PhysRevA.90.042102
21. C. Ciliberto, A. Rocchetto, A. Rudi, L. Wossnig, Phys. Rev. A **102**, 042414 (2020). https://doi.org/10.1103/PhysRevA.102.042414
22. S. Weinberg, *Lectures on Quantum Mechanics* (Cambridge University Press, Cambridge, 2013)
23. W.K. Wootters, W.H. Zurek, Nature **299**(5886), 802 (1982). https://doi.org/10.1038/299802a0
24. D. Dieks, Phys. Lett. A **92**(6), 271 (1982). https://doi.org/10.1016/0375-9601(82)90084-6
25. B. Schölkopf, A. Smola, *Learning with Kernels: Support Vector Machines, Regularization, Optimization, and Beyond* (The MIT Press, Cambridge, 2018)
26. J. Mercer, Philos. Trans. R. Soc. Lond. Series A **209**, 415 (1909). https://doi.org/10.1098/rsta.1909.0016
27. F. Pedregosa, G. Varoquaux, A. Gramfort, V. Michel, B. Thirion, O. Grisel, M. Blondel, P. Prettenhofer, R. Weiss, V. Dubourg, J. Vanderplas, A. Passos, D. Cournapeau, M. Brucher, M. Perrot, E. Duchesnay, J. Mach. Learn. Res. **12**, 2825 (2011)
28. L. Buitinck, G. Louppe, M. Blondel, F. Pedregosa, A. Mueller, O. Grisel, V. Niculae, P. Prettenhofer, A. Gramfort, J. Grobler, R. Layton, J. VanderPlas, A. Joly, B. Holt, G. Varoquaux, in *ECML PKDD Workshop: Languages for Data Mining and Machine Learning* (2013), pp. 108–122
29. N.C. Thompson, K. Greenewald, K. Lee, G.F. Manso. The computational limits of deep learning (2022). https://arxiv.org/abs/2007.05558
30. P. Rebentrost, M. Mohseni, S. Lloyd, Phys. Rev. Lett. **113**, 130503 (2014). https://doi.org/10.1103/PhysRevLett.113.130503
31. S. Shalev-Shwartz, S. Ben-David, *Understanding Machine Learning: From Theory to Algorithms* (Cambridge University Press, Cambridge, 2014)

Chapter 2
Kernelizing Quantum Mechanics

We live in a classical world.

Abstract We show how to use quantum states, as coherent and squeezed states for a kernel machine.

2.1 Introduction

In this chapter, we show the way quantum states and controllable quantum circuits realize quantum classifiers. The simplest and "nearly classical" states are coherent states, which can implement Gaussian kernel machines. When we depart these states and enter the quantum world deeply, we realize unconventional kernel machines, for example, when using squeezed states. However, we experience an augmented computational effort when simulating quantum machines with classical computers. On the other hand, open issues exist in the quantum circuits that compute the kernel as scalar products. In the following, we discuss in detail the links between kernel methods and quantum circuits with examples.

2.2 The Computational Model by Quantum Feature Maps and Quantum Kernels

The first step to use a quantum computer is transferring our classical input to the quantum computer, which works on quantum data (see Fig. 1.2).

In the context of quantum machine learning, the transfer is indicated as a *quantum feature map* [1]. One takes a classical datum as a vector $\mathbf{X}$ and obtains a quantum register $|\phi(\mathbf{X})\rangle$. In a quantum computer, this encoding may be realized by acting on the parameters of quantum gates with auxiliary qubits ("ancillas") as inputs. For example, a qubit in vacuum state $|0\rangle$ is rotated by angle θ around an axis $\hat{n}$, such that

C. Conti, *Quantum Machine Learning*, Quantum Science and Technology,
https://doi.org/10.1007/978-3-031-44226-1_2

the feature map is the rotation operator and encodes a single scalar value $X^0 = \theta$. The classical datum is the rotation angle.

The task of the quantum processor determines the choice of the feature map. For quantum machine learning, we need to find a feature map that enables to represent and train arbitrary functions. We also need to extract a classical output from the evolution of the quantum device.

A simple approach to extract a classical output is measuring the probability of finding the system in a given *weight state* $|w\rangle$, that is, measuring $|\langle w|\phi(x)\rangle|^2$. Choosing the vector $|w\rangle$ determines the class of mathematical functions the QML device implements.

Having in mind our discussion of the quantum kernel machine in the previous chapter, we can think to map each datum $\mathbf{X}^{(i)}$ to a state $|\psi(\mathbf{X}^{(i)})\rangle$ and assign the corresponding label y_i when we observe such a state. This means that our observable is

$$\hat{O} = \sum_{i=0}^{N_X-1} y_i |\psi(\mathbf{X}^{(i)})\rangle\langle\psi(\mathbf{X}^{(i)})| \,. \tag{2.1}$$

In general, the states $|\psi(\mathbf{X}^{(i)})\rangle$, with $i = 0, 1, \ldots, N_X - 1$, are not orthogonal. Even if the system is in a state $|\psi(\mathbf{X}^{(j)})\rangle$ with $j \neq i$, we have a contribution from all the label states to the output. Thus, we introduce arbitrary weights α_j in (2.1) to improve the accuracy:

$$\hat{O} = \sum_{i=0}^{N_X-1} \alpha_i y_i |\psi(\mathbf{X}^{(i)})\rangle\langle\psi(\mathbf{X}^{(i)})| \,. \tag{2.2}$$

The model for our target function f reads

$$f(\mathbf{X}) = \langle\psi(\mathbf{X})|\hat{O}|\psi(\mathbf{X})\rangle = \sum_{i=0}^{N_X-1} \alpha_i y_i |\langle\phi(\mathbf{X}^{(i)})|\phi(\mathbf{X})\rangle|^2 \tag{2.3}$$

letting

$$K(\mathbf{X}, \mathbf{Y}) = |\langle\phi(\mathbf{X})|\phi(\mathbf{Y})\rangle|^2 \,. \tag{2.4}$$

To fit a target function $f_T(\mathbf{X})$ known in N_X points, we consider the linear system

$$f(\mathbf{X}^{(j)}) = \sum_{i=0}^{N_{X-1}} \theta_i |\langle\phi(\mathbf{X}^{(i)})|\phi(\mathbf{X}^{(j)})\rangle|^2 = \sum_{i=0}^{N_X-1} G_{ij}\theta_i \tag{2.5}$$

where $\theta_j = \alpha_j y_j$ and the elements

$$G_{ij} = |\langle \phi(\mathbf{X}^{(i)}) | \phi(\mathbf{X}^{(j)}) \rangle|^2 \tag{2.6}$$

form the Gram matrix. Equation (2.5) can be solved by direct inversion, or by linear regression, either on a classical or a quantum computer.

With reference to Eq. (2.5), for the density matrix representation, we have that the data $\mathbf{X}^{(i)}$ and $\mathbf{X}^{(j)}$ map to $\rho[\mathbf{X}^{(i)}]$ and $\rho[\mathbf{X}^{(j)}]$ and the scalar product is

$$K[\mathbf{X}^{(i)}, \mathbf{X}^{(j)}] = \langle \rho[\mathbf{X}^{(i)}], \rho[\mathbf{X}^{(j)}] \rangle = \text{Tr}\left\{ \rho[\mathbf{X}^{(i)}] \rho[\mathbf{X}^{(j)}] \right\} . \tag{2.7}$$

In Eq. (2.7) we used the scalar product defined in the density matrix feature space, which is the trace of the product. Thus Eq. (2.5) is equivalent to

$$f(\mathbf{X}^{(j)}) = \sum_{i=0}^{N_X - 1} \theta_i \, \text{Tr}\left\{ \rho[\mathbf{X}^{(i)}] \rho[\mathbf{X}^{(j)}] \right\} . \tag{2.8}$$

Equation (2.8) shows the advantage in working in the density matrix feature space. The corresponding representation contains directly measurable real-valued scalar products, and no modulus square appears.

The emerging computational model is the following:

- Map all the points in the datasets in quantum states.
- Compute the scalar products by using the quantum device.
- Invert the Gram matrix to determine the weights.

This model is advantageous if the computation of the scalar products is fast on the quantum device. Also, the inversion of the Gram matrix can be either done classically or using a quantum computer. In the latter case, if there is not overhead in mapping the Gram matrix to a quantum state, one gains a further advantage.

We detail in the next sections examples of feature mapping. The resulting method is the quantum version of the classical kernel [2].

2.3 Quantum Feature Map by Coherent States

Assuming one has a vector $\mathbf{X}$ as input, the simplest mapping is encoding each single component X_p to the amplitude of a coherent state.

Consider, for example, the case $D = 1$ with a scalar real input X_0; we build the coherent state

$$|\alpha_0\rangle = |\alpha = X_0\rangle = \hat{\mathcal{D}}(X_0)|0\rangle . \tag{2.9}$$

In this feature mapping, the scalar product between two different values $\alpha_0 = X_0$ and $\alpha_0' = X_0'$, is

$$\langle \alpha_0 | \alpha_0' \rangle = e^{-\frac{1}{2}|\alpha_0 - \alpha_0'|^2} . \tag{2.10}$$

The kernel function is the squared modulus of (2.10), that is,

$$K(X_0, X_0') = |\langle \alpha_0 | \alpha_0' \rangle|^2 = e^{-|X_0 - X_0'|^2} . \tag{2.11}$$

For D modes, we build the product state

$$|\boldsymbol{\alpha}\rangle = |\alpha_0\rangle \otimes |\alpha_1 \otimes\rangle \ldots \otimes |\alpha_{D-1}\rangle = |\alpha_0, \alpha_1, \ldots, \alpha_{D-1}\rangle \tag{2.12}$$

with $\alpha_j = X_j$, and $j = 0, 1, \ldots, D-1$.

By using the Gaussian kernels, we represent any real-valued function $f(\mathbf{X})$ by a superposition of Gaussian functions:

$$f(\mathbf{X}) = \sum_{i=0}^{N_X - 1} \theta_i e^{-|\mathbf{X} - \mathbf{X}^{(i)}|^2} , \tag{2.13}$$

where $\mathbf{X}^{(i)}$ are the N_X fitting point and θ_i the fitting parameters (weights).

2.4 Quantum Classifier by Coherent States

A specific application of quantum feature maps is quantum classifiers. A classifier is a device that discriminates ensembles of points generated by different rules, which is a typical application of support vector machines. Here we are revisiting the idea in a quantum context to make a specific example showing that the two computing models coincide when using coherent states and Gaussian kernels.

In support vector machines, one wants to determine the decision boundary between two datasets. The following code generates the *moons dataset*, i.e., two interleaved and independent sets of points forming semicircles in a two-dimensional space. The dataset includes a function $f_T(X_0, X_1)$ that returns 1 or 0 if the input belongs to one or the other of the two half-moons (see Fig. 1.2). The code below generates and plots the dataset by the `make_moons` function of the `Python` `scikit-learn` package (3-Clause BSD licence).[1]

```
figure = plt.figure(figsize=(18, 9))
cm = plt.cm.RdBu
cm_bright = ListedColormap(["#FF0000", "#0000FF"])
```

[1] scikit-learn.org.

```
dsplot = make_moons(noise=0.0, random_state=0)
# X, y tensors with the dataset
X, y = dsplot
ax = plt.subplot(1,2,1)
# Plot the testing points
ax.scatter(
        X[:, 0], X[:, 1], c=y, cmap=cm_bright, alpha=0.6,
        ↪ edgecolors="k"
    )
ax1 = plt.subplot(1,2,2)
dsplot = make_moons(noise=0.1, random_state=0)
X, y = dsplot
# Plot the testing points
ax1.scatter(
        X[:, 0], X[:, 1], c=y, cmap=cm_bright, alpha=0.6,
        ↪ edgecolors="k"
    )
x_min, x_max = X[:, 0].min() - 0.5, X[:, 0].max() + 0.5
y_min, y_max = X[:, 1].min() - 0.5, X[:, 1].max() + 0.5
xx, yy = np.meshgrid(np.arange(-2, 2),
                         np.arange(2, 2, h));
ax1.set_xlim(-1.5, 2.5)
ax1.set_ylim(-1.5, 2.5)
ax.set_xlim(-1.5, 2.5)
ax.set_ylim(-1.5, 2.5)
ax.set_xlabel('$x_0$')
ax.set_ylabel('$x_1$')
ax1.set_ylabel('$x_1$')
ax1.set_xlabel('$x_0$')
```

We generate two datasets with different amounts of noise. Figure 2.1 shows the figure generated by the previous code with two noises.[2] In the context of classical support vector machines, using a kernel like in Eq. (2.13) is a well-known strategy with the name of *Gaussian radial basis functions*, and many packages exist to perform the classification (Fig. 2.1).

Thus, feature mapping with coherent states and support vector machines with Gaussian radial basis functions (RBFs) are precisely the same computational model from a mathematical perspective. The considered classifier is also denoted as *support vector classifier* (SVC).

T make a classification, we follow the `Python` package `scikit-learn` by using the code below. We do not need to code the kernel function as it is included in `scikit-learn`. In `scikit-learn` the adopted RBF is $K(\mathbf{X}, \mathbf{Y}) = e^{-\gamma|\mathbf{X}-\mathbf{Y}|^2}$. For the scalar product in Eq. (2.11), we have to set `gamma=1.0` ($\gamma = 1.0$).

[2] The code is in `jupyternotebooks/quantumfeaturemap/coherentstate.ipynb`.

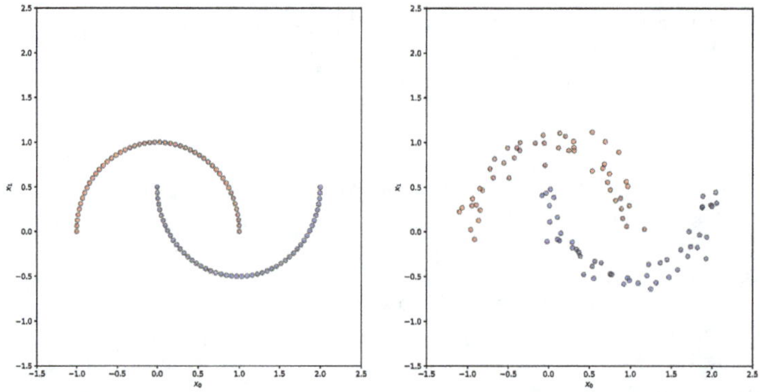

Fig. 2.1 Examples of the moons dataset (left panel) for a classification problem. We report the dataset including noise (right panel)

Training the classifier means determining the values of the parameters $\boldsymbol{\theta}$, as typically done by a linear regression that is present in the `fit` function, as in the following code.

```
# generate the dataset
X, y = make_moons(noise=0.2)
# preprocess dataset,remove mean and scale to unit variance
X = StandardScaler().fit_transform(X)
#  split into training and test part
X_train, X_test, y_train, y_test = train_test_split(
        X, y, test_size=0.4, random_state=42)
# define the classifier, support vector machine classifier with
↪  RBF
clf=SVC(gamma=1)
# fit the training set
clf.fit(X_train, y_train)
# compute the score on the test data
score=clf.score(X_test,Y_test)
```

Figure 2.2 shows the result for the classifications for two different datasets: the moons dataset and a dataset with data distributed in concentric circles obtained by `get_circles`. The noise is added to both the datasets.[3]

[3] The code for these examples is in `jupyternotebooks/quantumfeaturemap/coherentstate.ipynb`.

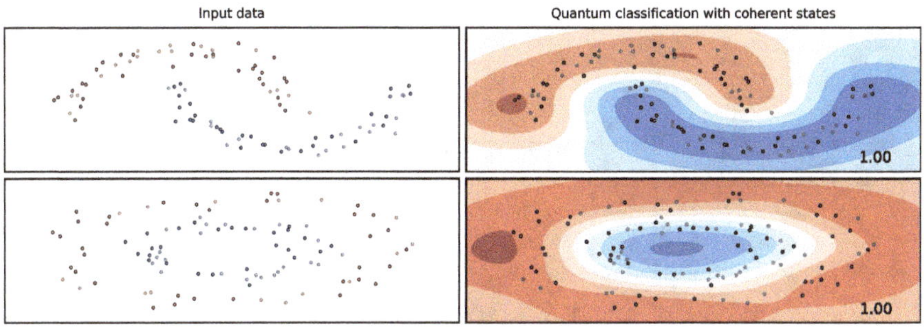

Fig. 2.2 Classification of two datasets (left panels) by using the quantum kernel method with coherent states. The score of the classification is indicated in the right panels. The classification boundary is the boundary in the colorized regions

2.5 How to Measure Scalar Products with Coherent States

Having stated the equivalence of the support vector machines and quantum kernel methods, one can argue about the advantage of using a quantum protocol. In kernel methods, one needs the Gram matrix, which takes $O(N^2)$ operations. Hence, one significant aspect is speeding up the computation of the scalar products, which are the elements of the Gram matrix.

For the Gaussian kernels, one has efficient approaches in conventional computers, because numerically evaluating a Gaussian function is not a demanding task. We do not expect to have a substantial advantage in adopting a quantum computer as far as dealing with very large scales (at which quantum computers are not available at the moment), and we miss an efficient way to measure the scalar product.

For coherent states, a simple strategy for the scalar product

$$|\langle\alpha|\beta\rangle|^2 = e^{-|\alpha-\beta|^2} \tag{2.14}$$

is observing that a coherent state in terms of the number states $|n\rangle$ reads

$$|\alpha\rangle = \mathcal{D}(\alpha)|0\rangle = e^{-\frac{1}{2}|\alpha|^2}\sum_n \frac{\alpha^n}{\sqrt{n!}}|n\rangle\,. \tag{2.15}$$

The probability of zero particles is hence

$$\Pr(n=0) = \text{Tr}(\rho|0\rangle\langle 0|) = |\langle 0|\alpha\rangle|^2 = e^{-|\alpha|^2}\,. \tag{2.16}$$

Equation (2.16) tells us that the zero-particle event corresponds to computing Gaussian functions. This is the first step we need for Eq. (2.14). To retrieve a scalar product, we mix two Gaussian states by a beam splitter as in Fig. 2.3. This is the simplest approach in quantum optics, where laser beams readily provide coherent

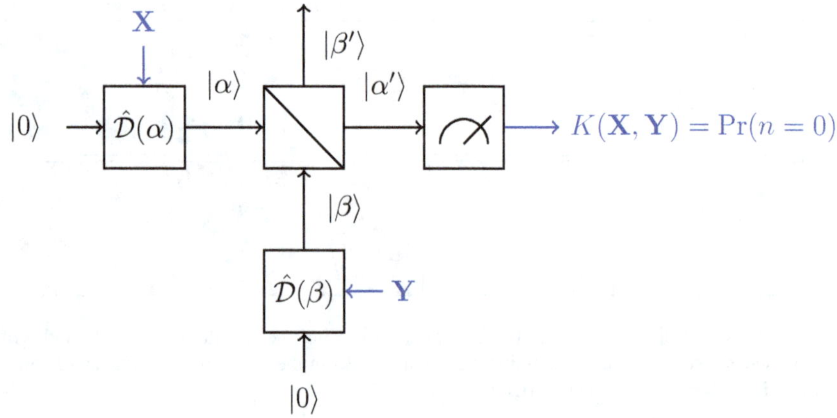

Fig. 2.3 Beam splitting and photon counting to compute scalar products and the quantum kernel with coherent states as feature maps. The classical inputs are **X** and **Y**, and $K(\mathbf{X}, \mathbf{Y})$ is the classical output

states and are combined by beam splitters. Two different coherent states $|\alpha\rangle$ and $|\beta\rangle$ are mixed by a $50 : 50$ beam splitter, such that the new states emerging are still coherent states $|\alpha'\rangle$ and $|\beta'\rangle$ with

$$\alpha' = \frac{1}{\sqrt{2}}(\alpha - \beta) ,$$

$$\beta' = \frac{1}{\sqrt{2}}(\alpha + \beta) .$$

If we measure $\Pr(n = 0)$ on $|\alpha'\rangle$, we get

$$\Pr(n = 0) = e^{-|\alpha'|^2} = e^{-\frac{1}{2}|\alpha-\beta|^2} , \tag{2.17}$$

which is the scalar product in Eq. (2.10). If we want to get rid of the factor $1/2$ and have the scalar product in Eq. (2.11), we have to use the states $|\sqrt{2}\alpha\rangle$ and $|\sqrt{2}\beta\rangle$ at the entrance of the beam splitter, which is a simple scaling in the dataset.

Summarizing, photon counting furnishes the scalar product and enables the use of a quantum circuit made by a beam splitter and coherent states to perform quantum machine learning. To compute the Gram matrix, we need to loop over all the input vectors in the dataset.

If we have a large register with D coherent states in a product state, we need either D distinct beam splitters or to combine the input coherent states in a single beam splitter. As we are dealing with a product state with $\boldsymbol{\alpha} = (\alpha_0, \alpha_1, \ldots \alpha_{D-1})$, the probability is the product of probabilities. The photon counting furnishes the vectorial scalar product

$$\Pr(n=0) = e^{-|\alpha'|^2} = e^{-\frac{1}{2}|\boldsymbol{\alpha}-\boldsymbol{\beta}|^2} . \tag{2.18}$$

We can doubt that using quantum hardware with product states provides an effective advantage. Computing Gaussian functions is not a demanding task for a classical computer. Other quantum states provide more elaborated feature maps, as in the following example.

2.6 Quantum Feature Map by Squeezed Vacuum

We consider a single-mode squeezed vacuum [3]

$$|\zeta\rangle = \mathcal{S}(\zeta)|0\rangle = \frac{1}{\sqrt{\cosh(r)}} \sum_{n=0}^{\infty} \frac{\sqrt{(2n)!}}{2^n\, n!} \left[-e^{\mathrm{i}\varphi} \tanh(r)\right]^n |2\,n\rangle , \tag{2.19}$$

where $z = r\,e^{\mathrm{i}\varphi}$ is the complex squeezing parameters. The squeezed vacuum depends on the two parameters r and φ. For the sake of definiteness, we use the phase φ and fix the amplitude to a constant value $r = c$ (one can equivalently use the amplitudes to encode the data). Assuming that we have a single real value X_0 scaled in the range $[0, 2\pi)$ in our dataset with $D = 1$, we can simply state

$$\varphi = x_0 . \tag{2.20}$$

We will retain c as a hyperparameter. The scalar product of the squeezed states is [4]

$$\langle c, X|c, Y\rangle = \frac{\operatorname{sech}(c)}{\left[1 - e^{\mathrm{i}(X-Y)} \tanh(c)^2\right]^{1/2}} . \tag{2.21}$$

As in the case of the coherent states, the squared modulus gives the kernel function (see Fig. 2.4)

$$K(X, Y) = |\langle c, X|c, Y\rangle|^2 . \tag{2.22}$$

We generalize to the case with D squeezed state and use as a quantum feature map the product state (see Fig. 2.4)

$$|\boldsymbol{\varphi}\rangle = |c, \varphi_0\rangle \otimes |c, \varphi_1\rangle \otimes \ldots \otimes |c, \varphi_{D-1}\rangle , \tag{2.23}$$

where $\boldsymbol{\varphi} = \mathbf{X} = (X_0, X_1, \ldots, X_{D-1})$ is the input datum from the dataset. As we use a product state for the feature mapping, the kernel function is the product of the kernel functions for the single modes

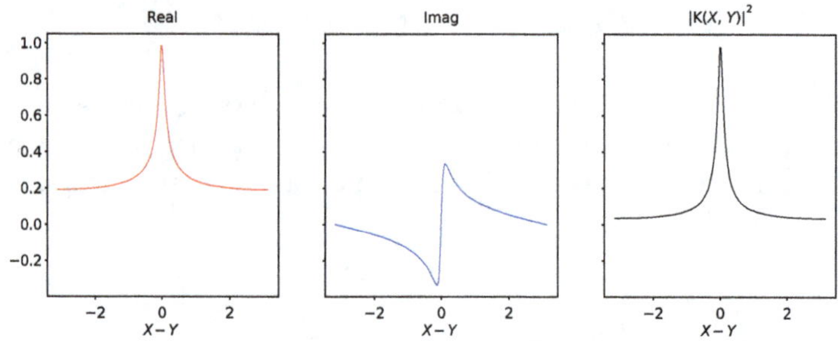

Fig. 2.4 Kernel function for one-mode register with squeezed states. We show the real part, the imaginary part, and the squared modulus for $c = 2$

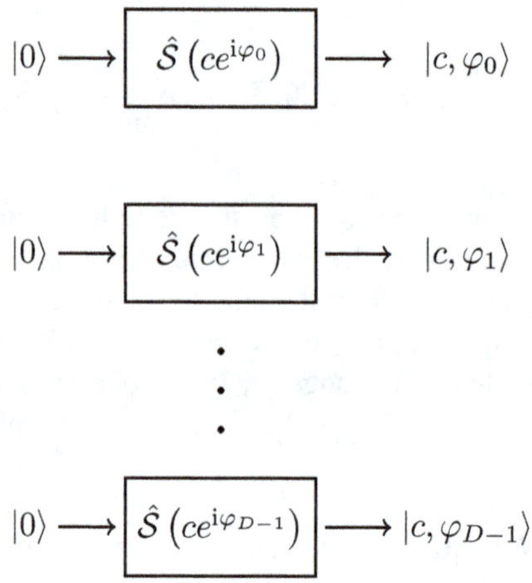

Fig. 2.5 Representation of a quantum register for feature mapping the input vector $\mathbf{X}$ into the phases of D squeezing operators. We take equal amplitudes with $r = c$

$$K(\mathbf{X}, \mathbf{Y}) = \prod_{j=0}^{D-1} K\left(X_j, Y_j\right) \, . \tag{2.24}$$

We consider a specific example with $D = 2$ (Fig. 2.5).

2.7 A Quantum Classifier by Squeezed Vacuum

We repeat the example in Sec. 2.4 using squeezed vacuum states. For a dataset with $D = 2$ as in Fig. 2.1, we consider the following quantum feature map

$$|\phi(\boldsymbol{x})\rangle = |c, X_0\rangle \otimes |c, X_1\rangle \tag{2.25}$$

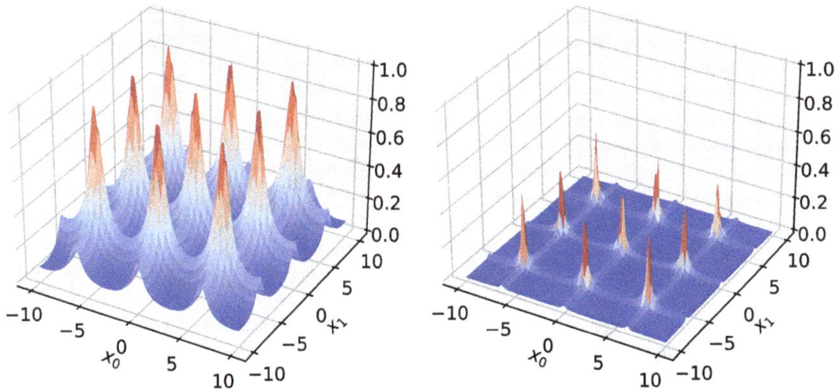

Fig. 2.6 Kernel function for the squeezed vacuum feature mapping for $c = 1$ (left panel) and $c = 2$ (right panel). We plot the kernel as a function of X_0 and X_1 when $\mathbf{Y} = 0$

with $D = 2$ coherent states. We use as kernel the squared modulus

$$K(\mathbf{X}, \mathbf{Y}) = |\langle c, X_0|c, Y_0\rangle|^2 |\langle c, X_1|c, Y_1\rangle|^2 . \tag{2.26}$$

Figure 2.6 shows the kernel function for different values of c.[4] Note that the map is periodical, as we use the phase φ to encode the data. Thus, we have to scale in the input dataset in the range $[0, 2\pi)$. As for coherent state feature mapping, we adopt the `Python scikit-learn` package for support vector machines. However, the kernel function for the squeezed vacuum does not correspond to conventional kernels. We define a custom kernel for the classification that computes the squared modulus of Eqs. (2.21) and (2.26) for $D = 2$. This is done in the following code.

```
def squeezing_kernel(X, Y):
    """ squeezing kernel for D=2 """
    nb1, nx=X.shape
    nb2, ny=Y.shape
    out = np.zeros((nb1,nb2))
    s0 = np.cosh(csqueeze)*np.cosh(csqueeze)
    s1 = np.cosh(csqueeze)*np.cosh(csqueeze)
    for ib1 in range(nb1):
        for ib2 in range(nb2):
            d0 = 1-np.exp(1j*(X[ib1,0]-Y[ib2,0])) \
                 *np.tanh(csqueeze)*np.tanh(csqueeze)
            d1 = 1-np.exp(1j*(X[ib1,1]-Y[ib2,1])) \
                 *np.tanh(csqueeze)*np.tanh(csqueeze)
            out[ib1,ib2]=np.abs(np.sqrt(1.0/(s0*d0*s1*d1)))**2
    return out
```

[4] The code is in `jupyternotebooks/quantumfeaturemap/squeezedvacuum.ipynb`.

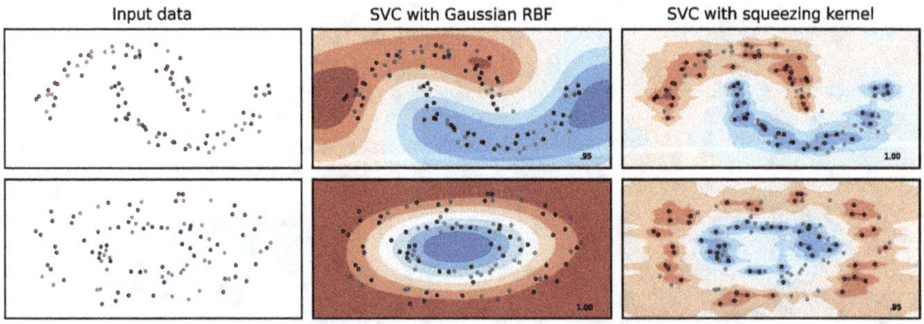

Fig. 2.7 Comparison of quantum classifier performances. In this example, the coherent states work better than the squeezed vacuum for the moons dataset. The opposite holds for the concentric circles. The mapping is universal and reaches a score equal to one (reported in the panel); the data are separable in the feature space

---

We recall that for a coherent state, the classifier is

```
coherent_state_classifier= SVC(gamma=1.0)
```

For the squeezed vacuum feature map, we specify the custom kernel function as

---

```
squeezed_state_classifier= SVC(kernel=squeezing_kernel)
```

We perform the classification as in Sec. 2.4 and compare the coherent state and the squeezed state quantum classifiers in Fig. 2.7. The two quantum feature maps perform in a comparable way. In applications one has to choose the best classifier. The choice depends on the specific dataset and the complexity of the quantum device. However, we need quantum hardware for the scalar products of squeezed states, which we choose as feature mapping.

2.8 Measuring Scalar Products with General States: The SWAP Test

At variance with coherent state, measuring the scalar product between squeezed states cannot be done by simple photon counting efficiently. A general approach to compute the scalar product of two states $|\psi\rangle$ and $|\varphi\rangle$ is using a large Hilbert space formed by a tripartite system including the two states and an ancilla qubit.

The latter is an additional qubit used in a projection measurement. This approach is the so-called SWAP test [1].

We start considering a single qubit (Fig. 2.8). Assuming the qubit in the ground state $|0\rangle$, the Hadamard gate acts as in Fig. 2.9 to obtain a superposition of $|0\rangle$ and $|1\rangle$, as follows:

$$\hat{\mathrm{H}}|0\rangle = \frac{1}{\sqrt{2}}\hat{\mathrm{H}}|0\rangle + \frac{1}{\sqrt{2}}\hat{\mathrm{H}}|1\rangle \ . \tag{2.27}$$

Then we consider a SWAP gate that acts on the two states $|\psi\rangle$ and $|\varphi\rangle$ as in Fig. 2.10, i.e.,

$$\widehat{\mathrm{SWAP}}|\psi\rangle|\phi\rangle = |\phi\rangle|\psi\rangle \ ; \tag{2.28}$$

and the Fredkin gate [5], which is a controlled swap such that it swaps the states only when the control qubit $|c\rangle$ is $|1\rangle$, as in Fig. 2.11.

Committing all these gates together, we have the SWAP test in Fig. 2.12, summarized as follows:

1. We consider the initial product state with the ancilla qubit

$$|0\rangle|\psi\rangle|\varphi\rangle \ . \tag{2.29}$$

2. The Hadamard gate acts on the ancilla; we get

$$\frac{1}{\sqrt{2}}\left(|0\rangle + |1\rangle\right)|\psi\rangle|\varphi\rangle \ . \tag{2.30}$$

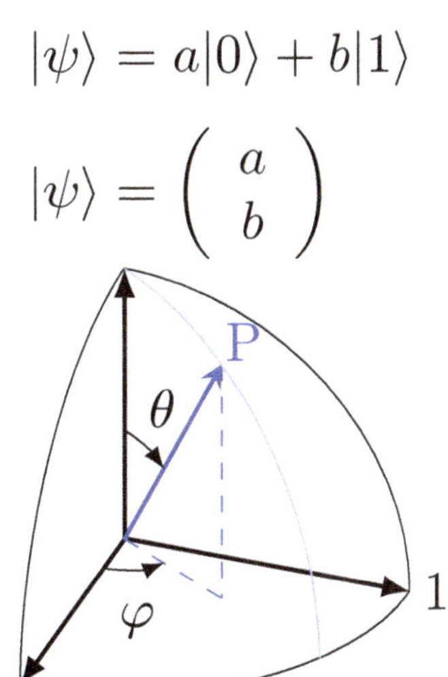

Fig. 2.8 A single qubit: a superposition of two levels $|0\rangle$ and $|1\rangle$ is represented as a vector of two complex numbers or as a point in the Bloch sphere. It is used as ancilla qubit in the SWAP operation

Fig. 2.9 Hadamard gate acting on a qubit

$$|\psi\rangle \longrightarrow \boxed{\hat{\mathrm{H}}} \longrightarrow |\phi\rangle$$

$$|\phi\rangle = \hat{\mathrm{H}}|\psi\rangle$$

$$\hat{\mathrm{H}} = \tfrac{1}{\sqrt{2}}\begin{pmatrix} 1 & 1 \\ 1 & -1 \end{pmatrix}$$

$$\hat{\mathrm{H}}|0\rangle = \hat{\mathrm{H}}\begin{pmatrix} 1 \\ 0 \end{pmatrix} = \tfrac{1}{\sqrt{2}}\begin{pmatrix} 1 \\ 1 \end{pmatrix}$$

$$\hat{\mathrm{H}}|1\rangle = \hat{\mathrm{H}}\begin{pmatrix} 0 \\ 1 \end{pmatrix} = \tfrac{1}{\sqrt{2}}\begin{pmatrix} 1 \\ -1 \end{pmatrix}$$

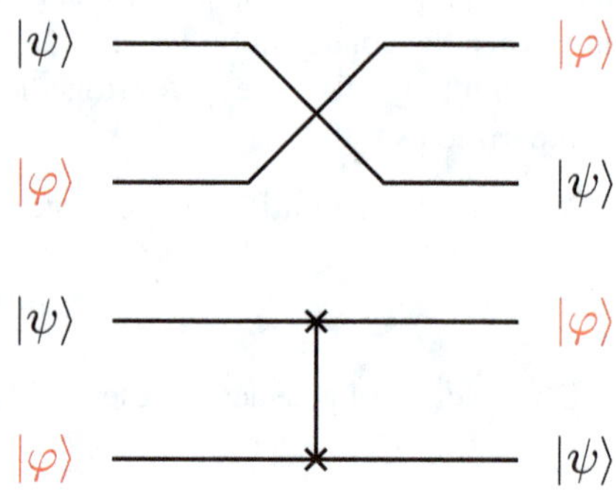

Fig. 2.10 SWAP gate on the two states $|\psi\rangle$ and $|\varphi\rangle$. We show equivalent symbols for the SWAP gate

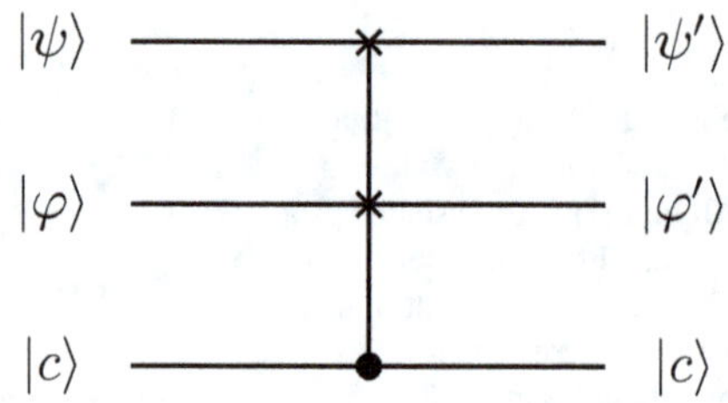

Fig. 2.11 Controlled SWAP gate or Fredkin gate

3. After the Fredkin gate, we have

$$\frac{1}{\sqrt{2}}|0\rangle|\psi\rangle|\varphi\rangle + \frac{1}{\sqrt{2}}|1\rangle|\varphi\rangle|\psi\rangle \ . \tag{2.31}$$

4. After the second Hadamard gate on the ancilla, we have the final state

$$|\Phi\rangle = \frac{1}{2}\left(|0\rangle + |1\rangle\right)|\psi\rangle|\varphi\rangle + \frac{1}{2}\left(|0\rangle - |1\rangle\right)|\varphi\rangle|\psi\rangle$$

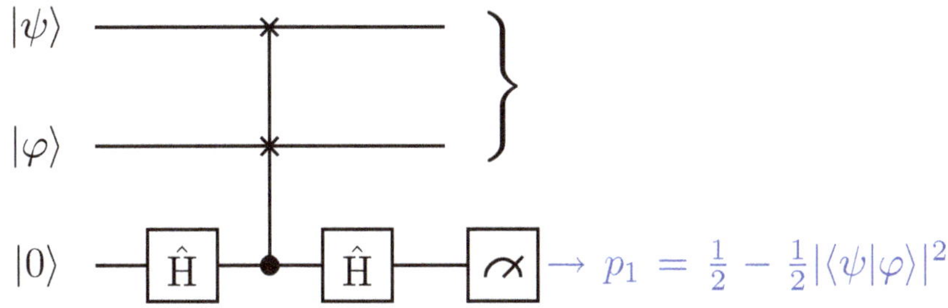

Fig. 2.12 The SWAP test quantum circuit to compute the scalar product. After the gate, the states are in entangled superposition [see Eq. (2.32)]

$$= \frac{1}{2}|0\rangle\left(|\psi\rangle|\varphi\rangle + |\varphi\rangle|\psi\rangle\right) + \frac{1}{2}|1\rangle\left(|\psi\rangle|\varphi\rangle - |\varphi\rangle|\psi\rangle\right) . \tag{2.32}$$

5. The probability p_1 of counting one particle in the ancilla qubit is

$$p_1 = \mathrm{Tr}\left(\rho|1\rangle\langle 1| \otimes \mathbf{I} \otimes \mathbf{I}\right) = \frac{1}{4}\left(\langle\psi|\langle\varphi| - \langle\varphi|\langle\psi|\right)\left(|\psi\rangle|\varphi\rangle - |\varphi\rangle|\psi\rangle\right) . \tag{2.33}$$

6. From Eq. (2.33) we get

$$|\langle\psi|\varphi\rangle|^2 = 1 - 2p_1 , \tag{2.34}$$

which shows that we obtain the modulus square of the scalar product $\langle\psi|\varphi\rangle$ by the probability of observing one particle in the ancilla qubit.

Hence, measuring the kernel function heavily uses entanglement and requires enlarging the Hilbert space by an ancilla qubit.

If we need the value of the complex scalar product, a strategy is further augmenting the size of the Hilbert space by increasing the number of dimensions of $|\psi\rangle$ and $|\varphi\rangle$ [1]. We consider the new states

$$|\psi'\rangle = \frac{1}{\sqrt{2}}\begin{pmatrix}|\psi\rangle\\|1\rangle\end{pmatrix} \tag{2.35}$$

and

$$|\varphi'\rangle = \frac{1}{\sqrt{2}}\begin{pmatrix}|\varphi\rangle\\|1\rangle\end{pmatrix} , \tag{2.36}$$

such that their scalar product is

$$\langle\varphi'|\psi'\rangle = 1 + \langle\varphi|\psi\rangle . \tag{2.37}$$

Using the new states in the SWAP test, we have [see Eq. (2.34)]

$$|\langle\psi'|\varphi'\rangle|^2 = 1 - 2p_1 \ . \tag{2.38}$$

After Eq. (2.37),

$$|\langle\psi|\varphi\rangle + 1| = \sqrt{1 - 2p_1} \ . \tag{2.39}$$

$\langle\psi|\varphi\rangle$ is a complex quantity not determined uniquely by Eq. (2.39). However, if we know that the scalar product is real-valued, we can obtain its sign from Eq. (2.39). This is a simple example of a *phase retrieval* technique. It shows that complex-valued scalar products introduce an additional level of complexity in the quantum processor. In addition, the measurement of phases is typically subject to experimental errors and indeterminacy.

2.9 Boson Sampling and Quantum Kernels

In the quantum kernel method, we need a strategy to encode a classical datum in a quantum state. We also need a quantum device to compute the kernel function. When numerically determining the Gram matrix is efficient in a classical computer, we do not expect a substantial advantage in using a quantum processor.

We have considered continuous variable states as coherent and squeezed states. These states belong to the class of Gaussian states, which plays a significant role in quantum machine learning.

Another interesting aspect is that measuring kernel function requires the probability of observing one particle in a given state. This means that we need to sample the probability distributions of all the possible outputs at specific events. For example, we need to find the probability of observing the ground state $|000\rangle$.

The samples of the probability distribution give the labels of our classifier. This kind of measurement [6] is named "boson sampling." In recent years, boson sampling (BS) gained a special role in quantum computing [7].

If one considers a linear circuit made of squeezers and interferometers (detailed in later chapters), it turns out that numerically predicting the probability of particle patterns is a demanding task for classical computers [6].

In other words, if we want to use a conventional laptop to compute the probability of observing the output state

$$|\mathbf{n}\rangle = |0\rangle \otimes |2\rangle \otimes |1\rangle \otimes |4\rangle, \tag{2.40}$$

from the output channels of a quantum circuit, we have to expect that it will be time-consuming. In (2.40) $\mathbf{n} = (0, 2, 1, 4)$ is the BS pattern.

One can reverse the argument and state that a dedicated quantum circuit has an advantage in computing BS. The quantum circuit will return such a probability in a human-scale time, while the laptop (or a powerful computer) will take thousands of years.

The previous argument allows claiming that the quantum computer has an advantage over the classical computer for BS, and hence BS is a task that demonstrates the quantum advantage. Experiments demonstrated computing BS probabilities with dedicated quantum circuits with orders of magnitude acceleration with respect to classical computers [8–11].

Historically, these experiments are a critical achievement, as they are the first that demonstrate a computational task, which is advantageous on real quantum computers. However, the argument is not free from controversy in the sense that the competition may not be posed in a fair way.

Putting BS from a different perspective, we have a quantum experiment, namely, bosons propagating in a circuit, and we have an outcome that we obtain by some statistical estimation that returns a probability. Then, we claim that the quantum experiment is a computation. We have a theory that predicts the outcome of the quantum experiment. But predicting the outcome requires a lot of computation. For this reason, performing the experiment is faster than making the computation on a classical computer. Specialized quantum processors for boson sampling have been reported at a scale reaching $n > 50$ [7], which is quite challenging for a classical processor [12].

Hence, the argument for the quantum computational advantage is the upside down of the typical attitude in experimental physics. For example, assume you want to predict the waves in the sea when you launch a stone. Performing the experiment is easy: you need the sea, a stone, and a camera to take a picture and count the ripples on the water surface. Predicting the number of waves by a classical computer is demanding. It is even more demanding if you account for the sea motion before launching the stone (i.e., you do not assume that the sea has a flat surface, or that the stone is pointwise). Following the same argument in BS, we can affirm that in this specialized computational task (i.e., counting the waves generated by a stone), the sea is a computer that has a computational advantage with respect to conventional computers!

One can find other examples, such as predicting the gravitational waves emitted by black-hole mergers. This requires big computers for solving Einstein's general relativity equations ab initio. Real emitted gravitational waves require few seconds. Therefore, we claim that a black-hole merger is a computer device that has an advantage with respect to the classical computers in the specialized task of counting the ripples of space-time!

A counter-argument is that once we have a large-scale quantum computer that exploits entanglement and error correction, we compute both the ripples in the sea and the space-time more efficiently than a classical computer. Hence quantum computers will prevail. In simple words, to realize a quantum computer, one needs the same effort to build "the sea" or a "black-hole merger" in the laboratory, which is not so far from reality.

In general terms, we expect an advantage of quantum computers in tasks that are hard for classical computers, such as boson sampling.

A quantum computer will be more efficient in realizing quantum feature mapping and classifications based on boson sampling, a task probably prohibitive for classical computing.

However, there is also another exit strategy to this situation to defend the usefulness of BS quantum circuits, which is outlining their potential use in quantum machine learning and quantum feature mapping.

Computing the BS probability corresponds to computing the probability

$$\Pr(\mathbf{n}) = |\langle \mathbf{n}|\psi\rangle|^2 \tag{2.41}$$

where $\mathbf{n}$ is a particle pattern and $|\psi\rangle$ is the output state of our quantum device. If the evolution of the quantum state depends on a number of parameters $\boldsymbol{\theta}$, the probability $\Pr(\mathbf{n}, \boldsymbol{\theta})$ will also be a function of the parameters. We can select the parameters $\boldsymbol{\theta}$ to fit a certain dataset.

We remark, however, that this gives an argument in favor of quantum machine learning only if the performance in the specific task of the quantum feature mapping is better (in terms of speed or energy consumption) than any other classical computer.

2.10 Universality of Quantum Feature Maps and the Reproducing Kernel Theorem*

By extending the arguments for the support vector machines, we understand the class of functions that we represent by a quantum computer.

Given a dataset with inputs and labels $\left[\mathbf{X}^{(j)}, y^j\right]$, with $j = 0, 1, \ldots, N_X - 1$, we approximate the target function as a linear combination of specific kernel functions

$$f(\mathbf{X}) = \sum_{j=0}^{N_X-1} \theta_j K\left[\mathbf{X}, \mathbf{X}^{(j)}\right] . \tag{2.42}$$

Even though we know this is a reasonable ansatz and we know that the training of the weights θ_j converges, we do not know which is the resulting function after the optimization. We do not know the space of functions given by Eq. (2.42).

Theorems in the *kernel theory* furnish the answer [2]. Assuming N_X observations, one has a kernel function $K(\mathbf{X}, \mathbf{Y})$ if the *Gram matrix* $\boldsymbol{K}$ with elements

$$K_{ij} = K(\mathbf{X}^{(i)}, \mathbf{X}^{(j)}) \tag{2.43}$$

is positive semidefinite, that is, $(N_X \geq 2)$

$$\sum_{i,j=0}^{N_X-1} c_i c_j K_{ij} \geq 0 , \tag{2.44}$$

with c_i real-valued coefficients.
Letting $\mathbf{c} = \{c_0, c_1, \ldots, c_{N_X-1}\}$, (2.44) also reads

$$\mathbf{c} \cdot \boldsymbol{K} \cdot \mathbf{c}^\dagger \geq 0 . \tag{2.45}$$

Given a feature map $\mathcal{F}$, and its feature vectors $\Phi(\mathbf{X})$, the inner products of two feature vectors define a kernel [1]

$$K(\mathbf{X}, \mathbf{Y}) = \langle \Phi(\mathbf{Y}), \Phi(\mathbf{Y}) \rangle , \tag{2.46}$$

with the scalar product in the feature space. For example, in the space of the density matrices with $\Phi(\mathbf{X}) = \rho(\mathbf{X})$, we have

$$K(\mathbf{X}, \mathbf{Y}) = \langle \Phi(\mathbf{X}), \Phi(\mathbf{Y}) \rangle = \mathrm{Tr}\left[\rho(\mathbf{X})\rho(\mathbf{Y})\right] . \tag{2.47}$$

Once we have a kernel function and a dataset of observation, we can consider the space of functions that can be written as

$$f(\mathbf{X}) = \sum_{j=0}^{N_X-1} \theta_j K\left[\mathbf{X}, \mathbf{X}^{(j)}\right] . \tag{2.48}$$

In other words, the dataset and the kernel generate a space of functions. The *representer theorem* states that the space of functions is the so-called reproducing kernel Hilbert space (RKHS).

To understand what the RKHS is, we have to consider the fact that not only the feature map defines a kernel function as the scalar product in (2.46), but the following theorem holds: *any kernel function can be written as* (2.46). So there is a one-to-one correspondence between the relevant space of functions (the RKHS) and the kernel.

Assume we do not know the feature mapping, but we start with the knowledge of the kernel function $K(\mathbf{X}, \mathbf{Y})$. Given the kernel function K, we can indeed choose an arbitrary set of M observations $\mathbf{X}^{(j)}$ and the functions of the $\mathbf{X}$ variables

$$K_j(\mathbf{X}) \equiv K(\mathbf{X}, \mathbf{X}^{(j)}) . \tag{2.49}$$

Then we consider the space spanned by linear combinations of these functions, i.e., all the functions written as

$$f(\mathbf{X}) = \sum_{j=0}^{M-1} w_j K_j(\mathbf{X}) \ . \tag{2.50}$$

Given two real functions f and g in this space, with

$$g(\mathbf{X}) = \sum_{j=0}^{M-1} v_j K_j(\mathbf{X}) \ , \tag{2.51}$$

we introduce a scalar product by the following formula:

$$\langle f(\mathbf{X}), g(\mathbf{X}) \rangle = \sum_{ij} w_i v_j K_{ij} \ . \tag{2.52}$$

As we are dealing with a kernel function, we have

$$\langle f(\mathbf{X}), f(\mathbf{X}) \rangle \geq 0 \tag{2.53}$$

as requested for a scalar product. In particular, given an integer $h \in [0, \ldots, M-1]$ if we consider the function $f(\mathbf{X}) = K_h(\mathbf{X})$, we have that its components are $w_j = \delta_{jh}$ for the chosen dataset of observations. We also have

$$\langle K_j(\mathbf{X}), K_h(\mathbf{X}) \rangle = K_{jh} \ . \tag{2.54}$$

This follows from the definition of the scalar product as in (2.52).

So the space of the linear combinations of the function $K_j(\boldsymbol{x})$ is a linear space equipped with a scalar product, and we can consider its closure, including all the limits of the Cauchy series. The resulting Hilbert space is the RKHS.

The reason for this fancy name RKHS is the following property, which arises from the definition of the scalar product:

$$\langle f(\mathbf{X}), K_h(\mathbf{X}) \rangle = \sum_j w_j \langle K_j(\mathbf{X}), K_h(\mathbf{X}) \rangle = \sum_j w_j K_{jh} = f(\mathbf{X}_h) \ . \tag{2.55}$$

Each sample of a function in the RKHS is a scalar product.

Having stated that a scalar product in the feature Hilbert space defines a kernel function, and – reciprocally – any kernel function defines a Hilbert space with a scalar product. We have finally the conclusion concerning the universality, that is, *having chosen a feature space, all functions in the RKHS are represented, and any represented function belongs to the RKHS*. Thus the RKHS, which is the space defined by the specific quantum processor, is the universality class of the adopted feature mapping.

Thus care must be adopted when building a quantum circuit to have the most general universality class. If the quantum circuit does not converge with acceptable performance, it implies that it cannot reproduce the proper space of functions for the given task.

For example, above we considered the functions represented by coherent states. All the functions that can be expanded in Gaussians are represented. These functions are those fast decreasing and norm-integrable.

In general terms, once we have chosen a kernel function, we define a space of functions, and the representer theorem states that the optimization with an ansatz like (2.55) produces elements of the RKHS.

2.11 Further Reading

- The inductive bias of quantum kernels: In [13], a detailed analysis of the spectral properties of quantum kernels shows that we can expect a realistic advantage on classical kernel machines only in specialized cases. For continuous functions, also in infinite-dimensional spaces, classical systems are much more reliable. Limitations are also outlined for the exploitation of the hypothetical quantum advantage in linear algebra.
- On the statistical limits of quantum supervised limits: In [14], authors outline that the quantum speedup can be only polynomial with respect to classical systems. Indeed, one has to take into account the measurement process. Repeated measurements are needed to achieve a given precision, e.g., in computing the kernel function. Superpolynomial advantage must be expected only for algorithms for which no efficient classical counterpart is known.
- Quantum-assisted machine learning: In [15], a quantum-assisted machine learning has been demonstrated by using ion-trap devices. A hybrid algorithm with a quantum-computer accelerator demonstrated high performance in generating high-quality data.

References

1. M. Schuld, F. Petruccione, *Supervised Learning with Quantum Computers* (Springer, Berlin, 2018)
2. B. Schölkopf, A. Smola, *Learning with Kernels: Support Vector Machines, Regularization, Optimization, and Beyond* (The MIT Press, Cambridge, 2018)
3. M. Schuld, N. Killoran, Phys. Rev. Lett. **122**, 040504 (2019). https://doi.org/10.1103/PhysRevLett.122.040504
4. S.M. Barnett, P.M. Radmore, *Methods in Theoretical Quantum Optics* (Oxford University Press, New York, 1997)
5. E. Fredkin, T. Toffoli, Int. J. Theor. Phys. **21**(3–4), 219 (1982). https://doi.org/10.1007/BF01857727
6. S. Aaronson, A. Arkhipov, Theory Comput. **9**(4), 143 (2013). https://doi.org/10.4086/toc.2013.v009a004

7. Y.H. Deng, S.Q. Gong, Y.C. Gu, Z.J. Zhang, H.L. Liu, H. Su, H.Y. Tang, J.M. Xu, M.H. Jia, M.C. Chen, H.S. Zhong, H. Wang, J. Yan, Y. Hu, J. Huang, W.J. Zhang, H. Li, X. Jiang, L. You, Z. Wang, L. Li, N.L. Liu, C.Y. Lu, J.W. Pan, Phys. Rev. Lett. **130**, 190601 (2023). https://doi.org/10.1103/PhysRevLett.130.190601
8. H. Wang, J. Qin, X. Ding, M.C. Chen, S. Chen, X. You, Y.M. He, X. Jiang, L. You, Z. Wang, C. Schneider, J.J. Renema, S. Höfling, C.Y. Lu, J.W. Pan, Phys. Rev. Lett. **123**, 250503 (2019). https://doi.org/10.1103/PhysRevLett.123.250503
9. H.S. Zhong, H. Wang, Y.H. Deng, M.C. Chen, L.C. Peng, Y.H. Luo, J. Qin, D. Wu, X. Ding, Y. Hu, P. Hu, X.Y. Yang, W.J. Zhang, H. Li, Y. Li, X. Jiang, L. Gan, G. Yang, L. You, Z. Wang, L. Li, N.L. Liu, C.Y. Lu, J.W. Pan, Science **370**, 1460 (2020)
10. Y. Li, M. Chen, Y. Chen, H. Lu, L. Gan, C. Lu, J. Pan, H. Fu, G. Yang (2020)
11. L.S. Madsen, F. Laudenbach, M.F. Askarani, F. Rortais, T. Vincent, J.F.F. Bulmer, F.M. Miatto, L. Neuhaus, L.G. Helt, M.J. Collins, A.E. Lita, T. Gerrits, S.W. Nam, V.D. Vaidya, M. Menotti, I. Dhand, Z. Vernon, N. Quesada, J. Lavoie, Nature **606**(7912), 75 (2022). https://doi.org/10.1038/s41586-022-04725-x
12. Y. Li, M. Chen, Y. Chen, H. Lu, L. Gan, C. Lu, J. Pan, H. Fu, G. Yang. Benchmarking 50-photon gaussian boson sampling on the sunway taihulight (2020). https://doi.org/10.48550/ARXIV.2009.01177
13. J.M. Kübler, S. Buchholz, B. Schölkopf, CoRR (2021). http://arxiv.org/abs/2106.03747v2
14. C. Ciliberto, A. Rocchetto, A. Rudi, L. Wossnig, Phys. Rev. A **102**, 042414 (2020). https://doi.org/10.1103/PhysRevA.102.042414
15. M.S. Rudolph, N.B. Toussaint, A. Katabarwa, S. Johri, B. Peropadre, A. Perdomo-Ortiz, Phys. Rev. X **12**, 031010 (2022). https://doi.org/10.1103/PhysRevX.12.031010

Chapter 3
Qubit Maps

Don't reinvent the wheel...

Abstract We consider quantum feature mapping in qubit states with entanglement. We introduce tensor notations for qubits in `TensorFlow`. We review an experiment with IBM quantum computers with memristors. We deepen the mathematical aspects of quantum feature maps and support vector machines.

3.1 Introduction

Feature mapping using product states does not exploit the potential of quantum computing. Also, most quantum computers deal with qubits and not with continuous variable states. Here we discuss a feature mapping originally used in IBM systems based on superconducting qubits [1].

In our epoch, we have evidence of quantum advantage (albeit limited to specialized demonstrations such as boson sampling). We have to figure out some guidelines for the design quantum systems to support this evidence. One strategy is implementing circuits hard to simulate on classical devices. However, when facing real applications (e.g., natural language processing), the resulting quantum processor will not necessarily be more efficient than the classical processors. But quantum computers will allow new algorithms whose outcomes we cannot imagine at the moment.

3.2 Feature Maps with Qubits

In the previous chapter, we have seen examples of feature mapping with continuous variables as with coherent and squeezed states. In the most studied quantum computers, one adopts entangled qubits, which also provide a route to implement new feature mappings.

C. Conti, *Quantum Machine Learning*, Quantum Science and Technology,
https://doi.org/10.1007/978-3-031-44226-1_3

One can start from a qubit register as

$$|\mathbf{0}\rangle = |0\rangle \otimes |0\rangle \otimes |0\rangle \tag{3.1}$$

and then introduce a feature map as a unitary operator dependent on the input dataset

$$|\Phi(\mathbf{X})\rangle = \hat{U}_{\Phi}(\mathbf{X})|\mathbf{0}\rangle \ . \tag{3.2}$$

As qubits are vectors in the Bloch sphere, the operators may be rotations with the dataset encoded in angles. The generators of rotations are the Pauli matrices.

$$\hat{\mathrm{I}} = \begin{pmatrix} 1 & 0 \\ 0 & 1 \end{pmatrix} \quad \hat{\mathrm{X}} = \begin{pmatrix} 0 & 1 \\ 1 & 0 \end{pmatrix} \quad \hat{\mathrm{Y}} = \begin{pmatrix} 0 & -\mathrm{i} \\ \mathrm{i} & 0 \end{pmatrix} \quad \hat{\mathrm{Z}} = \begin{pmatrix} 1 & 0 \\ 0 & -1 \end{pmatrix} \ . \tag{3.3}$$

Here we show how to model these gates by using `TensorFlow`[1]. Interestingly, we see how the computational time on a classical simulator of the quantum computers rapidly grows when we represent multidimensional entangled states.

3.3 Using Tensors with Single Qubits

We write a single qubit in the basis $|z\rangle$, with $z \in \{0, 1\}$ and components f^z,

$$|\psi\rangle = f^0|0\rangle + f^1|1\rangle = \sum_{z\in\{0,1\}} f^z|z\rangle = f^z\mathbf{e}_z \ . \tag{3.4}$$

where we also introduced the notation for the basis vectors with lower index

$$|z\rangle = \mathbf{e}_z \ , \tag{3.5}$$

to remove the summation symbol, when we have repeated indices.

In `TensorFlow`, we represent $|\psi\rangle$ by a tensor of two complex numbers `[f0,f1]`. We use "contravariant" notation [2] for the qubit components (i.e., upper indices) f^z, as it happens for a generic tensor in a basis $\mathbf{e}_0$ and $\mathbf{e}_1$, which correspond to $|0\rangle$ and $|1\rangle$, respectively. The basis vectors $|0\rangle$ and $|1\rangle$ are defined by the code[2]

```
qubit0=tf.constant([1,0],dtype=tf.complex64)
qubit1=tf.constant([0,1],dtype=tf.complex64)
# Let us print the qubit tensors
print(qubit0)
# Returns
```

[1] tensorflow is a library available under the Apache 2.0 license. Website https://www.tensorflow.org

[2] Details can be found in the notebook `jupyternotebooks/quantumfeaturemap/QubitsMap.ipynb`.

```
# tf.Tensor([1.+0.j 0.+0.j], shape=(2,), dtype=complex64)
print(qubit1)
# Returns
# tf.Tensor([0.+0.j 1.+0.j], shape=(2,), dtype=complex64)
```

We denote a qubit by two complex numbers f^z with $z \in \{0, 1\}$ with `shape=(2,)` in `TensorFlow`.

We print the shape and the components of the qubits as follows.

```
print(qubit0.shape)
# Returns
# (2,)
print(qubit0[0])
# Returns
# tf.Tensor((1+0j), shape=(), dtype=complex64)
print(qubit0[1])
# Returns
tf.Tensor(0j, shape=(), dtype=complex64)
# Returns
# tf.Tensor((1+0j), shape=(), dtype=complex64)
# Or we can print real and imag by tf.print
tf.print(tf.math.real(qubit0))
# Returns
# [1 0]
tf.print(tf.math.imag(qubit0))
# Returns
# [0 0]
# Note that tf.print does not work with complex tensors
# in the TensorFlow version
tf.print(qubit0)
# Returns
# [? ?]
```

3.3.1 One-Qubit Gate as a Tensor

A gate on the single qubit $|\psi\rangle$ corresponds to an operator

$$\hat{G}|\psi\rangle = \sum_{z\in\{0,1\}} f^z \hat{G}|z\rangle \ . \tag{3.6}$$

We have

$$\hat{G}|z\rangle = \sum_{w\in\{0,1\}} G^w{}_z|w\rangle \ , \tag{3.7}$$

i.e., the action of the gate operator on the basis vector is

$$\hat{G}|0\rangle = G^0{}_0|0\rangle + G^1{}_0|1\rangle \tag{3.8}$$

$$\hat{G}|1\rangle = G^0{}_1|0\rangle + G^1{}_1|1\rangle \,. \tag{3.9}$$

We have from Eq. (3.6)

$$\hat{G}|\psi\rangle = \sum_{z,w\in\{0,1\}} f^z G^w{}_z|w\rangle = \sum_{z\in\{0,1\}} F^z|z\rangle \,, \tag{3.10}$$

with

$$F^z = G^z{}_w f^w \tag{3.11}$$

after exchanging z and w indices and omitting the sum symbol over repeated indices. According to Eq. (3.11), the action of a gate is a 2×2 complex matrix $G^z{}_w$ on the complex vector f^w. This is also described as the contraction of the tensor $G^u{}_w f^v$ with respect to the lower index of $G^u{}_w$ and the upper index v of f^v.

3.4 More on Contravariant and Covariant Tensors in `TensorFlow`*

A reader with a background in general relativity may argue about the connections of `tensors` in `TensorFlow` and tensors in physics, with interest in the covariant and contravariant notation.

A `tensor` in `TensorFlow` is just a list of lists; hence it is not a tensor, as we intend in physics, for example, an element in the product spaces of tangent and cotangent spaces (vector and forms) [2], as $\mathbf{G} = G^{ij}{}_k \mathbf{e}_i \otimes \mathbf{e}_j \otimes \boldsymbol{\omega}^k$. This difference emerges when we try to represent a "physical" tensor with many indices by a `tf.tensor` with a given `shape`.

Tensors have upper and lower indices, e.g., $G^{ij}{}_k$ is a rank-3 tensor (i.e., *three* indices) of type $\binom{2}{1}$ (i.e., *two* upper indices and *one* lower index). Such a tensor, when represented in `TensorFlow`, will have a `shape` given by a list of *three* elements, corresponding to the number of elements in each dimension.

For example, we may decide that $G^{ij}{}_k$ has `shape=(3,4,3)`, indicating that $i \in \{0, 1, 2\}$, $j \in \{0, 1, 2\}$, and $k \in \{0, 1, 2, 3\}$. As another example, we consider $M^j{}_k$ with `shape=(1,2)`, such that $j \in \{0\}$ and $k \in \{0, 1\}$.

Note that – in our chosen notation – in the list `shape`, elements at even positions as `shape[0]=3` and `shape[2]=3` in `shape=(3,4,3)` correspond to the upper (contravariant) indices i and j and elements at the odd position as `shape[1]=4` to the lower (covariant) indices as k.

We must be careful in this mapping, because in physics G^{ij} differs from $G^{i}_{\ j}$. In our approach, we represent by a `tf.constant` $G^{i}_{\ j}$ with `shape=(N,M)`, but not G^{ij}.

A possibility is using `tf.constant` with shape with size 1 in some directions. A dimension with size 1 has only one possible value for the corresponding index; hence it is redundant (as an alternative, we could also use dimensions with `size=None`).

Thus the tensor G^{ij} with rank 2 and type $\binom{2}{0}$ will make to a rank-3 `tf.constant` with `shape=(N,1,N)` with $N > 1$ (if $N = 1$ we have a scalar). For vectors, this implies that V^j is the same as $V^j{}_k$ with size 1 dimension for k; the corresponding index is $k = 0$.

Notably enough, if we use dimensions with size 1, we can represent a tensor by the `shape`. In other words, `shape=(1,N,N,1,1,N)` identify (with $N > 1$) a tensor G_{ijk}.

Interestingly, the `tf.squeeze` operation removes dimensions with size 1. `tf.squeezing` a tensor with `shape=(1,N,M)` with $N > 1$ and $M > 1$, i.e., $G_i{}^j$, will hence map into `shape=(N,M)` corresponding to $G^i{}_j$. Also, `tf.squeezing` `shape=(1,N)` maps V_j into V^j. We must be careful, however, as this lowering and rising indices by squeezing do not involve the metric. Hence, it is not equivalent to general relativity and does not hold in Riemannian geometry.

In this book, we can forget about these subtle aspects and map a tensor in a `tf.tensor` in a self-explanatory way as in the following examples.

3.5 Hadamard Gate as a Tensor

We consider a one-qubit Hadamard gate $\hat{\mathrm{H}}$

$$\hat{\mathrm{H}} = \begin{pmatrix} \mathrm{H}^0{}_0 & \mathrm{H}^0{}_1 \\ \mathrm{H}^1{}_0 & \mathrm{H}^1{}_1 \end{pmatrix} = \frac{1}{\sqrt{2}} \begin{pmatrix} 1 & 1 \\ 1 & -1 \end{pmatrix} . \tag{3.12}$$

We define the corresponding tensor as

```
Hsingle=tf.constant([[1/np.sqrt(2), 1/np.sqrt(2)],
                     [1/np.sqrt(2), -1/np.sqrt(2)]],
                     ↪  dytpe=tf.complex64)
```

where `Hsingle` indicates a gate on the single qubit. The element $\mathrm{H}^0{}_0$ is `Hsingle[0,0]`, $\mathrm{H}^0{}_1$ is `Hsingle[0,1]`, $\mathrm{H}^1{}_0$ is `Hsingle[1,0]`, and $\mathrm{H}^1{}_1$ is `Hsingle[1,1]`.

The tensor `Hsingle` is a list of two vectors corresponding to (i) `Hsingle[0]`, the first row of $\hat{\mathrm{H}}$, with elements `Hsingle[0,0]` and `Hsingle[0,1]`, and

(ii) `Hsingle[1]`, the second row of $\hat{\text{H}}$, with elements `Hsingle[1,0]` and `Hsingle[1,1]`. Hence, when we contract the list of two columns `Hsingle` with the two-component vector `qubit`, we are building a new list with two elements,

`Hsingle[0]qubit=Hsingle[0,0]qubit[0]+Hsingle[0,1] qubit[1],`

and

`Hsingle[1]qubit=Hsingle[1,0]qubit[0]+Hsingle[1,1] qubit[1].`

The action of a gate on f^z is a contraction of the rank-3 tensor $G^u{}_w f^v$ with respect to the indices (w, v) with $u, v, w \in \{0, 1\}$. The contraction is obtained in `TensorFlow` by the operation `tf.tensordot`. This corresponds to

```
tf.tensordot(Hsingle,qubit0,axes=[[1],[0]] )
```

where the list `axes=[[1],[0]]` denotes the index 1 for `Hsingle` and the index `0` for `qubit0`. Note that the index `[[1]]`, which is odd, corresponds to the covariant (i.e., lower) index w, and the index `[[0]]` corresponds to the contravariant (i.e., upper) index u.

In this case, the contraction is the conventional matrix-vector product and can be abbreviated as

```
tf.tensordot(Hsingle,qubit0,axes=1)
# which returns
# <tf.Tensor: shape=(2,), dtype=complex64,
# numpy=array([0.70710677+0.j, 0.70710677+0.j], dtype=complex64)>
```

that is,

$$\hat{\text{H}}|0\rangle = \frac{1}{\sqrt{2}}\left(|0\rangle + |1\rangle\right) . \tag{3.13}$$

3.6 Recap on Vectors and Matrices in `TensorFlow`*

Let us review the way `TensorFlow` handles vector and matrices.[3] A column vector is just a list of numbers. Thus, for example,[4]

[3] Details can be found in the notebook `jupyternotebooks/quantumfeaturemap/TensorsAndVectors.ipynb`.

[4] We use real numbers (i.e., `float`) as elements not to confuse them with integer value for indices, but we could consider complex or integer elements, as well.

$$V = \begin{pmatrix} V^0 \\ V^1 \end{pmatrix} = \begin{pmatrix} 1.0 \\ 2.0 \end{pmatrix} \tag{3.14}$$

is defined as

```
# We define a vector as a list of comma-separated number
V=tf.constant([1.0, 2.0])
print(V)
# Returns
# tf.Tensor([1. 2.], shape=(2,), dtype=float32)
# V[0] and V[1] are the elements of the list
print(V[0])
# returns a scalar with empty shape
# tf.Tensor(1.0, shape=(), dtype=float32)
# and seemingly
print(V[1])
# returns a scalar with empty shape
# tf.Tensor(2.0, shape=(), dtype=float32)
```

A matrix can be defined as a list of row vectors. For example, a 2×2 matrix is a tensor with `shape=(2,2)`, which means a list of two tensors with `shape=(1,2)`. We consider

$$M = \begin{pmatrix} M^0{}_0 \; M^0{}_1 \\ M^1{}_0 \; M^1{}_1 \end{pmatrix} = \begin{pmatrix} 3.0 \; 4.0 \\ 5.0 \; 6.0 \end{pmatrix} , \tag{3.15}$$

which we code as

```
M=tf.constant([[3.0, 4.0],[5.0, 6.0]])
print(M)
#returns
# tf.Tensor(
#[[3. 4.]
# [5. 6.]], shape=(2, 2), dtype=float32)
# The elements M[0] and M[1] are the two sublists that
↪   corresponds to the rows
print(M[0])
# returns
tf.Tensor([3. 4.], shape=(2,), dtype=float32)
print(M[1])
# returns
tf.Tensor([3. 4.], shape=(2,), dtype=float32)
# note that the shape is (2,) so they correspons to column
↪   vectors,
# even if they are the rows of the matrix
# Seemingly we can extract the columns of the matrix as
print(M[:,0])
```

```
# returns
# tf.Tensor([3. 5.], shape=(2,), dtype=float32)
print(M[:,1])
# returns
# tf.Tensor([4. 6.], shape=(2,), dtype=float32)
# Columns have also shape=(2,) when extracted
```

3.6.1 Matrix-Vector Multiplication as a Contraction

Matrix-Vector multiplication is built in tensor notation as the vector with components $M^i{}_k V^k$, which is the contraction of the tensor $M^i{}_j V^k$ with respect to the indices j and k. The tensor $M^i{}_j V^k$, which results from the outer product of M and V, has rank 3 (i.e., has three indices) of type $\binom{2}{1}$ tensor, with `shape=(2,2,2)`. The tensor has *two* upper indices and *one* lower index. We build the outer product in `TensorFlow` as

```
# outer product, axes=0 as no contraction is done
outerMV=tf.tensordot(M,V,axes=0)
print(outerMV)
# returns
# tf.Tensor: shape=(2, 2, 2), dtype=float32, numpy=
# array([[[ 3.,  6.],
#         [ 4.,  8.]],
#
#        [[ 5., 10.],
#         [ 6., 12.]]], dtype=float32)
# The shape has three indices $(2,2,2)$, each with two possible
↪   values
# For example, we can access the element
print(outerMV[0,0,1])
# returns
# tf.Tensor(6.0, shape=(), dtype=float32)
```

Given the matrix $M^i{}_j$, and the corresponding `tensor M` in `TensorFlow`, the index j is the index `1` of `M`, and the index k is the index `0` of `V`. We can build the matrix-vector multiplication by calling `tf.tensordot` and contracting the proper indices.

```
# inner product, axes=[1,0] says we contract index 1 of M and
↪   index 0 of V
innerMV=tf.tensordot(M,V,axes=[1,0])
print(innerMV)
# returns
```

```
tf.Tensor([11. 17.], shape=(2,), dtype=float32)
# The shape corresponds to a column vector as expected
```

The list of indices in the inner product can be abbreviated by `axes=1` for conventional matrix-vector multiplication, as follows.

```
# inner product, axes=1 is conventional matrix multiplication
innerMV=tf.tensordot(M,V,axes=1)
print(innerMV)
# returns
tf.Tensor([11. 17.], shape=(2,), dtype=float32)
# The shape corresponds to a column vector as expected
```

3.6.2 Lists, Tensors, Row, and Column Vectors

Note that in the example above, all lists of numbers are interpreted as column vectors if we do not specify the shape. If we need a row vector, we can explicitly indicate a tensor with `shape=(1,2)`.

```
# define a row vector with explicit shape
VR=tf.constant([1.0,2.0],shape=(1,2))
print(VR)
#returns
#tf.Tensor([[1. 2.]], shape=(1, 2), dtype=float32)
```

But this is actually a list of *one* list of two numbers, which we can access as follows.

```
print(VR[0])
# returns
tf.Tensor: shape=(2,), dtype=float32, numpy=array([1., 2.],
↪  dtype=float32)
# the elements are
print(VR[0,0])
# tf.Tensor(1.0, shape=(), dtype=float32)
print(VR[0,1])
# tf.Tensor(2.0, shape=(), dtype=float32)
```

Seemingly, we can handle a column vector by explicitly stating its `shape=(2,1)`.

```
# define a row vector with explicit shape
VC=tf.constant([1.0,2.0],shape=(2,1))
print(VC)
# returns
# tf.Tensor([[1. 2.]], shape=(2, 1), dtype=float32)
print(VC[0])
# returns (note the shape is 1)
# tf.Tensor([1.], shape=(1,), dtype=float32)
print(VC[1])
# returns (note the shape is 1)
# tf.Tensor([2], shape=(1,), dtype=float32)
# the elements are
print(VC[0,0])
# tf.Tensor(1.0, shape=(), dtype=float32)
print(VC[1,0])
# tf.Tensor(2.0, shape=(), dtype=float32)
# Remark: accessing VC[0,1] and VC[1,1] returns an error as
# as VC is a list of two one-number lists
```

We can also make matrix-vector multiplications with `(2,2)` and `(2,1)` tensors as

```
tf.tensordot(M,VC,axes=[1,0])
# returns
# <tf.Tensor: shape=(2, 1), dtype=float32, numpy=
# array([[11.],
#        [17.]], dtype=float32)>
# note the shape is (2,1)
# The same can be done as
tf.tensordot(M,VC,axes=1)
```

On the contrary, the following commands return an error if we try to make the inner product with `(2,2)` and `(1,2)` tensors.

```
tf.tensordot(M,VR,axes=[1,0])
tf.tensordot(M,VR,axes=1)
```

However, by carefully indicating the contraction indices as `axes=[1,1]` in `tf.tensordot`, we compute the product of the matrix with the transposed row vector VR, which is VC, that is,

```
tf.tensordot(M,VR,axes=[1,1])
# returns
# <tf.Tensor: shape=(2, 1), dtype=float32, numpy=
```

```
# array([[11.],
#        [17.]], dtype=float32)>
```

This can be understood by observing that VR is not a row vector, but a tensor of rank 2 with two indices, as it holds for VC. For VR the first index has values `[0,1]` and the second index has one single value `[0]`; the opposite holds for VC.

In other words, we have two possible ways to define column vectors, either with shape (2, 1) or with shape (2,); the first type is a tensor of rank 2 (i.e., with two indices), and the second is a tensor with rank 1 (i.e., one index). The rank-2 tensor VC is a list of *two* lists, `VC[0]` and `VC[1]`, each with *one* number, `VC[0,0]` for the list `VC[0]` and `VC[1,0]` for the list `VC[1]`. The tensor with rank 1, V, is a list of *two* numbers, `V[0]` and `V[1]`. Tensors are just lists of lists of numbers.

As a final remark, there is no need to consider a list (2,) as a column vector or a row vector, but it is just a list of two numbers with no orientation. But, to fix the ideas, we will conventionally retain a vector with `shape=(N,)` as a column vector with dimension N, i.e., a tensor with rank 1.

3.7 One-Qubit Gates

We define here the tensors corresponding to the simple one-qubit gates in Eq. (3.3).[5] These tensors are combined to generated multi-qubit operations.
The identity matrix $\hat{\text{I}}$ is

```
# single qubit identity
Isingle = tf.constant([[1, 0], [0, 1]], dtype=tf.complex64)
```

For the X gate $\hat{\text{X}}$, we have

```
# X-gate for the single qbit
Xsingle = tf.constant([[0, 1], [1, 0]], dtype=tf.complex64)
```

For the Y gate $\hat{\text{Y}}$, we have

```
# Y gate for the single qbit
Ysingle = tf.constant(np.array(
    [[0.0, -1j], [1j, 0.0]],
    dtype=np.complex64), dtype=tf.complex64)
```

[5] The qubit functions are in the module `thqml/quantummap.py`.

For the Z gate $\hat{Z}$, we have

```
# Single qubit Z gate
Zsingle = tf.constant([[1, 0], [0, -1]], dtype=tf.complex64)
```

We can combine these gates one after another. For example, to generate the single qubit state

$$|\psi\rangle = \hat{Z}\hat{H}|0\rangle = \frac{1}{\sqrt{2}}(|0\rangle - |1\rangle) \ , \tag{3.16}$$

we have[6]

```
tf.tensordot(Zsingle,tf.tensordot(Hsingle,qubit0)axes=1)
# returns<tf.Tensor: shape=(2,), dtype=complex64,
# numpy=array([ 0.70710677+0.j, -0.70710677+0.j],
# dtype=complex64)>
```

3.8 Scalar Product

We define a function that returns the scalar product of qubit tensors. Given two qubits

$$|\psi\rangle = \sum_{z\in\{0,1\}} \psi^z|z\rangle \tag{3.17}$$

and

$$|\phi\rangle = \sum_{z\in\{0,1\}} \phi^z|z\rangle \ , \tag{3.18}$$

their scalar product is the complex number

$$\langle\psi|\phi\rangle = \sum_{z\in\{0,1\}} (\psi^z)^*\phi^z \ . \tag{3.19}$$

[6] Other examples are found in the notebook `jupyternotebooks/quantumfeaturemap/QubitsMap.ipynb`.

This is the contraction of the rank-2 tensor $(\psi^z)^*\phi^w$ and can be done as follows.

```
@ tf.function # this decorator speeds up computing
def Scalar(a, b):
    # we reshape the vector a column vectors
    # (this reshaping is not needed for a single qubit
    #  but it is useful for multi-qubit)
    a1 = tf.reshape(a, (tf.size(a), 1))
    b1 = tf.reshape(b, (tf.size(b), 1))
    # scalar multiplication after conjugation
    sc = tf.tensordot(tf.transpose(a1, conjugate=True), b1,
                      axes=1)  # must be a tensor not a scalar
    return sc
```

In the `Scalar` function, we reshape the vectors as column vectors, such that this function can work at any number of qubits; a feature will be handful below.

3.9 Two-Qubit Tensors

We now consider a two-qubit state

$$|\psi\rangle = \psi^{00}|0\rangle \otimes |0\rangle + \psi^{01}|0\rangle \otimes |1\rangle + \psi^{10}|1\rangle \otimes |0\rangle + \psi^{11}|1\rangle \otimes |1\rangle \quad (3.20)$$

$$= \sum_{z,w\in\{0,1\}^2} \psi^{zq}|z\rangle \otimes |w\rangle\ . \quad (3.21)$$

For example, letting $|\varphi\rangle = \sum_z f^z|z\rangle$ and $|\phi\rangle = \sum_w g^w|w\rangle$, we have

$$|\psi\rangle = |\varphi\rangle \otimes |\phi\rangle = \sum_{z,w\in\{0,1\}^2} \psi^{zq}|z\rangle \otimes |w\rangle, \quad (3.22)$$

with the following components

$$\psi^{zw} = f^z g^w\ , \quad (3.23)$$

which are the outer product of the two tensors f^z and g^w, i.e., a product state. Let us consider the two-qubit state $|0\rangle|0\rangle$; the corresponding tensor is built as

```
# build a two qubit state as an outer product,
# obtained by axes=0 in tf.tensordot
q00 = tf.tensordot(qubit0,qubit0,axes=0)
# print(q00) returns
#<tf.Tensor: shape=(2, 2), dtype=complex64, numpy=
```

```
# array([[1.+0.j, 0.+0.j],
#        [0.+0.j, 0.+0.j]], dtype=complex64)>
```

The two-qubit state `q00` is a tensor with `shape=(2,2)`; it has two indices $z, w \in \{0, 1\}$. Seemingly, we define the other tensors that form the two-qubit basis.

```
q01 = tf.tensordot(qubit0,qubit1,axes=0)
# returns
#tf.Tensor(
#[[0.+0.j 1.+0.j]
# [0.+0.j 0.+0.j]], shape=(2, 2), dtype=complex64)
q10 = tf.tensordot(qubit1,qubit0,axes=0)
# returns
#tf.Tensor(
#[[0.+0.j 0.+0.j]
# [1.+0.j 0.+0.j]], shape=(2, 2), dtype=complex64)
q11 = tf.tensordot(qubit1,qubit1,axes=0)
#tf.Tensor(
#[[0.+0.j 0.+0.j]
# [0.+0.j 1.+0.j]], shape=(2, 2), dtype=complex64)
```

Once we have defined the basis tensors, we can use linear combinations to build other states.

```
q10+q01
# returns
#<tf.Tensor: shape=(2, 2), dtype=complex64, numpy=
#array([[0.+0.j, 1.+0.j],
#        [1.+0.j, 0.+0.j]], dtype=complex64)>
q00+3j*q11
#<tf.Tensor: shape=(2, 2), dtype=complex64, numpy=
#array([[1.+0.j, 0.+0.j],
#        [0.+0.j, 0.+3.j]], dtype=complex64)>
```

In general, the resulting state is not normalized, as we will detail below.

3.9.1 Further Remarks on the Tensor Indices

The two-qubit tensors have shape `(2,2)`; hence they are lists of two tensors with *two* elements. To fix the ideas, let us consider a two-qubit tensor built as an outer product of the two one-qubit (unnormalized) states

$$|\psi\rangle = 3|0\rangle + 2\mathrm{i}|1\rangle \ , \qquad (3.24)$$

$$|\phi\rangle = 3\mathrm{i}|0\rangle + 1.5|1\rangle\ . \tag{3.25}$$

We build the state

$$|\Psi\rangle = |\psi\rangle \otimes |\phi\rangle = 9\mathrm{i}|00\rangle + 4.5|01\rangle - 6|10\rangle + 3\mathrm{i}|11\rangle\ . \tag{3.26}$$

In terms of `TensorFlow` tensors, we have[7]

```
psi = 3.0*qubit0+2j*qubit1
phi = 3j*qubit0+1.5*qubit1
# outer product of the two tensors
Psi = tf.tensor(psi,phi,axes=0)
# Let us print the tensors
print(psi)
#tf.Tensor([3.+0.j 0.+2.j], shape=(2,), dtype=complex64)
print(phi)
#tf.Tensor([0. +3.j 1.5+0.j], shape=(2,), dtype=complex64)
print(Psi)
#  tf.Tensor(
# [[ 0. +9.j  4.5+0.j]
# [-6. +0.j  0. +3.j]], shape=(2, 2), dtype=complex64)
```

The tensor `Psi` has shape `(2, 2)`, which is a list of *two* tensors with shape `(2,)`, namely, `Psi[0]` and `Psi[1]`. Let us check these two tensors in detail

```
# we print the tensors composing Psi
print(Psi[0])
#<tf.Tensor: shape=(2,), dtype=complex64,
# numpy=array([0. +9.j, 4.5+0.j], dtype=complex64)>
print(Psi[1])
#<tf.Tensor: shape=(2,), dtype=complex64,
# numpy=array([-6.+0.j,  0.+3.j], dtype=complex64)>
```

Thus `Psi[0]` corresponds to the components of $|00\rangle$ and $|01\rangle$, and `Psi[1]` are the components $|10\rangle$ and $|11\rangle$.
If we represent $|\Psi\rangle$ as

$$|\Psi\rangle = \sum_{z,w\in\{0,1\}^2} \Psi^{zw}|z\rangle \otimes |w\rangle \tag{3.27}$$

we have that `Psi[0]` is the tensor Ψ^{0w} and `Psi[1]` is Ψ^{1w}.

[7] Details can be found in the notebook `jupyternotebooks/quantumfeaturemap/QubitsGym.ipynb`.

`Psi[0,0]` is a scalar with `shape=()` corresponding to Ψ^{00}, and `Psi[0,1]` is Ψ^{01}. Seemingly we have for `Psi[1,0]` and `Psi[1,1]`

```
print(Psi[0,0])
# tf.Tensor(9j, shape=(), dtype=complex64)
print(Psi[0,1])
# tf.Tensor((4.5+0j), shape=(), dtype=complex64)
print(Psi[1,0])
#tf.Tensor((-6+0j), shape=(), dtype=complex64)
print(Psi[1,1])
#tf.Tensor(3j, shape=(), dtype=complex64)
```

Note that the order of the tensors in the outer product is relevant, as it determines the order of the contravariant indices (z, w) in Ψ^{zw}, i.e.,

$$\Psi^{zw} \neq \Psi^{wz} \; . \tag{3.28}$$

3.10 Two-Qubit Gates

Let us consider a gate on a two-qubit state, for example, $\hat{\mathrm{H}}^{\otimes 2} = \hat{\mathrm{H}} \otimes \hat{\mathrm{H}}$

$$\begin{aligned} |\psi\rangle &= \hat{\mathrm{H}}^{\otimes 2}|\varphi\rangle \otimes |\phi\rangle = \hat{\mathrm{H}}|\varphi\rangle \otimes \hat{\mathrm{H}}|\phi\rangle = \\ &= \textstyle\sum_{z,w\in\{0,1\}^2} \varphi^x \mathrm{H}^z{}_x |z\rangle \otimes \phi^y \mathrm{H}^w{}_y |w\rangle \\ &= \textstyle\sum_{z,w\in\{0,1\}^2} \psi^{zw} |z\rangle \otimes |w\rangle \; , \end{aligned} \tag{3.29}$$

where sum over repeated indices x and y is implicit and we used

$$|\varphi\rangle = \sum_{z\in\{0,1\}} \varphi^z |z\rangle \tag{3.30}$$

$$|\phi\rangle = \sum_{z\in\{0,1\}} \phi^z |z\rangle \tag{3.31}$$

$$\hat{\mathrm{H}}|z\rangle = \sum_{w\in\{0,1\}} \mathrm{H}^z{}_w |w\rangle \; . \tag{3.32}$$

For the components, we have

$$\psi^{zw} = \mathrm{H}^z{}_x \mathrm{H}^w{}_y \varphi^x \phi^y \tag{3.33}$$

with $x, y, z, w, v \in \{0, 1\}$. Hence the Hadamard gate for two qubits is represented by the rank-4 tensor

$$H^{z}{}_{x}{}^{w}{}_{y} = H^{z}{}_{x} H^{w}{}_{y} . \tag{3.34}$$

The gate acts on the two-qubit state

$$|\Psi\rangle = \sum_{x,y \in \{0,1\}^2} \Psi^{xy} |x\rangle \otimes |y\rangle \tag{3.35}$$

as follows:

$$\hat{H}^{\otimes 2} |\Psi\rangle = \sum_{z,w \in \{0,1\}^2} H^{z}{}_{x}{}^{w}{}_{y} \Psi^{xy} |z\rangle \otimes |w\rangle , \tag{3.36}$$

contracting the lower (covariant) indices with the state indices x and y.

We build $H^{z}{}_{x}{}^{w}{}_{y}$ by `tf.tensordot` as an outer tensor product using `Hsingle`.

```
HH=tf.tensordot(Hsingle,Hsingle,axes=0)
# the output is a 4 index tensor
# with shape=(2,2,2,2)
# contravariant indices are in the even positions [0] and [2]
# covariant indices are in the odd positions [1] and [3]
print(HH)
# returns
#tf.Tensor(
#[[[[ 0.49999997+0.j   0.49999997+0.j]
#    [ 0.49999997+0.j  -0.49999997+0.j]]
#
#   [[ 0.49999997+0.j   0.49999997+0.j]
#    [ 0.49999997+0.j  -0.49999997+0.j]]]
#
#
#  [[[ 0.49999997+0.j   0.49999997+0.j]
#    [ 0.49999997+0.j  -0.49999997+0.j]]
#
#   [[-0.49999997+0.j  -0.49999997+0.j]
#    [-0.49999997+0.j   0.49999997+0.j]]]],
# shape=(2, 2, 2, 2),
# dtype=complex64)
```

The outer product returns rank-4 tensors with shape (2, 2, 2, 2). The element [0, 0] is $H^{0}{}_{0} H^{x}{}_{y}$, that is, the matrix corresponding to $\hat{H}$ times the element $H^{0}{}_{0}$.

Hence, `HH[1,1]` is a `shape=(2,2)` tensor corresponding to $H^{1}{}_{1} H^{x}{}_{y}$.

```
print(HH[1,1])
#tf.Tensor(
#[[-0.49999997+0.j -0.49999997+0.j]
```

```
# [-0.49999997+0.j  0.49999997+0.j]], shape=(2, 2),
↪  dtype=complex64)
# Element H^0_0 H^1_1
print(H[0,0,1,1,])
#tf.Tensor((-0.49999997+0j), shape=(), dtype=complex64)
```

We remark that the order from left to right of the four indices in $\mathrm{H}^{z}{}_{w}{}^{x}_{y}$ corresponds to the order in the `tensor` indices in `HH[z,w,x,y]`, which justifies the adopted notation for $\mathrm{H}^{z}{}_{x}{}^{w}{}_{y}$ instead of H^{zw}_{xy}. In general, a gate $\hat{G}$ is linear operation on the state tensor ψ^{yz} and is a tensor with two upper and two lower indices

$$\hat{G}|\psi\rangle = \sum_{x,y\in\{0,1\}^2} \mathrm{G}^{x}{}_{z}{}^{y}{}_{w}\psi^{zw}|x\rangle \otimes |y\rangle \ . \tag{3.37}$$

where we have

$$\hat{G}|x\rangle \otimes |y\rangle = \sum_{z,w\in\{0,1\}^2} \mathrm{G}^{z}{}_{x}{}^{w}{}_{y}|z\rangle \otimes |w\rangle \ . \tag{3.38}$$

$\hat{G}$ is a rank-4 tensor with `shape=(2,2,2,2)` and contravariant indices corresponding to odd indices. So, for example, the element $G^{0}{}_{0}{}^{1}{}_{0}$ corresponds to `indices=[0,0,1,0]`.
The action of the gate on the state ψ^{xy} is a contraction over the gate odd (i.e., lower) indices, as – for example – in

```
tf.tensordot(HH,q00, axes=[[1,3],[0, 1]]),
```

where `axes` refers to a list of indices for the two tensors, with [1, 3] the odd covariant indices of $\mathrm{H}^{z}{}_{x}{}^{w}{}_{y}$, i.e., xy, and [0, 1] are the two upper indices of ψ^{xy}.

In the following, we detail useful gates. We test the gates on different states. It is convenient defining a function `gate` to simplify the application of a gate to a state, as follows.

```
@tf.function
def gate(A,psi):
    """ A tensorflow function for applying a gate to a state """
    return tf.tensordot(A,psi, axes=[[1,3],[0,1]])
```

The function `gate` prevents us repeating the lengthy `tensordot` command when we want to apply a gate to a state.

3.10.1 Coding Two-Qubit Gates

In the following, we denote the $\hat{\mathrm{H}} \otimes \hat{\mathrm{H}} = \hat{\mathrm{H}}^{\otimes 2}$, as the HH gate, that is, a Hadamard gate acting on each of the two qubits. The corresponding tensors are built as the outer product of two Hsingle gates, as follows.

```
HH=tf.tensordot(Hsingle,Hsingle,axes=0)
```

We also build other two-qubit gates as $\hat{\mathrm{H}} \otimes \hat{\mathrm{I}}$ or $\hat{\mathrm{I}} \otimes \hat{\mathrm{H}}$, and others by the one-qubit operations Xsingle, Ysingle, and Zsingle. We report in the following some of the corresponding tensors, to be used in a later section.[8]

```
# Hadamard gate for the two qubits
HH = tf.tensordot(Hsingle, Hsingle, axes=0)

# Hadamard gate for the second qubit
IH = tf.tensordot(Isingle, Hsingle, axes=0)

# Hadamard gate for the first qubit
HI = tf.tensordot(Hsingle, Isingle, axes=0)

# $X\otimes X$ gate as a tensor product
XX = tf.tensordot(Xsingle, Xsingle, axes=0)

# Operator $X\otimes I$ ( X on first qubit)
XI = tf.tensordot(Xsingle, Isingle, axes=0)

# Operator $I\otimes X$ (X on the second qbuit)
IX = tf.tensordot(Isingle, Xsingle, axes=0)

# ## $Y\otimes Y$ gate as a tensor product
YY = tf.tensordot(Ysingle, Ysingle, axes=0)

# Operator $Y\otimes I$ ( Y on first qubit)
YI = tf.tensordot(Ysingle, Isingle, axes=0)

# Operator $I\otimes Y$ (Y on the second qbuit)
IY = tf.tensordot(Isingle, Ysingle, axes=0)

# Operator $Z_0\otimes Z_1$ gate as a tensor product
ZZ = tf.tensordot(Zsingle, Zsingle, axes=0)

# Operator $Z_0\otimes I_1$ ( Z on first qubit)
ZI = tf.tensordot(Zsingle, Isingle, axes=0)
```

[8] These and other qubit tensor operations are defined in the Python module thqml/quantummap.py.

```
# Operator $I_0\otimes Z_1$ (Z on the second qbuit)
IZ = tf.tensordot(Isingle, Zsingle, axes=0)
```

3.10.2 CNOT Gate

A CNOT is an entangling gate: it introduces correlations between two qubits. One qubit acts as a control and is unaltered by the gate. The other qubit is affected by the NOT operation if the control qubit is in the $|1\rangle$ state or unaltered if the control is in the $|0\rangle$ state. The CNOT 4×4 matrix is

$$\widehat{\text{CNOT}} = \begin{pmatrix} 1 & 0 & 0 & 0 \\ 0 & 1 & 0 & 0 \\ 0 & 0 & 0 & 1 \\ 0 & 0 & 1 & 0 \end{pmatrix} . \tag{3.39}$$

To represent this matrix as a (2, 2, 2, 2) tensor, we consider its block representation

$$\widehat{\text{CNOT}} = \begin{pmatrix} \mathbf{G}^0{}_0 & \mathbf{G}^0{}_1 \\ \mathbf{G}^1{}_0 & \mathbf{G}^1{}_1 \end{pmatrix} . \tag{3.40}$$

where $\mathbf{G}^j{}_k$ are 2×2 matrices with $j, k \in \{0, 1\}$

$$\mathbf{G}^0{}_0 = \begin{pmatrix} 1 & 0 \\ 0 & 1 \end{pmatrix} , \mathbf{G}^0{}_1 = \begin{pmatrix} 0 & 0 \\ 0 & 0 \end{pmatrix} , \tag{3.41}$$

$$\mathbf{G}^1{}_0 = \begin{pmatrix} 0 & 0 \\ 0 & 0 \end{pmatrix} , \mathbf{G}^1{}_1 = \begin{pmatrix} 0 & 1 \\ 1 & 0 \end{pmatrix} . \tag{3.42}$$

In the `TensorFlow` tensor `CNOT[0,0]` corresponds to the $\mathbf{G}^0{}_0$ matrix, `CNOT[0,1]` to $\mathbf{G}^0{}_1$, and so forth. The element $(1, 1)$ of $\mathbf{G}^1{}_0$ will be `CNOT[1,0,1,1]`.

A two-qubit state is a complex 4×1 vector

$$\Psi = \psi^0\phi^0|00\rangle + \psi^0\phi^1|01\rangle + \psi^1\phi^0|10\rangle + \psi^1\phi^1|11\rangle = \begin{pmatrix} \psi^0\phi^0 \\ \psi^0\phi^1 \\ \psi^1\phi^0 \\ \psi^1\phi^1 \end{pmatrix} , \tag{3.43}$$

which in blocks is

$$\Psi = \begin{pmatrix} \psi^0 \boldsymbol{\phi} \\ \psi^1 \boldsymbol{\phi} \end{pmatrix} . \tag{3.44}$$

The multiplication in block notation times **G** returns the two-qubit state

$$\begin{pmatrix} \psi^0 \mathbf{G}^0{}_0 \boldsymbol{\phi} + \psi^1 \mathbf{G}^0_1 \boldsymbol{\phi} \\ \psi^0 \mathbf{G}^1{}_0 \boldsymbol{\phi} + \psi^1 \mathbf{G}^1{}_1 \boldsymbol{\phi} \end{pmatrix} = \begin{pmatrix} \psi^0 \mathbf{G}^0{}_0 \boldsymbol{\phi} \\ \psi^1 \mathbf{G}^1{}_1 \boldsymbol{\phi} \end{pmatrix} , \tag{3.45}$$

where we used $\mathbf{G}^0{}_1 = \mathbf{G}^1{}_0 = 0$. As anticipated, Eq. (3.45) shows that the first qubit is not affected by the gate. Also, according to (3.45), if the control qubit is $|0\rangle$, i.e., $\psi^0 = 1$ and $\psi^1 = 0$, $\mathbf{G}^0{}_0$ is the operator acting on the second qubit; as $\mathbf{G}^0{}_0 = \hat{\mathrm{I}}$ is the identity, the second qubit is unaltered. On the contrary, if the control qubit is $|1\rangle$, i.e., $\psi^0 = 0$ and $\psi^1 = 1$, $\mathbf{G}^1{}_1$ is the operator acting on the second qubit; as $\mathbf{G}^1{}_1 = \hat{\mathrm{X}}$, the second qubit is subject to X gate.

In block representation, the CNOT gate also reads

$$\mathbf{G} = \begin{pmatrix} \hat{\mathrm{I}} & \mathbf{0} \\ \mathbf{0} & \hat{\mathrm{X}} \end{pmatrix} , \tag{3.46}$$

with $\mathbf{0}$ denoting a 2×2 matrix of zeros.

We build the CNOT rank-4 tensor as follows.

```
# we first define the numpy matrix
# (direct access to tensor is tensorflow is not as simple as in
↪   numpy)
CNOT_np=np.zeros((2,2,2,2))
CNOT_np[0,0,0,0]=1
CNOT_np[0,0,1,1]=1
CNOT_np[1,1,0,1]=1
CNOT_np[1,1,1,0]=1
# then we introduce a tf.tensor with the CNOT matrix
CNOT = tf.constant(CNOT_np,dtype=tf.complex64)
```

We test the CNOT gate acting on different qubits

```
gate(CNOT, q00)
# returns
#<tf.Tensor: shape=(2, 2), dtype=complex64, numpy=
#array([[0.9999999+0.j, 0.        +0.j],
#       [0.       +0.j, 0.        +0.j]], dtype=complex64)>
gate(CNOT, q11)
# returns
#<tf.Tensor: shape=(2, 2), dtype=complex64, numpy=
#array([[0.+0.j, 0.+0.j],
#       [1.+0.j, 0.+0.j]], dtype=complex64)>
```

```
gate(CNOT, q01)
#<tf.Tensor: shape=(2, 2), dtype=complex64, numpy=
#array([[0.+0.j, 1.+0.j],
#        [0.+0.j, 0.+0.j]], dtype=complex64)>
gate(CNOT, q10)
#<tf.Tensor: shape=(2, 2), dtype=complex64, numpy=
#array([[0.+0.j, 0.+0.j],
#        [0.+0.j, 1.+0.j]], dtype=complex64)>
gate(CNOT, q10+q01)
#<tf.Tensor: shape=(2, 2), dtype=complex64, numpy=
#array([[0.+0.j, 1.+0.j],
#        [0.+0.j, 1.+0.j]], dtype=complex64)>
```

3.10.3 CZ Gate

The controlled-Z gate 4 × 4 matrix is

$$\widehat{\mathrm{CZ}} = \begin{pmatrix} 1 & 0 & 0 & 0 \\ 0 & 1 & 0 & 0 \\ 0 & 0 & 1 & 0 \\ 0 & 0 & 0 & -1 \end{pmatrix} . \tag{3.47}$$

The CZ gate applies the Z−gate if the control qubit is $|1\rangle$. Its block representation is

$$\widehat{\mathrm{CZ}} = \begin{pmatrix} \hat{\mathrm{I}} & \mathbf{0} \\ \mathbf{0} & \hat{\mathrm{Z}} \end{pmatrix} . \tag{3.48}$$

For defining the CZ gate in `TensorFlow`, as for the CNOT gate, we start from the 4 × 4 matrix representation and write down the corresponding tensor as follows.

```
# CZ gate
# first define the tensor as np array
CZ_np = np.zeros((2, 2, 2, 2))
CZ_np[0, 0, 0, 0] = 1
CZ_np[0, 0, 1, 1] = 1
CZ_np[1, 1, 0, 0] = 1
CZ_np[1, 1, 1, 1] = -1
# then define a tf.constant
CZ = tf.constant(CZ_np, dtype=tf.complex64)
```

If we apply the CZ gate to the two-qubit basis, we have

$$\widehat{\mathrm{CZ}}\,|00\rangle = |00\rangle\ , \tag{3.49}$$

$$\widehat{\mathrm{CZ}}\,|01\rangle = |01\rangle\ , \tag{3.50}$$

$$\widehat{\mathrm{CZ}}\,|10\rangle = |01\rangle\ , \tag{3.51}$$

$$\widehat{\mathrm{CZ}}\,|11\rangle = -|11\rangle\ . \tag{3.52}$$

This can be checked by the following code.

```
print(Gate(CZ, q00))
print(Gate(CZ, q01))
print(Gate(CZ, q10))
print(Gate(CZ, q11))
#tf.Tensor(
#[[1.+0.j 0.+0.j]
# [0.+0.j 0.+0.j]], shape=(2, 2), dtype=complex64)
#tf.Tensor(
#[[0.+0.j 1.+0.j]
# [0.+0.j 0.+0.j]], shape=(2, 2), dtype=complex64)
#tf.Tensor(
#[[0.+0.j 0.+0.j]
# [1.+0.j 0.+0.j]], shape=(2, 2), dtype=complex64)
#tf.Tensor(
#[[ 0.+0.j  0.+0.j]
# [ 0.+0.j -1.+0.j]], shape=(2, 2), dtype=complex64)
```

By cascading multiple gates, we have

$$\widehat{\mathrm{CZ}}\,\hat{\mathrm{H}}^{\otimes 2}|00\rangle = \frac{1}{2}\left(|00\rangle + |01\rangle + |10\rangle - |11\rangle\right)\ , \tag{3.53}$$

$$\widehat{\mathrm{CZ}}\,\hat{\mathrm{H}}^{\otimes 2}|01\rangle = \frac{1}{2}\left(|00\rangle - |01\rangle + |10\rangle + |11\rangle\right)\ , \tag{3.54}$$

$$\widehat{\mathrm{CZ}}\,\hat{\mathrm{H}}^{\otimes 2}|10\rangle = \frac{1}{2}\left(|00\rangle + |01\rangle - |10\rangle + |11\rangle\right)\ , \tag{3.55}$$

$$\widehat{\mathrm{CZ}}\,\hat{\mathrm{H}}^{\otimes 2}|11\rangle = \frac{1}{2}\left(|00\rangle - |01\rangle - |10\rangle - |11\rangle\right)\ . \tag{3.56}$$

These are obtained by the following code.

```
# set output precision in printing
np.set_printoptions(precision=4)
# compute output states
print(Gate(CZ, Gate(HH, q00)))
print(Gate(CZ, Gate(HH, q01)))
print(Gate(CZ, Gate(HH, q10)))
print(Gate(CZ, Gate(HH, q11)))
```

```
# returns
# tf.Tensor(
# [[ 0.5+0.j  0.5+0.j]
#   [ 0.5+0.j -0.5+0.j]], shape=(2, 2), dtype=complex64)
# tf.Tensor(
# [[ 0.5+0.j -0.5+0.j]
# [ 0.5+0.j  0.5+0.j]], shape=(2, 2), dtype=complex64)
# tf.Tensor(
# [[ 0.5+0.j  0.5+0.j]
# [-0.5+0.j  0.5+0.j]], shape=(2, 2), dtype=complex64)
# tf.Tensor(
# [[ 0.5+0.j -0.5+0.j]
#   [-0.5+0.j -0.5+0.j]], shape=(2, 2), dtype=complex64)
```

3.11 Quantum Feature Maps with a Single Qubit

We encode a classical datum in qubit quantum states by parametric gates. We investigate a single qubit and rotation gates as those implemented in the quantum hardware with superconducting qubits [1].

We start considering a qubit in the ground state $|0\rangle$. The first step is the Hadamard gate to excite both $|0\rangle$ and $|1\rangle$ of the Z-basis. Then, we encode some datum in their superposition. The simplest way is to perform a rotation by using the angles as variables for the feature map.

For example, we consider a rotation around the axis Y, whose generator is the Pauli matrix,

$$\hat{\mathrm{Y}} = \begin{pmatrix} 0 & -\mathrm{i} \\ \mathrm{i} & 0 \end{pmatrix} . \tag{3.57}$$

If we have one single scalar classical datum $\mathbf{X} = X_0$, we encode it in the angle of the rotation θ_Y letting $\theta_Y = X_0$. Our feature map is (including the Hadamard gate $\hat{\mathrm{H}}$)

$$|\phi(X_0)\rangle = e^{\mathrm{i}X_0\hat{\mathrm{Y}}}\hat{\mathrm{H}}|0\rangle . \tag{3.58}$$

We may design a more complex encoding by other rotations. For example,

$$|\phi(X_0)\rangle = e^{\mathrm{i}\theta_Y(X_0)Y}e^{\mathrm{i}\theta_Z(X_0)Z}\hat{\mathrm{H}}|0\rangle , \tag{3.59}$$

where we use multiple angles θ_Y and θ_Z and nonlinear maps from the angles and the input datum X_0 as $\theta_Z = X_0^2$. These maps are implemented by the circuit that performs the rotations and, more specifically, by the input system that converts the dataset into the controls of the quantum gates.

3.11.1 Rotation Gates

We define an operator representing a rotation in the Z-direction with angle θ as

$$\hat{Z}_\theta = e^{i\theta\hat{Z}} = \cos(\theta)\hat{I} + i\sin(\theta)\hat{Z}\,. \tag{3.60}$$

The corresponding tensor is obtained by a `tf.function` with the angle θ as input and returning the gate tensor.[9]

```
@tf.function
def EZ(theta):
    """ return exp(1j theta Z) """
    ct = tf.cast(tf.math.cos(theta), dtype=tf.complex64)
    st = tf.cast(tf.math.sin(theta), dtype=tf.fcomplex64)
    EZS = ct*Isingle+1j*st*Zsingle
    return EZS
```

Note that we cast (by `tf.cast`) the cosine and the sine of the input angle as a complex `tf.complex64`, as this is needed for the complex tensor `EZS`.

Seemingly, we define the rotations with respect to X and Y:

$$\hat{X}_\theta = e^{i\theta\hat{X}}\,,$$

$$\hat{Y}_\theta = e^{i\theta\hat{Y}}\,.$$

The corresponding code is as follows.

```
@tf.function
def EX(theta):
    """ return exp(1j theta X) """
    ct = tf.cast(tf.math.cos(theta), dtype=tf.complex64)
    st = tf.cast(tf.math.sin(theta), dtype=tf.complex64)
    EXS = ct*Isingle+1j*st*Xsingle
    return EXS

@tf.function
def EY(theta):
    """ return exp(1j theta Y) """
    ct = tf.cast(tf.math.cos(theta), dtype=tf.complex64)
    st = tf.cast(tf.math.sin(theta), dtype=tf.complex64)
    EYS = ct*Isingle+1j*st*Ysingle
    return EYS
```

[9] See the jupyternotebooks/quantumfeaturemap/QubitsMap.ipynb.

However, as we deal with a single qubit ($D = 1$), we have few possibilities for encoding. Things become more interesting when we consider two qubits.

3.12 Quantum Feature Maps with Two Qubits

We consider two qubits ($D = 2$), denoted as qubit "0" and qubit "1." We have $2^D = 4$ basis vectors, and we use entangled states for the feature map.

Entanglement appears when introducing controlled gates, for example, rotating qubit 1 if the qubit 0 is in a specific state.

Let us consider the operator

$$e^{\mathrm{i}\phi\hat{Z}\otimes\hat{Z}} = \cos(\phi)\hat{I}\otimes\hat{I} + \mathrm{i}\sin(\phi)\hat{Z}\otimes\hat{Z} \tag{3.61}$$

applied to a two-qubit state $|\psi_0\rangle \otimes |\psi_1\rangle$. When the qubit 0 is in a vector of the basis $|\psi_0\rangle = |z_0\rangle$ with $z_0 \in \{0, 1\}$, we have

$$e^{\mathrm{i}\phi\hat{Z}\otimes\hat{Z}}|z_0\rangle|\psi_1\rangle = e^{\mathrm{i}\phi(-1)^{z_0}\hat{I}\otimes\hat{Z}}|z_0\rangle|\psi_1\rangle\ , \tag{3.62}$$

which corresponds to a Z-rotation of angle ϕ for $|\psi_1\rangle$ if $z_0 = 0$, or angle $-\phi$ if $z_0 = 1$. If the control qubit 0 is $|0\rangle$, the controlled qubit 1 undergoes a ϕ Z-rotation; if the control qubit 0 is $|1\rangle$, the rotation angle is $-\phi$.

The operator in Eq. (3.61) corresponds to a sequence of a CNOT, a rotation for qubit 1, and a further CNOT. If the qubit 0 is in the ground state $|0\rangle$, the qubit 1 is rotated by angle ϕ. On the contrary, if A is $|1\rangle$, B rotates about Z by $-\theta$. The second CNOT gate is needed to restore the original basis in the B.

The final state is entangled, and we use the entanglement to encode some binary combination of the input data. If our datum is a couple of real variables X_0 and X_1, such that $\mathbf{X} = (X_0, X_1)$, we use the controlled gate to store information about combinations of X_0 and X_1. For example, by letting

$$\phi_{01} = (\pi - X_0)(\pi - X_1)\ , \tag{3.63}$$

authors in [1] use the operator

$$e^{\mathrm{i}\phi_{01}\hat{Z}\otimes\hat{Z}}\ . \tag{3.64}$$

The rationale behind this approach is that we want to represent a function of two variables $f(X_0, X_1)$, which can be expanded into powers of X_0 and X_1 and combinations, e.g.,

$$f(X_0, X_1) = X_0^2 + X_0X_1\ . \tag{3.65}$$

We use the one-qubit gates for the part of this expansion that only depends on X_0 or on X_1, while we use entanglement gates for the terms containing combinations like X_0X_1. This is not the only strategy but looks to be the first we can reasonably conceive to exploit entanglement.

The resulting feature map operator $\hat{U}_\phi$ is

$$\hat{U}_\phi = \left[e^{i\phi_0(X_0)\hat{Z}} \otimes \hat{I}\right]\left[\hat{I} \otimes e^{i\phi_1(X_1)\hat{Z}}\right] e^{i\phi_{01}\hat{Z}\otimes\hat{Z}} . \tag{3.66}$$

We observe that

$$\begin{aligned} e^{i\phi_0\hat{Z}} \otimes \hat{I} &= [\cos(\phi_0)I + i\sin(\phi_0)Z] \otimes \hat{I} = \cos(\phi_0)I \otimes I + i\sin(\phi_0)Z \otimes \hat{I} \\ &= e^{i\phi_0\hat{Z}\otimes\hat{I}} , \end{aligned} \tag{3.67}$$

and also

$$\hat{I} \otimes e^{i\phi_1\hat{Z}} = e^{i\phi_1\hat{I}\otimes\hat{Z}} . \tag{3.68}$$

Introducing

$$\begin{aligned} \hat{Z}_0 &= \hat{Z} \otimes \hat{I} \\ \hat{Z}_1 &= \hat{I} \otimes \hat{Z} , \end{aligned} \tag{3.69}$$

such that as $\hat{Z}_j$ is the Z gate acting only the qubit j, with $j = 0, 1, \dots D - 1$, we simplify the notation as

$$\hat{U}_\phi = e^{i\phi_0(X_0)\hat{Z}_0} e^{i\phi_1(X_1)\hat{Z}_1} e^{i\phi_{01}\hat{Z}\otimes\hat{Z}} . \tag{3.70}$$

We encode the two real inputs as follows:

$$\phi_0 = X_0 \tag{3.71}$$

$$\phi_1 = X_1 \tag{3.72}$$

$$\phi_{10} = (\pi - X_0)(\pi - X_1) , \tag{3.73}$$

which is the feature map studied experimentally in [1]. The input dataset is scaled such that $X_{0,1} \in (0, 2\pi]$.

One can imagine more complex encoding by cascading the quantum circuit, which increases the degree of entanglement between the qubits.

3.13 Coding the Entangled Feature Map

To implement the entangled feature map in `TensorFlow`, we need to realize the rotation operators, like $e^{\mathrm{i}\theta\hat{\mathrm{Z}}}$. In computing the exponential function, we use

$$e^{\mathrm{i}\theta\hat{\mathrm{A}}} = \cos(\theta)\hat{\mathrm{I}} + \mathrm{i}\sin(\theta)\hat{\mathrm{A}}\ , \tag{3.74}$$

which holds for $\hat{A} \in \{\hat{\mathrm{I}}, \hat{\mathrm{X}}, \hat{\mathrm{Y}}, \hat{\mathrm{Z}}\}$ and for any operator such that $\hat{A}^2 = \hat{\mathrm{I}}$. For two qubits, we define rotations that act only on the first or the second qubit, as $\hat{\mathrm{I}} \otimes e^{\mathrm{i}\theta\hat{\mathrm{Z}}}$ or $e^{\mathrm{i}\theta\hat{\mathrm{Z}}} \otimes \hat{\mathrm{I}}$. The tensors corresponding to these operators are outer products with `Isingle` as follows.

```
@tf.function
def EZI(theta):
    """ return exp(theta Z0)  """
    ct = tf.cast(tf.math.cos(theta), dtype=tfcomplex)
    st = tf.cast(tf.math.sin(theta), dtype=tfcomplex)
    EZS = ct*Isingle+1j*st*Zsingle
    EZ0 = tf.tensordot(EZS, Isingle, axes=0)
    return EZ0

@tf.function
def EIZ(theta):
    """ return  exp(theta Z1) """
    ct = tf.cast(tf.math.cos(theta), dtype=tfcomplex)
    st = tf.cast(tf.math.sin(theta), dtype=tfcomplex)
    EZS = ct*Isingle+1j*st*Zsingle
    EZ0 = tf.tensordot(Isingle, EZS, axes=0)
    return EZ0
```

Equation (3.74) also holds for two-qubit gates as `ZZ`. The operator

$$e^{\mathrm{i}\theta\hat{\mathrm{Z}}\otimes\hat{\mathrm{Z}}} = \cos(\theta)\hat{\mathrm{I}}^{\otimes 2} + \mathrm{i}\sin(\theta)\hat{\mathrm{Z}}^{\otimes 2} \tag{3.75}$$

is a controlled rotation such that, if the first qubit is 1, the rotation is θ, and $-\theta$ otherwise. This operator is coded using the two-qubit `tf.function`.

```
@tf.function
def EZZ(theta):
    """ return exp(theta Z0 Z1) """
    ct = tf.cast(tf.math.cos(theta), dtype=tfcomplex)
    st = tf.cast(tf.math.sin(theta), dtype=tfcomplex)
    EZZ = ct*II+1j*st*ZZ
    return EZZ
```

3.13.1 The Single Pass Feature Map

We define a function with input parameters θ_0, θ_1, and θ_{01} and return the image state as follows:

$$|\Phi(\mathbf{X})\rangle = \hat{U}_\Phi|00\rangle = e^{i\theta_0\hat{Z}_0}e^{i\theta_1\hat{Z}_1}e^{i\theta_{01}\hat{Z}\otimes\hat{Z}}\hat{H}^{\otimes 2}|00\rangle\ . \tag{3.76}$$

Equation (3.76) is defined as a `@tf.function` function with three real numbers `theta0`, `theta1`, and `theta01` as inputs.

```
@tf.function
def featuremapU(phi, theta0, theta1, theta01):
    """ given a state phi, return the state after a feature map
↪   """
    phi1=gate(H,phi)
    phi2=gate(expZ0(theta0),phi1)
    phi3=gate(expZ1(theta1),phi2)
    phi4=gate(expZZ(theta01), phi3)
    return phi4
```

3.13.2 The Iterated Feature Map

Once we have a function for the feature map, we iterate the mapping to create a more complex representation of the data. For example, the authors of [1] use a double iteration as

$$|\Phi(\mathbf{X})\rangle = \mathcal{U}_\Phi(\mathbf{X}) = \hat{U}_\Phi(\mathbf{X})\hat{U}_\Phi(\mathbf{X})|00\rangle\ , \tag{3.77}$$

being $\boldsymbol{X} = (\theta_0, \theta_1, \theta_{01})$. Equation (3.77) in `TensorFlow` reads

```
@tf.function
def featuremapUU(phi, theta0, theta1, theta01):
    """ iterated feature map """
    phi1 = featuremapU(phi, theta0, theta1, theta01)
    phi2 = featuremapU(phi1, theta0, theta1, theta01)
    return phi2
```

The reason to use a double iteration is that one obtains maps that are computationally demanding to simulate in classical computing. Thus the resulting model is advantageous for a quantum computer and facilitates to claim quantum advantage for these specific feature mappings. The final goal is to demonstrate that certain tasks can only be done on quantum computers in a human-scale time, a challenge that is still hard to undertake.

3.14 The Entangled Kernel and Its Measurement with Two Qubits

We can now use the previous map to introduce a kernel as follows:

$$K(\mathbf{X}, \mathbf{Y}) = |\langle \Phi(\mathbf{X})|\Phi(\mathbf{Y})\rangle|^2 = |\langle 00|\mathcal{U}_\Phi(\mathbf{X})^\dagger \mathcal{U}_\Phi(\mathbf{Y})|00\rangle|^2 . \tag{3.78}$$

The kernel in Eq. (3.78) corresponds to the probability of finding the state

$$\mathcal{U}_\Phi(\mathbf{X})^\dagger \mathcal{U}_\Phi(\mathbf{Y})|00\rangle \tag{3.79}$$

in the state $|00\rangle$. It is the probability of counting 0 particles in each mode, i.e., sampling the probability of the output at a specific pattern.

Therefore, we find a connection between the boson sampling and the quantum kernel mapping.

We define the following protocol:

1. Build a tunable quantum computer that implements the feature map.
2. For each couple of inputs, $\mathbf{X}^{(i)}$ and $\mathbf{X}^{(j)}$, set the parameters to implement the feature map.
3. Perform runs of the quantum computer to determine the probability of observing the output in the ground state $|00\rangle$.
4. The resulting sampled probability distribution is the value of the kernel $K\left[\mathbf{X}^{(i)}, \mathbf{X}^{(j)}\right]$.
5. Train the kernel machine with the available dataset and labels.
6. Use the quantum computer to test the prediction with the test dataset.

3.15 Classical Simulation of the Entangled Kernel

We simulate the protocol by computing the $\Pr(\mathbf{n} = \{0, 0\})$ using `scikit-learn`, by following the same strategy as in Chap. 1.[10]

First, we define a `numpy` function for the kernel by the mapping in Eq. (3.73).

[10] Details in the notebook `jupyternotebooks/quantumfeaturemap/QuantumKernelMachineQubits.ipynb`.

```
# define a map from the input dataset to the angles
def MapTheta(x):
    """ take a two vector and return a theta vector """
    theta0=x[0]
    theta1=x[1]
    theta01=(np.pi-x[0])*(np.pi-x[1])
    return [theta0, theta1,theta01]
# Define a kernel function in numpy for the scikit-learn.
# We must define the kernel in a way such that it can handle
# a multidimensional input (a batch of input data)
# and return a numpy function
def Kernel_np(xA_np,xB_np):
    """ compute the Kernel function """
    nb1, nx= xA_np.shape
    nb2, ny= xB_np.shape
    out = np.zeros((nb1,nb2))
    # loop of the data to have
    # the proper size in the output
    for i1 in range(nb1):
        for i2 in range(nb2):
            xB=[tf.constant(xB_np[i2,0]),tf.constant(xB_np
              [i2,1])]
            xA=[tf.constant(xA_np[i1,0]),tf.constant(xA_np
              [i1,1])]
            thetaA=MapTheta(xA)
            thetaB=MapTheta(xB)
            psiA=FeatureMapUU(qm.q00,thetaA)
            psiB=FeatureMapUU(qm.q00,thetaB)
            sc=qm.Scalar(psiA,psiB)
            # modulus square
            abs2 = tf.square(tf.abs(sc))
            # convert the output in numpy
            out[i1,i2]=abs2.numpy()
    return out
```

We use the function to plot the kernel map, as in Fig. 3.1.

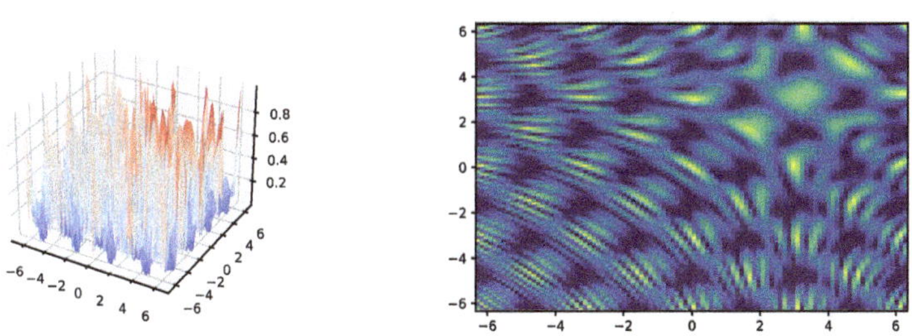

Fig. 3.1 Entangled quantum map with two-dimensional data (kernel function $K(\mathbf{X}), \mathbf{Y})$ at $\mathbf{Y} = \mathbf{0}$ as a function of X_0 and X_1)

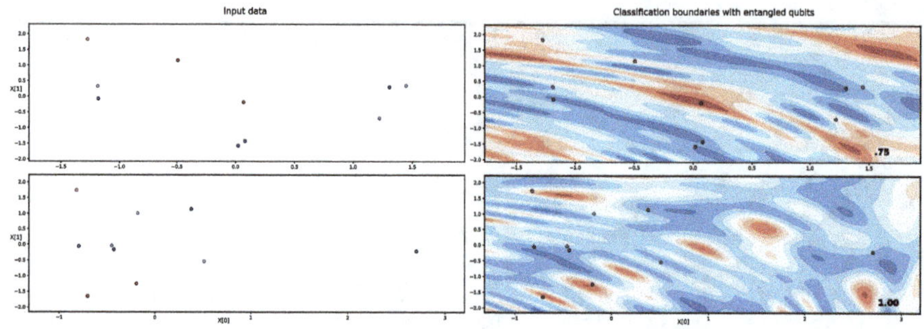

Fig. 3.2 Result of two classifications with the entangled qubit kernel. The right panels also show the classification score

Note that the kernel profile is much more complex than the kernels in Chap. 1.

We perform the training and classification by using the `fit` function of the `scikit-learn` package. We show in Fig. 3.2 the results for a typical dataset.

The classifier provides a good score for the example reported, which we have limited to a few points for obtaining the results in a small amount of time.

3.15.1 A Remark on the Computational Time

For the qubit classifier, the computing time is much longer than for the squeezing/coherent case in the previous chapter. This difference outlines how the kernel function is computationally hard. Thus, in principle, we could have a quantum advantage in implementing this model by a real quantum computer. However, for all the inputs in the dataset, we must perform boson sampling, which is demanding if we want a good estimate of the means and standard deviations. We have evidence that an entangled map is demanding with classical computers. The existence of a quantum advantage for kernel classifiers is an open issue.

3.16 Further Reading

- The quantum machine learning paper by Gambetta: The first demonstration of quantum machine learning [1].
- Tensor notation for qubits: Penrose in [3] describes different notations for tensors including qubits.

References

1. V. Havlíček, A.D. Córcoles, K. Temme, A.W. Harrow, A. Kandala, J.M. Chow, J.M. Gambetta, **567**, 209 (2019). https://doi.org/10.1038/s41586-019-0980-2
2. C.W. Wisner, K.S. Thorne, J.A. Wheeler, *Gravitation* (W. H. Freeman, 1973)
3. R. Penrose, *The Road to Reality* (Jonathan Cape, 2007)

Chapter 4
One-Qubit Transverse-Field Ising Model and Variational Quantum Algorithms

Two-qubit or not two-qubit.

Abstract We give a first introduction to combinatorial optimization with the Ising model and its quantum counterpart. We start investigating the basics of the transverse-field Ising model in the case of one qubit. We give an example of a neural network variational ansatz for the ground state.

4.1 Introduction

The Ising model dates back to more than one century ago with the work of Lenz and Ising [1]. The model was introduced to study the interactions of spins arranged in a solid. How do tiny magnets align when they are in regular or disordered configurations? What is their ground state? Despite the strong link with condensed matter physics, the range of applications of this model increased impressively in the last few years. Nowadays, we can identify the Ising model as the prototype of any optimization problem. A significant aspect is the link between a spin glass (a generic name for systems of interacting spins with disordered couplings) and a neural network. The link is so fundamental in a way such that we can retain spin glasses as one leading theoretical paradigm in machine learning [2].

Classically, spins are vectorial degrees of freedom represented by arrows that arrange in any possible direction in space. In a simplified picture, each spin has only two directions tagged by a binary variable $\sigma = \pm 1$. A spin may represent the status of different physical systems, such as a firing neuron or a traffic light.

With the classical Ising model, we denote the function

$$H = J_{12}\sigma_1\sigma_2 + J_{13}\sigma_1\sigma_3 + \ldots + J_{23}\sigma_2\sigma_3 + \ldots = \sum_{i<j} J_{ij}\sigma_i\sigma_j \ , \tag{4.1}$$

with i and j in the range $[0, N-1]$. Equation (4.1) is the Hamiltonian of a spin system coupling N binary variables $\sigma_0, \sigma_1, \ldots \sigma_{N-1}$ through the interaction

C. Conti, *Quantum Machine Learning*, Quantum Science and Technology,
https://doi.org/10.1007/978-3-031-44226-1_4

coefficients J_{ij}. Conventionally, one often defines the Ising Hamiltonian with a minus sign [2], as in[1]

$$H = -\sum_{i<j} J_{ij}\sigma_i\sigma_j \,. \tag{4.2}$$

4.2 Mapping Combinatorial Optimization to the Ising Model

Before considering the quantum counterpart, we give two examples of the connection between the Ising model and combinatorial optimization. Indeed, the modern importance of minimizing the Ising Hamiltonian is in that it is a universal task encompassing many combinatorial problems [3].

4.2.1 Number Partitioning

As shown in Fig. 4.1, we consider a set of real numbers $(\xi_0, \xi_1, \xi_2, \ldots \xi_{N-1})$, and we want to split into two balanced ensembles A and B such that

$$\sum_{\xi_j\in A} \xi_j \simeq \sum_{\xi_j\in B} \xi_j \,, \tag{4.3}$$

which corresponds to minimizing the quantity

$$H = \frac{1}{2}\left(\sum_{\xi_j\in A} \xi_j - \sum_{\xi_j\in B} \xi_j\right)^2 \tag{4.4}$$

by a proper choice of the partitions of the ξs in the two subsets A and B. The factor 1/2 in (4.4) is helpful below. This problem maps to the Ising model by introducing the spin variables σ_j, with one spin for each number ξ_j.

If $\xi_j \in A$, we have $\sigma_j = 1$; if $\xi_j \in B$, we have $\sigma_j = -1$. Different configurations of spins correspond to different partitions in the ensembles A and B. Minimizing (4.4) means finding a proper spin state $\sigma_0, \sigma_1, \ldots, \sigma_{N-1}$. Indeed, we write (4.4) as

[1] As a matter of notation, the original literature denotes as "Ising model" only the simplest case with $J_{ij} = J$, corresponding to uniform coupling. On the other hand, when J_{ij} is a random variable, we have the "Sherrington-Kirkpatrick" or the "Edwards-Anderson" model, depending on the coupling ranges (see Section "Further Reading"). Therefore, we will loosely refer to the Ising model to simplify the scenario when considering a classical system of interacting binary spins with any coupling.

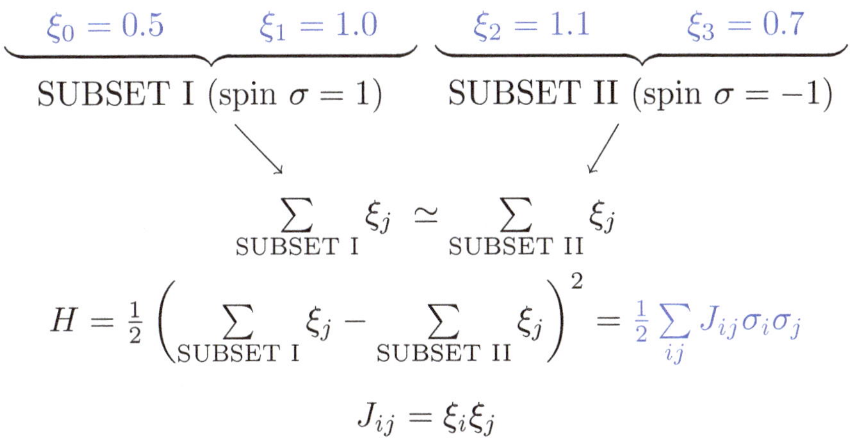

Fig. 4.1 In the number partitioning problem, one has to split a set of real numbers $\xi_0, \xi_1, \ldots, \xi_{N-1}$ into two balanced subsets. The problem can be solved by minimizing an Ising Hamiltonian (given in the figure in blue within an additive constant, as detailed in the text). The strategy is tagging each real number ξ_j with a binary spin σ_j

$$H = \frac{1}{2}\left(\sum_j \sigma_j\xi_j\right)^2 = \frac{1}{2}\sum_{ij} \xi_i\xi_j\sigma_i\sigma_j$$
$$= \frac{1}{2}\sum_{i<j} \xi_i\xi_j\sigma_i\sigma_j + \frac{1}{2}\sum_{i>j} \xi_i\xi_j\sigma_i\sigma_j + \frac{1}{2}\sum_i \xi_i^2\sigma_i^2 \,. \tag{4.5}$$

Being $\sigma_i^2 = 1$, and letting $\langle\xi^2\rangle = \sum_j \xi^2/N$, we have

$$H = \sum_{i<j} J_{ij}\sigma_i\sigma_j + \frac{N}{2}\langle\xi^2\rangle \,, \tag{4.6}$$

with

$$J_{ij} = \xi_i\xi_j \,. \tag{4.7}$$

As $\langle\xi^2\rangle$ is a constant that does not depend on the partition, minimizing (4.4) is equivalent to minimizing the Ising Hamiltonian in (4.1) with the couplings in (4.7) determined by our data ξ_j.

The former argument also shows that minimizing with the spin variables σ_j the Ising Hamiltonian (4.4) is equivalent to minimizing the function

$$H = \sum_{i<j} J_{ij}\sigma_i\sigma_j \,, \tag{4.8}$$

with $J_{ij} = J_{ji}$. Thus we will often write the Ising Hamiltonian as in (4.8).

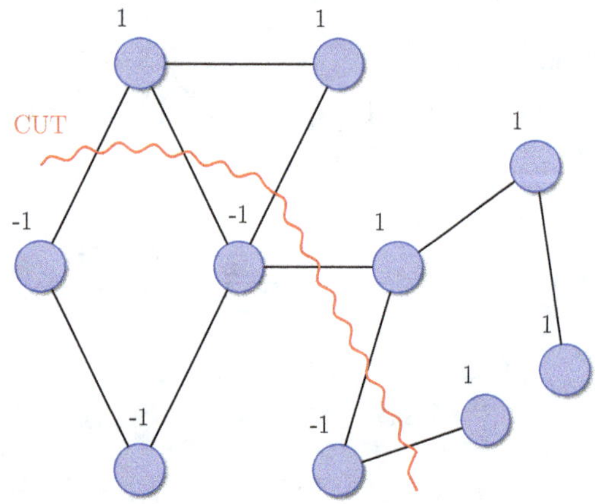

Fig. 4.2 A cut in a graph (E, V) made of edges E and vertices V splits the vertices into two sets, which we identify by tagging each vertex with a binary spin. In the (unweighted) max-cut problem, one has to find the cut that involves the largest number of edges. The max-cut is solved by minimizing an Ising Hamiltonian with respect to the binary spins

4.2.2 *Maximum Cut*

As a further example, we consider an important problem in graph theory, which is often used to benchmark devices for combinatorial optimization. Any Ising model as in (4.8), defined by the coupling J_{ij}, maps to a graph and to the problem of finding the max-cut of that graph, i.e., the largest cut involving the maximum number of edges (see Fig. 4.2). This is not too different from the number partitioning problem. The distinctive aspect is that we are dealing with edges of a graph. We attribute a spin $\sigma_j = 1$ to vertices in a partition A of the graph and $\sigma_j = -1$ to the vertices of the other partition B. An edge belongs to the cut if the two corresponding vertices have opposite spins.

The graph[2] $G = (V, E)$ is the set of vertices V and edges E. We have a total number N of vertices, each denoted by an index $j = 0, 1, \ldots, N-1$. Each edge is a couple of vertices identified by an unordered couple of indices (i, j).

The number of edges in the cut is computed as

$$N_{CUT} = \frac{1}{2} \sum_{(i,j)\in E} (1 - \sigma_i \sigma_j) , \tag{4.9}$$

which can be written as

$$N_{\mathrm{CUT}} = \frac{|E|}{2} - \frac{1}{2} \sum_{(i,j)\in E} \sigma_i \sigma_j , \tag{4.10}$$

where $|E|$ is the number of edges.

[2] We consider a simple undirected graph.

Thus the max-cut problem corresponds to maximize N_{CUT}, i.e., finding the minimum of (4.8) with $J_{ij} = 1$ with the sum extending over the edges of G. Minimizing the Ising Hamiltonian is equivalent to maximizing the cut.

An alternative formulation attributes a number $x_j = 0$ to the vertices in one cut and $x_j = 1$ to vertices in the other cut. The total number of cut edges is

$$N_{\text{CUT}} = \sum_{ij} x_i(1 - x_j) \,, \tag{4.11}$$

with i and j running over the vertices.

Equation (4.11) is equivalent to (4.10) by letting $x_j = (1 - \sigma_j)/2$, such that

$$\begin{aligned} N_{\text{CUT}} &= \sum_{ij} \frac{1}{4}(1 - \sigma_i)(1 + \sigma_j) = \frac{N^2}{4} - \sum_{ij} \frac{1}{4}\sigma_i\sigma_j \\ &= \frac{|E|}{2} - \frac{1}{2}\sum_{i<j} \sigma_i\sigma_j \,. \end{aligned} \tag{4.12}$$

The problem is generalized by considering a *weighted* graph such that we have a real number J_{ij} for each edge. The weighted max-cut corresponds to finding the cut that maximizes the sum of weights in the cut. Equation (4.9) becomes

$$N_{\text{CUT}} = \frac{1}{2} \sum_{(i,j)\in E} J_{ij}(1 - \sigma_i\sigma_j) \tag{4.13}$$

and the corresponding Ising Hamiltonian is (4.8).

Even though the mappings between the number partitioning and max-cut to the Ising model look similar, there is an important difference between the two optimizations. The complexity class of number partitioning is **P** while for the max-cut is **NP**; however, care is needed as the complexity depends on the specific graph topology [4].

4.3 Quadratic Unconstrained Binary Optimization (QUBO)

In the discussion of maximum cut, we have shown that one obtains an equivalent formulation either in terms of binary variables x_j with values in $\{0, 1\}$ or spins σ_j with values in $\{-1, 1\}$.

Optimization problems expressed with 0s and 1s are a large class, which includes the quadratic unconstrained binary optimization (QUBO). QUBO tasks are equivalent to the minimization of an Ising Hamiltonian and are solved by the most advanced quantum computers such as D-Wave.[3]

The cost function for a QUBO problem is [see Eq. (4.11)]

$$f_{\text{QUBO}}(\boldsymbol{x}) = \sum_{i=0}^{N-1} \sum_{j=0}^{N-1} Q_{ij} x_i x_j \tag{4.14}$$

with $x_i \in \{0, 1\}$. $f_{\text{QUBO}}(\boldsymbol{x})$ maps to an Ising model by letting $x_j = (1 + \sigma_j)/2$ with $\sigma_j \in \{-1, 1\}$.

4.4 Why Use a Quantum Computer for Combinatorial Optimization?

To understand the way a quantum computer solves a binary optimization task, we have to consider the quantum counterpart of Eq. (4.1), obtained by promoting the spins σ_j to operators $\hat{\sigma}_j$. As the spins are binary variables, they correspond to a two-level system, i.e., a qubit.

Conventionally, one attributes the state $|0\rangle$, or $|\downarrow\rangle$, to the spin with value $\sigma = -1$, pointing in the negative Z-direction, and the state $|1\rangle$, or $|\uparrow\rangle$, to $\sigma = 1$ in the positive Z-direction. Both states are eigenvectors of the magnetic momentum operator $\hat{\text{Z}} = \hat{\sigma}$.

The quantum operator (the quantum Ising model Hamiltonian) corresponding to (4.1) reads

$$\hat{\mathcal{H}} = \sum_{i<j} J_{ij} \hat{\text{Z}}_i \otimes \hat{\text{Z}}_j \ . \tag{4.15}$$

In a quantum system with Hamiltonian as in (4.15), the eigenstates are configurations of spins $|\boldsymbol{z}\rangle = |z_0, z_1, \ldots, z_{N-1}\rangle$, such that

$$\hat{\text{Z}}_j |z_j\rangle = (-1)^{z_j} |z_j\rangle = \sigma_j |z_j\rangle \ . \tag{4.16}$$

The energy of the ground state

$$\langle \boldsymbol{z}|\hat{\mathcal{H}}|\boldsymbol{z}\rangle = \sum_{i<j} J_{ij} \sigma_i \sigma_j \tag{4.17}$$

is the minimum energy of the classical model.

The rationales in using a quantum system for this minimization are as follows:

[3] https://www.dwavesys.com/.

Fig. 4.3 A quantum system with the transverse-field Ising Hamiltonian relaxes to the ground state, which corresponds to the minimum of the classical Ising model

1. The quantum system has a set of truly binary variables; hence it provides the minimum with binary variables, at variance with spins in classical systems, which are not quantized.
2. The quantum evolution is inherently parallel, governed by superposition, and one can expect a quantum advantage during minimization.
3. Minimization of complex functions with several minima may be improved by quantum tunnelling, which helps exiting the trap of local minima.

Overall, the strategy to solve a combinatorial optimization problem is first mapping to a QUBO or an Ising Hamiltonian and then using a quantum computer to find the ground state (see Fig. 4.3).

In the following, we will study how to design a quantum feature map to minimize the Ising Hamiltonian and the related variational algorithms.

4.5 The Transverse-Field Ising Model

The transverse-field Ising model is defined by the following Hamiltonian:

$$\hat{\mathcal{H}} = -\sum_{ij} J_{ij} \hat{\mathrm{Z}}_i \otimes \hat{\mathrm{Z}}_j - \sum_i h_i \hat{\mathrm{X}}_i \tag{4.18}$$

In Eq. (4.18), $\hat{\mathrm{Z}}_j$ and $\hat{\mathrm{X}}_j$ are the Pauli operators acting on the qubit with index j. The Hilbert space is the product of N qubit spaces with dimensions 2^N.

Note that Eq. (4.18) is a more general model than (4.15), as it involves a field term in the X-direction. The field h_i has the role of exciting superpositions of qubits to explore the phase space. In the absence of the field, a state like $|z_1 z_2\rangle$ is unperturbed during the evolution, as we will discuss below.

We simplify the notation by omitting the exterior product symbol, as follows:

$$\hat{\mathcal{H}} = -\sum_{ij} J_{ij} \hat{\mathrm{Z}}_i \hat{\mathrm{Z}}_j - \sum_i h_i \hat{\mathrm{X}}_i \ . \tag{4.19}$$

4.6 One-Qubit Transverse-Field Ising Model

With only one qubit, we have[4]

$$\hat{\mathcal{H}} = -J\hat{Z} - h\hat{X}\,, \tag{4.20}$$

where we assume $J \geq 0$ and $h \geq 0$. In matrix form, the Hamiltonian reads

$$\hat{\mathcal{H}} = \begin{pmatrix} -J & -h \\ -h & J \end{pmatrix}\,. \tag{4.21}$$

We are interested in finding the eigenstates

$$\hat{\mathcal{H}}|E\rangle = E|E\rangle\,. \tag{4.22}$$

The eigenvalues of $\hat{\mathcal{H}}$ are (see Fig. 4.4)

$$E_\pm = \pm\sqrt{J^2 + h^2} \tag{4.23}$$

with the corresponding normalized eigenvectors[5]

$$|E_-\rangle = \frac{1}{\sqrt{h^2 + (J + \sqrt{h^2 + J^2})^2}} \begin{pmatrix} J + \sqrt{h^2 + J^2} \\ h \end{pmatrix} \tag{4.24}$$

and

$$|E_+\rangle = \frac{1}{\sqrt{h^2 + (-J + \sqrt{h^2 + J^2})^2}} \begin{pmatrix} J - \sqrt{h^2 + J^2} \\ h \end{pmatrix}\,. \tag{4.25}$$

It is instructive to consider some particular cases.

4.6.1 $h = 0$

In this case the Hamiltonian is proportional to $\hat{Z}$; we have after (4.20)

$$\hat{\mathcal{H}} = -J\hat{Z} = \begin{pmatrix} -J & 0 \\ 0 & J \end{pmatrix} \tag{4.26}$$

---

[4] Details in the notebook `jupyternotebooks/quantumfeaturemap/SingleQubit TransverseFieldIsing.ipynb`.

[5] See the `Mathematica` file `mathematica/SingleQubitTransverseIsing.nb`.

such that the normalized eigenstates are the basis vectors $|E_-\rangle = |0\rangle = |\downarrow\rangle$ and $|E_+\rangle = |1\rangle = |\uparrow\rangle$ with eigenvalues

$$E_\pm = \pm J \; . \tag{4.27}$$

Thus the ground state is $|0\rangle$, with energy $E_- = -J < 0$, and the energy gap is $E_+ - E_- = 2J$ (see Fig. 4.4).

4.6.2 $J = 0$

The Hamiltonian is proportional to $\hat{\mathrm{X}}$,

$$\hat{\mathcal{H}} = -h\hat{\mathrm{X}} = \begin{pmatrix} 0 & -h \\ -h & 0 \end{pmatrix} \tag{4.28}$$

with normalized eigenvectors

$$|E_-\rangle = \frac{1}{\sqrt{2}}\begin{pmatrix} 1 \\ 1 \end{pmatrix} = \hat{\mathrm{H}}|0\rangle = |\nearrow\rangle \; , \tag{4.29}$$

and

$$|E_+\rangle = \frac{1}{\sqrt{2}}\begin{pmatrix} -1 \\ 1 \end{pmatrix} = \hat{\mathrm{H}}|1\rangle = |\nwarrow\rangle \; , \tag{4.30}$$

obtained by the Hadamard gate $\hat{\mathrm{H}}$ on the basis vectors.

The corresponding energies are

$$E_\pm = \pm h \; , \tag{4.31}$$

such that $2h > 0$ is the energy gap between the two levels (Fig. 4.4).

4.7 Coding the One-Qubit Transverse-Field Ising Hamiltonian

4.7.1 The Hamiltonian as a `tensor`

We start defining a variational quantum model to minimize the transverse-field Ising model Hamiltonian with a single qubit. We choose the ansatz represented in Fig. 4.5.

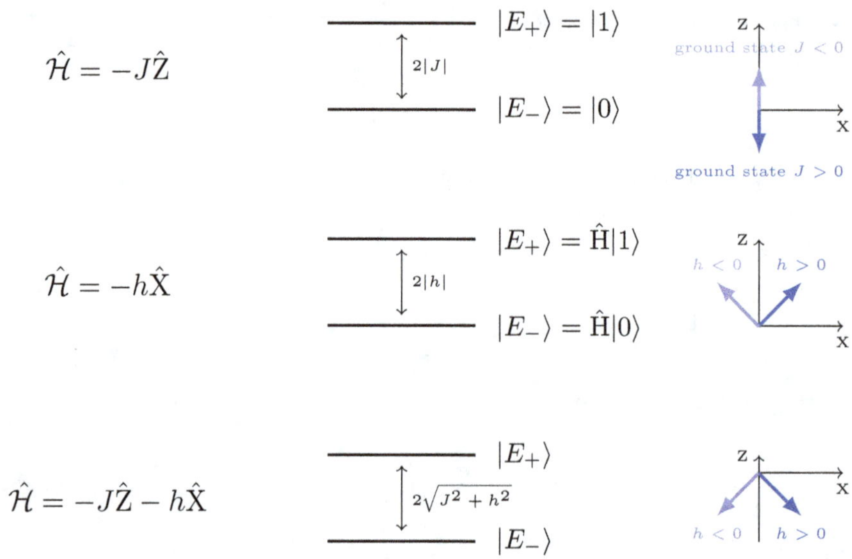

Fig. 4.4 Energy levels for the one-qubit transverse-field Ising Hamiltonian for different parameters. We also show the corresponding ground state Bloch vector in (X, Z) plane

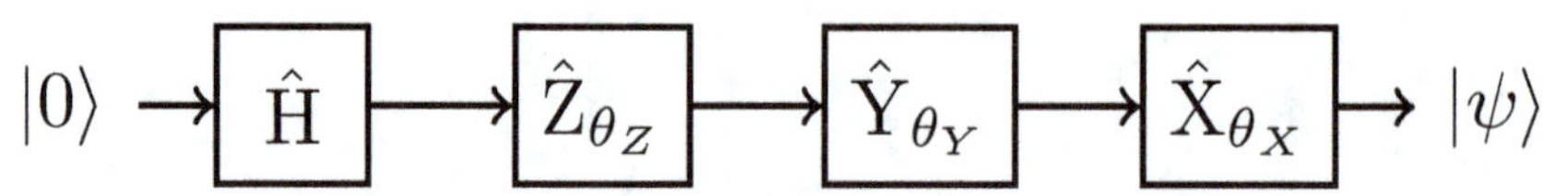

Fig. 4.5 Circuit for the quantum feature map for the one-qubit transverse-field Ising model

$$|\psi\rangle = e^{i\theta_X \hat{X}} e^{i\theta_Y \hat{Y}} e^{i\theta_Z \hat{Z}} \hat{H}|0\rangle \ . \tag{4.32}$$

The variational ansatz is determined by a unitary operator $\hat{U}(\theta_X, \theta_Y, \theta_Z)$, defined as

$$\hat{U} = e^{i\theta_X \hat{X}} e^{i\theta_Y \hat{Y}} e^{i\theta_Z \hat{Z}} \hat{H} \tag{4.33}$$

The parameters θ_X, θ_Y, and θ_Z are the parameter of the variational model, which we collect in a single vector $\boldsymbol{\theta} = (\theta_X, \theta_Y, \theta_Z)$.

We want to write a code that computes the Hamiltonian in (4.20) and find the parameters of (4.32), to minimize the expected value $\langle\psi|\hat{\mathcal{H}}|\psi\rangle$.[6]

[6] Details in the notebook `jupyternotebooks/quantumfeaturemap/SingleQubitTransverseFieldIsing.ipynb`.

We use the library `quantummap` in `thqml` that contains the various definitions of gates and related as detailed in the previous chapters. The Hamiltonian is a function returning a tensor.

```
# import the quantummap
from thqml import quantummap as quantummap
# recall the quantummap contains the definition
# of the relevant variables, as in previous chapters
# define the function returning the Hamiltonian
@tf.function
def Hamiltonian(J=1.0,h=0.0):
    """ single qubit tranverse Ising model Hamiltonian

        H=-J Z-h X

        Params
        ------
        h = field/gap (positive)
        J = Z coefficient (positive)
    """
    out =-J*quantummap.Zsingle -h*quantummap.Xsingle

    return out
```

Note that we use the `tf.tensors` `quantummap.Xsingle` and `quantummap.Zsingle` in the `Hamiltonian` function. We test `Hamiltonian` as follows.

```
H=Hamiltonian()
print(H)
# returns
tf.Tensor(
[[-1.+0.j -0.+0.j]
 [-0.+0.j  1.-0.j]], shape=(2, 2), dtype=complex64)
H=Hamiltonian(J=1.0,h=1.0)
print(H)
tf.Tensor(
[[-1.+0.j -1.+0.j]
 [-1.+0.j  1.-0.j]], shape=(2, 2), dtype=complex64)
```

4.7.2 Variational Ansatz as a `tensor`

We define the variational ansatz by

```
@tf.function
def FeatureMapU(theta, psi=quantummap.qubit0):
    """ Feature map with 1 qubits
        used as a variational ansatz for
        the one-qubit transverse field Ising H

        Hadamard and rotation

        Exp[theta(0)X]Exp[theta(1)*Y]Exp[theta(2)*Z]H|0\rangle

        Params
        ------
        theta[0] : rotation in the X
        theta[1] : rotation in the Y
        theta[2] : rotation in the Z

    """
    thetaX=theta[0]
    thetaY=theta[1]
    thetaZ=theta[2]
    psi=quantummap.Gate1(quantummap.Hsingle,psi)
    psi=quantummap.Gate1(quantummap.EX(thetaX),psi)
    psi=quantummap.Gate1(quantummap.EY(thetaY),psi)
    psi=quantummap.Gate1(quantummap.EZ(thetaZ),psi)

    return psi
```

The variational ansatz is a quantum feature map from the classical parameters $\boldsymbol{\theta}$ to a one-qubit Hilbert space. In the feature map, we use the exponential function of the rotations `quantummap.EX`, `quantummap.EY`, `quantummap.EZ`, and the gate function `quantummamp.Gate1`, which is the gate for one qubit.

```
# apply gate function on single qubit
@tf.function
def Gate1(A, psi):
    # here A is (2,2) tensor for a 2x2 matrix
    # psi is (2,) tensor for a single qubit
    # the gate is the matrix vector dot multiplication
    # as a contraction of index 1 for A
    # and index 0 for psi
    return tf.tensordot(A, psi, axes=[1, 0])
```

We test the quantum feature map as follows.

```
# Define a vector of parameters
theta=[2.0,0.0,1.0]
```

```
# Print the feature map at theta
print(FeatureMapU(theta))
# returns
#  tf.Tensor([-0.7    +0.0998j  0.3821+0.595j ], shape=(2,),
↪  dtype=complex64)
```

As expected, the variational ansatz returns a state corresponding a single qubit, i.e., a complex-valued `tensor` with `shape=(2,)`.

4.8 A Neural Network Layer for the Hamiltonian

We define a neural network model returning the expected value of the Hamiltonian on the variational ansatz $|\psi\rangle$. We use a custom layer for later integration in a trainable model. The layer is designed in a way it includes the parameters of the model, i.e., $\boldsymbol{\theta}$, and uses the `FeatureMapU` function.

```
class HamiltonianLayer(tf.keras.layers.Layer):
    # Define a layer returning the mean value of the Hamiltonian
    # on the variational ansatz
    #
    # The layer has a trainable parameters the
    # parameters theta of the trainable feature map
    # theta[0] as theta_x
    # theta[1] as theta_y
    # theta[2] as theta_z
    #
    # The Layer use the functions
    # Hamiltonian
    # FeatureMapU
    #
    # The constructor use parameters
    #
    #  Params
    #  ------
    #  J : positive Z coefficinet
    #  h : positive field
    #
    #  Example
    #  -------
    #  HL=HL(J=.5, h=1.2)
    #
    #  The call return the real part of
    #  <\psi|H\psi>
    #
    def __init__(self, J=1.0, h=0.0, **kwargs):
        super(HamiltonianLayer, self).__init__(**kwargs)
        # trainable parameter of the model
```

```
        # initially set as random (real variables)
        self.theta=tf.Variable(np.random.random(3),
                                 dtype=tf.float32,
                                name='theta')
        # gap coefficient
        self.J=J
        # field coefficient
        self.h=h
        # Hamiltonian
        self.H=Hamiltonian(J=self.J,h=self.h)
        # Feature map for the ground state
        self.FeatureMap=FeatureMapU
    def ground_state(self):
        """ Return the current ground state """
        psi=self.FeatureMap(self.theta,quantummap.qubit0)
        return psi
    def call(self,dummy):
        """ Return Re<H>
        Remark:  this layer has a dummy variable as input
        (required for compatibility with tensorflow call)
        """
        #psi=self.FeatureMap(self.theta,quantummap.qubit0)
        psi=self.ground_state()
        meanH=quantummap.Scalar(psi,
                                  quantummap.Gate1(self.H, psi))
        return tf.math.real(meanH)
```

We use the layer by creating an object belonging to the class `HamiltonianLayer` and then calling it with a dummy input, which returns the value of $\langle\hat{\mathcal{H}}\rangle$ at the randomly generated initial state.

```
# Create the layer with default parameters
HL=HamiltonianLayer()
# Print the parameters of the Hamiltonian
print([HL.J, HL.h])
# returns
#[1.0, 0.0]
# Display the Hamiltonian
print(HL.H)
# returns
#tf.Tensor(
#[[-1.+0.j -0.+0.j]
# [-0.+0.j  1.+0.j]], shape=(2, 2), dtype=complex64)
# Display the randomly generated ground_state
print(HL.ground_state)
# returns
# tf.Tensor([0.3151+0.7756j 0.5066-0.2064j],
# shape=(2,), dtype=complex64)
# Display the values of the random parameters
# corresponding to the ground_state
print(HL.theta)
```

```
# returns
# <tf.Variable 'theta:0' shape=(3,) dtype=float32,
#     numpy=array([0.399 , 0.2066, 0.7859], dtype=float32)>
#These parameters correspond to the trainable weights of the
↪  layer
tf.print(HL.weights)
# [[0.398956805 0.206608042 0.785909414]]
# Call the layer with a dummy input i.e. 3.1 (can be any number)
HL(3.1)
# returns a float corresponding to the
# current mean value -0.4016 on the ground_state
# <tf.Tensor: shape=(1, 1), dtype=float32,
# numpy=array([[-0.4016]], dtype=float32)>
```

We optimize the model to find out the parameters `HL.theta` that minimize the output of the `HamiltonianLayer`. First, we define a complete model with an input and an output layer.

```
# hyperparameter for the Hamiltonian
J=1.0
h=1.7
# Input layer (one dummy input)
xin = tf.keras.layers.Input(1);
# Hamiltonian layer
HL=HamiltonianLayer(J=J,h=h)
# output
meanH=HL(xin)
# trainable model returning meanH
Ising = tf.keras.Model(inputs = xin, outputs=meanH)
# test the model with dummy input 2.5
tf.print(Ising(2.5))
# returns (may vary with random initialization)
[[-0.122022316]]
```

We train the model by instructing `TensorFlow` to minimize a loss function, which is the mean value of the Hamiltonian. Hence, we add the loss function to the model with `tf.add_loss` corresponding to the tensor `meanH`, which is the output of the `HamiltonianLayer`.

```
Ising.add_loss(meanH)
```

Before training, we display the summary of the model.

```
Ising.summary()
# returns
```

```
Model: "model"
_________________________________________________________________
 Layer (type)                Output Shape              Param #
=================================================================
 input_1 (InputLayer)        [(None, 1)]               0

 hamiltonian_layer_1 (Hamilt  (1, 1)                   3
 onianLayer)

 add_loss (AddLoss)          (1, 1)                    0

=================================================================
Total params: 3
Trainable params: 3
Non-trainable params: 0
_________________________________________________________________
```

The output confirms that we have *three* parameters, as expected. We also plot the model by using `plot_model` as follows.

```
from keras.utils.vis_utils import plot_model
plot_model(Ising, to_file='singlequbitIsing_plot.png',
           show_shapes=True, show_layer_names=True)
```

The resulting plot is in Fig. 4.6 and shows the layers and the corresponding inputs and outputs. Note that we have an input layer with a dummy real-valued input variable to the `InputLayer`. The output of the `InputLayer` is also dummy and seeds the `HamiltonianLayer`, which returns a `shape=(1,1)` real-valued output. The shape of the output of the `HamiltonianLayer` is determined by the call to `tf.math.real` in its `call` function. The `HamiltonianLayer` seeds to the loss function to be minimized.

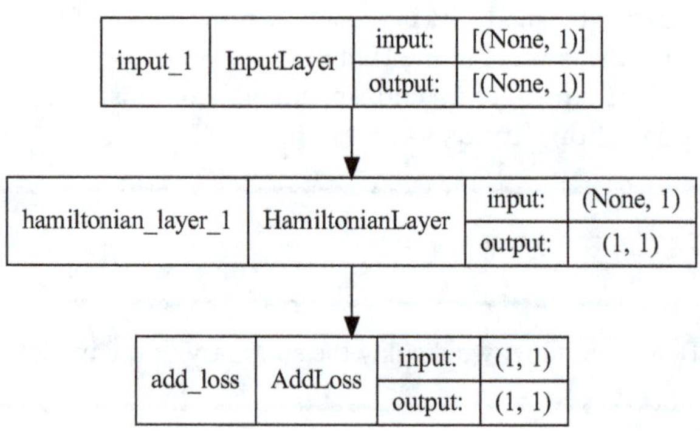

Fig. 4.6 Model graph of the one-qubit Ising Hamiltonian as obtained from the `plot_model` function

4.9 Training the One-Qubit Model

We train the model and compare the final state with (4.24) for different J and h.

We first consider the case $h = 0$ and $J = 1.0$, for which we expect the ground state to be $|E_-\rangle = |0\rangle = |\downarrow\rangle$, i.e., the eigenvector of $\hat{Z}$ with eigenvalue $E_- = -J = -1.0$.

First, we `compile` by assigning the Adam optimizer and then we use the `fit` method, as follows.[7]

```
# compile the model
# we choose a typical learning_rate 0.01
Ising.compile(optimizer=tf.keras.optimizers.Adam(learning_rate=
 0.01))
# then we train the model
history = Ising.fit(x=np.zeros(10,),epochs=1000, verbose=0)
```

Figure 4.7 shows the training history, revealing convergence after about 100 epochs. We show three cases for different parameters.

For $h = 0$ and $J = 1.0$, the asymptotic value is $\langle\hat{\mathcal{H}}\rangle = -1$, as expected. If we display the ground state after training, we have

```
# print the ground state after training
print(HL.ground_state())
# tf.Tensor([ 9.4154e-01+3.3690e-01j -1.5054e-07+6.4122e-09j],
#          shape=(2,), dtype=complex64)
# this state is tf.qubit0 within a phase factor
```

Note that this is not $|0\rangle$, as anticipated, but $e^{i\phi_0}|0\rangle$ with $\cos(\phi_0) = 0.94$ and $\sin(\phi_0) = 0.34$, i.e., $\phi_0 = 0.35$. This is not a mistake but reflects the fact that states are equivalent within an absolute phase factor ϕ_0, as one can check by looking at the density matrix. Hence the model returns one of the equivalent states with the same minimal expected energy.

For $h = 0.5$ and $J = 0$, the asymptotic value is $\langle\hat{\mathcal{H}}\rangle = -0.5$. This value corresponds to $E_- = -h$, as in (4.31), with eigenstate $|E_-\rangle = \hat{H}|0\rangle$ as in Eq. (4.29). If we display the ground state after training, we have

```
# print the ground state after training
# we convert to numpy for better printing
print(HL.ground_state().numpy())
# returns
```

[7] Details in the notebook `jupyternotebooks/quantumfeaturemap/SingleQubitTransverseFieldIsing.ipynb`.

```
# [-0.45+0.55j -0.45+0.55j]
# which corresponds to [1 1] within a phase factor
```

Hence, we obtain the expected $\hat{\mathrm{H}}|0\rangle$ within a phase factor.

Finally, for the case $h = 0.85$ and $h = 1$, the asymptotic value is $\langle\hat{\mathcal{H}}\rangle = -1.31$, which corresponds to E_- in (4.23). We display the ground state after training.

```
# print the ground state after training
print(HL.ground_state().numpy())
# returns
# [0.87+0.36j 0.32+0.13j]
```

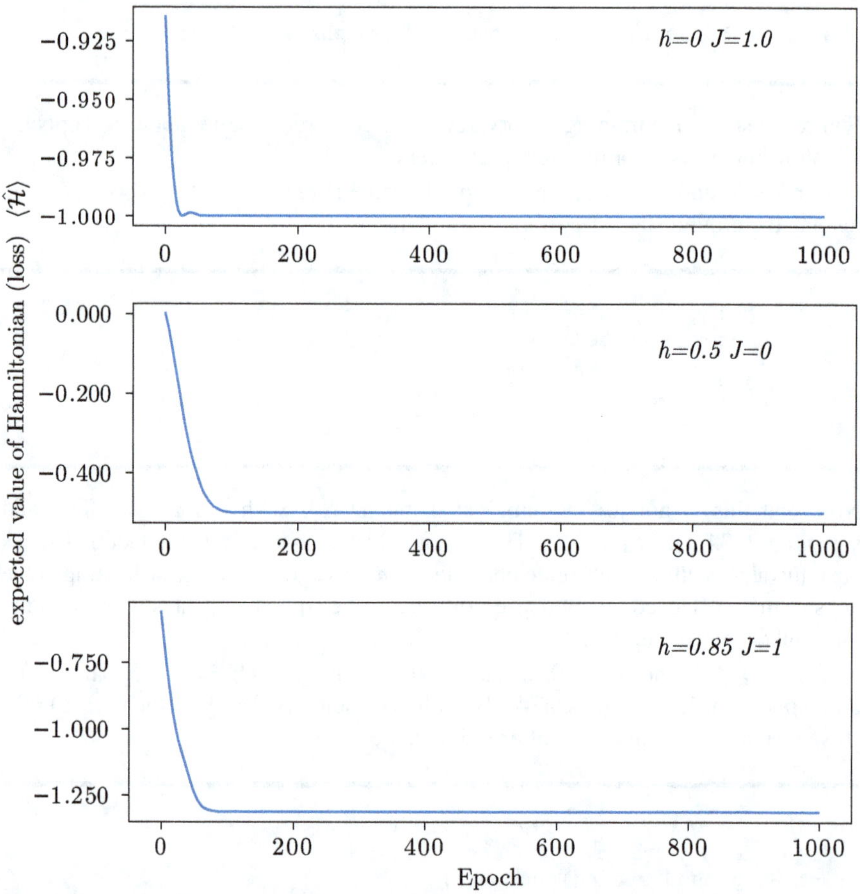

Fig. 4.7 Training history of the model in Fig. 4.6 for different values of J and h. The mean value $\langle\mathcal{H}\rangle$ converges to lowest energy eigenvalue E_-, and the network returns the lowest energy eigenvector

```
# if we scale by the second component
ev=HL.ground_state().numpy()
ev=ev/ev[1]
# returns
# array([2.72+0.j, 1.  +0.j], dtype=complex64)
```

The numerical value of the state is hence given by Eq. (4.24), within the arbitrary phase factor. The model converges to the ground state of the one-qubit transverse-field Ising Hamiltonian. In the next chapter, we consider the two-qubit case.

4.10 Further Reading

- Ising model and quantum software: There are many quantum libraries available online. They typically include examples in coding optimization problems. The most valuable resources are those in the documentation and tutorials with good introductions.
  - Qiskit, https://qiskit.org/documentation/optimization/tutorials/06_examples_max_cut_and_tsp.html.
  - PennyLane, https://pennylane.ai/qml/demos/tutorial_isingmodel_PyTorch.html.
  - D-Wave website furnishes many resources for learning – https://www.dwavesys.com/.
- Ising model: The Ising model is a celebrated paradigm in modern science. It has been the subject of intense investigations from several different perspectives. Most of the literature is focused on theoretical physics and statistical mechanics, while recent books put emphasis on link with combinatorial optimization.
  - A detailed introduction to the topic is the book by Nishimori [2], which covers many aspects, from the fundamentals to the neural networks.
  - A historical perspective is in the article by Brush [1].
- Quantum Ising model: The quantum version of the Ising model is as paradigmatic as its classical counterpart. The transverse-field Ising model enters many introductions to quantum thermodynamics and theoretical physics. In recent years, the emphasis shifted toward combinatorial optimization and quantum information.
  - The book by Suzuki, Inoue, and Chakrabarti is an advanced and comprehensive introduction with focus on phase transitions [5].
  - A renowned book is by Sachdev [6].
- Combinatorial optimization: The modern attention to the Ising model is due to its applications to combinatorial optimization and QUBO.

- The best introduction to combinatorial optimization is the famous book by Garey and Johnson [4].
- A must-read about the link between combinatorial optimization and the Ising model is the paper by Lucas [3].
- Introductory material about QUBO is in many websites and open software. A simple and clear tutorial is [7].

References

1. S.G. Brush, Rev. Mod. Phys. **39**, 883 (1967). https://doi.org/10.1103/RevModPhys.39.883
2. H. Nishimori, *Statistical Physics of Spin Glasses and Information Processing* (Oxford Science Publications, 2001)
3. A. Lucas, Front. Phys. **2**, 5 (2014). http://arxiv.org/abs/1302.5843
4. M.R. Garey, D.S. Johnson, *Computers and Intractability: A Guide to the Theory of NP-Completeness* (W. H. Freeman, 1979)
5. S. Suzuki, J. ichi Inoue, B.K. Chakrabarti, Quantum ising phases and transitions in transverse ising models (2013). https://doi.org/10.1007/978-3-642-33039-1
6. S. Sachdev, *Quantum Phase Transitions* (Cambridge University Press, Cambridge, 2011)
7. F. Glover, G. Kochenberger, Y. Du, arXiv:1811.11538 (2018). http://arxiv.org/abs/1811.11538

Chapter 5
Two-Qubit Transverse-Field Ising Model and Entanglement

Why do we always see the same side of the Moon?

Abstract We discuss the two-qubit transverse-field Ising model. We introduce entanglement measures. We show how to compute entanglement and unveil its significance in combinatorial optimization and variational quantum states.

5.1 Introduction

We now consider a two-qubit system that enables us to detail the concept of entanglement, which emerges when noticing that observations on one single qubit do not furnish all the information. The lack of information corresponds to having different states providing the same measured quantities, which means entropy. The density matrix furnishes the best approach to discovering and quantifying entanglement.

Notably enough, a certain amount of entanglement is needed to find the eigenstates of the Ising Hamiltonian, or their best approximations. The two-qubit case is the simplest one in which entanglement has such a significant role. Only in the presence of entanglement we can explore the Hilbert space and figure out the best variational ansatz.

5.2 Two-Qubit Transverse-Field Ising Model

The system we consider has two qubits. An example is the two-qubit superconducting circuit in [1].

The transverse-field Ising model (TIM) Hamiltonian reads

$$\hat{\mathcal{H}} = -J\,\hat{\mathrm{Z}}_0 \otimes \hat{\mathrm{Z}}_1 - h_0\hat{\mathrm{X}}_0 - h_1\hat{\mathrm{X}}_1 \tag{5.1}$$

C. Conti, *Quantum Machine Learning*, Quantum Science and Technology,
https://doi.org/10.1007/978-3-031-44226-1_5

The Hilbert space is spanned by the kets $|00\rangle$, $|01\rangle$, $|10\rangle$, and $|11\rangle$. We also use the notation with arrows $|\downarrow\downarrow\rangle$, $|\downarrow\uparrow\rangle$, $|\uparrow\downarrow\rangle$, and $|\uparrow\uparrow\rangle$.

As in Chap. 3, we use an entangled feature map as a variational ansatz for the ground state. Here, we consider in more detail the concept of entanglement and its presence in the ground state.

As the Hilbert space has dimension 4, we represent the Hamiltonian as 4×4 matrix that reads[1]

$$\hat{\mathcal{H}} = \begin{pmatrix} -J & -h_1 & -h_0 & 0 \\ -h_1 & J & 0 & -h_0 \\ -h_0 & 0 & J & -h_1 \\ 0 & -h_0 & -h_1 & -J \end{pmatrix} \tag{5.2}$$

To get an idea of the eigenvalues and eigenvectors, we start with the case without a transverse field.

5.2.1 $h_0 = h_1 = 0$

We have

$$\hat{\mathcal{H}} = -J\hat{Z}_0 \otimes \hat{Z}_1 \,, \tag{5.3}$$

which in matrix form reads

$$\hat{\mathcal{H}} = \begin{pmatrix} -J & 0 & 0 & 0 \\ 0 & J & 0 & 0 \\ 0 & 0 & J & 0 \\ 0 & 0 & 0 & -J \end{pmatrix} . \tag{5.4}$$

The eigenstates are the product states

$$|zw\rangle = |z\rangle \otimes |w\rangle \,, \tag{5.5}$$

with $z, w \in \{0, 1\}$. In matrix form the eigenstates are the basis vectors

$$|00\rangle = \begin{pmatrix} 1 \\ 0 \\ 0 \\ 0 \end{pmatrix} \; |01\rangle = \begin{pmatrix} 0 \\ 1 \\ 0 \\ 0 \end{pmatrix} \; |10\rangle = \begin{pmatrix} 0 \\ 0 \\ 1 \\ 0 \end{pmatrix} \; |11\rangle = \begin{pmatrix} 0 \\ 0 \\ 0 \\ 1 \end{pmatrix} . \tag{5.6}$$

[1] Details in the Mathematica file `TwoQubitTransverseIsing.nb`.

By using $\hat{Z}|z\rangle = (-1)^z|z\rangle$, we find that the eigenvalues are E:

$$E = -J(-1)^z(-1)^w \ . \tag{5.7}$$

We have four energy values $E \in \{-J, -J, J, J\}$ with two degenerate levels.

If $J > 0$, the ground state has energy $E_0 = -J$, and $z = w$, such that the spins are oriented in same direction. The two degenerate ground states are $|00\rangle = |\downarrow\downarrow\rangle$ and $|11\rangle = |\uparrow\uparrow\rangle$. This is the *ferromagnetic coupling*, or *ferromagnetic state*, which has aligned spins in the ground state.

If $J < 0$, the ground state has energy $E_0 = J$, and $z = -w$, such that the spins are oriented in opposite directions. The two degenerate states are $|01\rangle = |\downarrow\uparrow\rangle$ and $|10\rangle = |\uparrow\downarrow\rangle$. This is the *antiferromagnetic coupling*, or *antiferromagnetic state*, which has antiparallel spins in the ground state.

5.2.2 Entanglement in the Ground State with No External Field

Let us consider the case $J > 0$. The eigenstates are the product states $|0\rangle \otimes |0\rangle$ and $|1\rangle \otimes |1\rangle$ both with energy $E = -J$. These two states are not entangled. However, if we consider their linear combination

$$|\Phi^+\rangle = \frac{1}{\sqrt{2}}(|00\rangle + |11\rangle) \ , \tag{5.8}$$

which is a *maximally entangled two-qubit Bell state*, we also have an eigenstate of $\hat{\mathcal{H}}$. The degeneracy in the energy levels allows eigenstates with and without entanglement. This situation drastically changes if we include an external field.

To understand this fact, we first study $h \neq 0$; then we review the theory of two-qubit entanglement and its role in the two-quit transverse-field Ising Hamiltonian.

5.2.3 $h_0 = h_1 = h \neq 0$

Let us consider the case $h_0 = h_1 = h > 0$. The matrix Hamiltonian ($J > 0$) reads

$$\hat{\mathcal{H}} = \begin{pmatrix} -J & -h & -h & 0 \\ -h & J & 0 & -h \\ -h & 0 & J & -h \\ 0 & -h & -h & -J \end{pmatrix} \ . \tag{5.9}$$

The eigenvalues[2] E are

$$E_0 = -\sqrt{4h^2 + J^2} \tag{5.10}$$

$$E_1 = -J \tag{5.11}$$

$$E_2 = J \tag{5.12}$$

$$E_4 = \sqrt{4h^2 + J^2}\,. \tag{5.13}$$

The external field breaks the degeneracy and lowers the ground state energy E_0 with respect to the case $h_0 = h_1 = h = 0$. The unnormalized eigenstates are

$$|\psi_0\rangle = \begin{pmatrix} 1 \\ -\frac{J-\sqrt{4h^2+J^2}}{2h} \\ -\frac{J-\sqrt{4h^2+J^2}}{2h} \\ 1 \end{pmatrix}, |\psi_1\rangle = \begin{pmatrix} -1 \\ 0 \\ 0 \\ 1 \end{pmatrix}, |\psi_2\rangle = \begin{pmatrix} 0 \\ -1 \\ 1 \\ 0 \end{pmatrix},$$

$$|\psi_3\rangle = \begin{pmatrix} 1 \\ -\frac{J+\sqrt{4h^2+J^2}}{2h} \\ -\frac{J+\sqrt{4h^2+J^2}}{2h} \\ 1 \end{pmatrix}. \tag{5.14}$$

The states $|\psi_1\rangle$ and $|\psi_2\rangle$, after normalization and within a phase factor, are *maximally entangled Bell states*, namely,

$$|\psi_1\rangle = |\Phi^+\rangle = \frac{1}{\sqrt{2}}(|00\rangle - |11\rangle) \tag{5.15}$$

$$|\psi_2\rangle = |\Psi^-\rangle = \frac{1}{\sqrt{2}}(|01\rangle - |10\rangle)\,. \tag{5.16}$$

Some of the eigenstates are hence entangled. But we do not know if also the ground state is entangled. To clarify the situation, we first review the theory of entanglement for two qubits.

5.3 Entanglement and Mixtures

We introduce entanglement starting from the distinction between *pure states* and *mixtures*.

[2] See the `Mathematica` file `TwoQubitTransverseIsing.nb`.

We know that in quantum mechanics we have two possible systems: pure states and mixtures. A pure state is completely identified by the Hilbert space vector $|\psi\rangle$ within an arbitrary phase factor ϕ_0, such that $|\psi\rangle$ and $e^{i\phi_0}|\psi\rangle$ are the same pure state. The corresponding density matrix is

$$\rho = |\psi\rangle\langle\psi| , \tag{5.17}$$

which is independent of the phase factor ϕ_0.

When the vector in the Hilbert space does not contain the complete information for describing the system, we have a *mixture*, or a *mixed state*. A mixture occurs, for example, when there are a probability $p_0 < 1$ of observing the system in the state $|\psi_0\rangle$ and a probability $p_1 < 1$ of observing the system in the state $|\psi_1\rangle$. If no other states are observed, we have $p_0 + p_1 = 1$.

The complete description of a mixture is only given by the density matrix

$$\rho = p_0|\psi_0\rangle\langle\psi_0| + p_1|\psi_1\rangle\langle\psi_1| . \tag{5.18}$$

ρ contains all the relevant states and their occurrence probabilities.

If we consider an observable $\hat{A}$, the expected value for a pure state is

$$\langle\hat{A}\rangle = \mathrm{Tr}\left(\rho\hat{A}\right) = \langle\psi|\hat{A}|\psi\rangle . \tag{5.19}$$

and is entirely determined by the state vector $|\psi\rangle$. On the contrary, for a mixed state, the expected value is

$$\langle\hat{A}\rangle = \mathrm{Tr}\left(\rho\hat{A}\right) = p_0\langle\psi_0|\hat{A}|\psi_0\rangle + p_1\langle\psi_1|\hat{A}|\psi_1\rangle . \tag{5.20}$$

Equation (5.20) means that to predict the mean value of $\hat{A}$, we need to know not only $|\psi_0\rangle$ and $|\psi_1\rangle$ but also their probabilities p_0 and p_1, an information contained in (5.18). The knowledge of only $|\psi_0\rangle$ is a partial information, as it is the knowledge of only $|\psi_1\rangle$.

To describe the amount of information we have on a system, we use the concept of *entropy*. The more entropy, the less information we have when observing a single state.

A pure state has zero entropy in the sense that we know everything we can know about the system. In a mixed state, the entropy is greater than zero, because knowing that the system is the state $|\psi_0\rangle$ or in $|\psi_1\rangle$ does not provide all the information on possible observation outcomes. For example, if we perform a set of measurements of $\hat{A}$, and we find that the mean value is $\langle\psi_1|\hat{A}|\psi_1\rangle$, this may not be the case for the next set of measurements. After a certain number of observations, we find indeed that the behavior results from a mixture of $|\psi_0\rangle$ and $|\psi_1\rangle$ with specific probabilities – a situation that we describe stating that the system has a nonzero entropy.

Now consider a system composed of two subsystems, for example, two qubits A and B. Assume that we only do local observation, i.e., we measure only quantities on the qubit A. If the resulting observations are compatible with a pure state, the two qubits are independent. On the contrary, if knowing the state of the qubit A is not sufficient for making predictions, it means that we need to know also the state of B and other information to describe the system. In this case A is in a mixed state, and the two qubits are *entangled*. In other words, entanglement arises when the description of a local subsystem is the one of a mixed state. In this case, we need nonlocal information, i.e., information about both the subsystems A and B. In absence of entanglement, the two-qubit state is *separable*.

5.4 The Reduced Density Matrix

Entanglement is studied by the density matrix and the *reduced density matrix*. Consider a pure state of two qubits $|z\rangle$ and $|w\rangle$, with $z, w \in \{0, 1\}$,

$$|\psi\rangle = \sum_{z,w} \psi^{zw} |zw\rangle \,. \tag{5.21}$$

The corresponding density matrix, or density operator, is

$$\rho = \sum_{x,y,z,w} \psi^{xy} \psi_{zw} |xy\rangle\langle zw| \,. \tag{5.22}$$

In (5.22) the tensor with covariant indices $\psi_{zw} = (\psi^{zw})^*$ is a "covector" or "cotensor," i.e., the conjugate of the vector or tensor.

In terms of the tensor components of ρ, we have

$$\rho^{xy}{}_{zw} = \psi^{xy} \psi_{zw} \,, \tag{5.23}$$

which is the outer product of the tensor ψ^{xy} and its cotensor ψ_{zw}.

For later convenience, we first put the ket components with upper indices and then the bra components with lower indices. This follows the fact that in $\rho = |\psi\rangle\langle\psi|$, we first have the ket $|\psi\rangle$ and then the bra $\langle\psi|$.

Let us consider local observations on the qubit $|z\rangle$. Any measurement of a single-qubit operator $\hat{O}$ on A must be averaged over all the possible values of $|w\rangle$. The local measurement of $\hat{O}$ on A is the mean of all possible results of measurements for all the states of B. Entanglement occurs when the measurement on A depends on the state of B, a situation which is described as *non-locality*.

To determine the presence of entanglement, we observe that, when we measure $\hat{O}$ on the qubit A, the qubit B is not affected as we are considering a local operation. We are measuring the mean value of the two-qubit operator $\hat{O} \otimes \hat{I}_B$, with $\hat{I}_B$ the identity operator on B. The latter can be written as

$$\hat{I}_B = \sum_w |w\rangle_B \langle w|_B = |0\rangle_B \langle 0|_B + |1\rangle_B \langle 1|_B \,, \tag{5.24}$$

with obvious notation. Thus we have for the mean value of $\hat{O} \otimes \hat{I}_B$

$$\begin{aligned}\langle \hat{O}_A \otimes \hat{I}_B \rangle &= \mathrm{Tr}\,(\rho \hat{O}_A \otimes \hat{I}_B) \\ &= \sum_{uw} \langle uw|\rho \hat{O}_A \otimes \hat{I}_B |uw\rangle \\ &= \sum_u \langle u|_A \sum_w \langle w|_B \rho |w\rangle_B \hat{O}_A |u\rangle_A ,\end{aligned} \tag{5.25}$$

where we added the subscript A to $\hat{O}$ to emphasize that $\hat{O}$ acts on the qubit A.

Equation (5.25) is written as

$$\langle \hat{O}_A \otimes \hat{I}_B \rangle = \mathrm{Tr}(\rho_A \hat{O}) \tag{5.26}$$

where

$$\rho_A = \sum_w \langle w|_B \rho |w\rangle_B \tag{5.27}$$

is the density matrix of the system A averaged ("traced") over all possible states of B, i.e., the *reduced density matrix* of A. Thus, the mean value of the local operator $\hat{O}_A$ is obtained as the trace over the A part averaged over all the possible states B.

We remark that ρ_A is an operator on the Hilbert space $\mathcal{H}_A$ of A, while ρ is an operator on the product space $\mathcal{H} = \mathcal{H}_A \otimes \mathcal{H}_B$.

The partial trace in the A basis $|x\rangle_A \langle y|_A$ is

$$\rho_A = \sum_{x,y} \left(\sum_w \langle x|_A \langle w|_B \rho |y\rangle_A |w\rangle_B \right) |x\rangle_A \langle y|_A \,. \tag{5.28}$$

Assume we are able to compute ρ_A from ρ; if ρ_A looks like a pure state, say,

$$\rho_A = |\psi\rangle\langle\psi| \,, \tag{5.29}$$

it implies that the state $|\psi\rangle$ contains all the information about the subsystem A and the measurement is independent of the subsystem B. Local measurements on A are not affected by the absence of knowledge about B.

On the contrary, if ρ_A appears as a mixture, say,

$$\rho_A = p_0 |\psi_0\rangle\langle\psi_0| + p_1 |\psi_1\rangle\langle\psi_1| \,, \tag{5.30}$$

it means that the local measurement is not determined by one pure state. If we consider the mean value of an observable $\hat{O}_A$ on A, we have

$$\langle \hat{O}_A \rangle = p_0 \langle \psi_0 | \hat{O}_A | \psi_0 \rangle + p_1 \langle \psi_1 | \hat{O}_A | \psi_1 \rangle \, . \tag{5.31}$$

Different states of A, $|\psi_0\rangle$, and $|\psi_1\rangle$ contribute to the mean value; there is *entropy* or lack of information. We can find A in multiple states, and the knowledge of a single pure state is not enough to completely describe the system.

The knowledge of the state, $|\psi_0\rangle$ or $|\psi_1\rangle$, does not furnish all the information to predict the statistics of measurements. We also need the probabilities p_0 and p_1. The states and the probabilities depend on B. This circumstance is interpreted as the fact that the system B is nonlocally correlated with A. Local measurements on A depend on B, a situation denoted as entanglement.

Before continuing the discussion, to fix these ideas, let us deepen our math about the reduced density matrix.

5.5 The Partial Trace

Consider the operator $\hat{O}_A$ on the single qubit A; its matrix representation is

$$\hat{O}_A = \begin{pmatrix} O^0{}_0 & O^0{}_1 \\ O^1{}_0 & O^1{}_1 \end{pmatrix} = \sum_{x,y \in \{0,1\}} O^x{}_y |x\rangle_A \langle y|_A \, . \tag{5.32}$$

The identity operator $\hat{I}_B$ for the qubit B is

$$\hat{I}_B = \sum_{w,z \in \{0,1\}} \delta^w{}_z |w\rangle_B \langle z|_B \, . \tag{5.33}$$

Here we add the subscripts A, B to $\hat{O}$ and $\hat{I}$ to clarify the notation; we will also omit the variable values {0, 1} in the sum symbol.

The operator $\hat{O}_A \otimes \hat{I}_B$ is

$$\hat{O}_A \otimes \hat{I}_B = \sum_{x,y,z,w} O^x{}_y \delta^z{}_w |x\rangle_A |z\rangle_B \langle y|_A \langle w|_B \, . \tag{5.34}$$

We have

$$\rho = \sum_{s,p,q,r} \rho^{sp}{}_{qr} |s\rangle_A |p\rangle_B \langle q|_A \langle r|_B ; \tag{5.35}$$

we explicitly indicate A and the B in the bra and ket for better clarity.

We have

$$\rho\,\hat{O}_A \otimes \hat{I}_B = \sum_{s,p,y,w} \rho^{sp}{}_{xz} O^x{}_y \delta^z{}_w |s\rangle_A |p\rangle_B \langle y|_A \langle w|_B \,, \tag{5.36}$$

where we used

$$\langle q|_A \langle r|_B |x\rangle_A |z\rangle_B = \delta^q{}_x \delta^r{}_z \,. \tag{5.37}$$

From (5.36), we have

$$\rho\,\hat{O}_A \otimes \hat{I}_B = \sum_{s,p,y,w} \rho^{sp}{}_{xw} O^x{}_y |s\rangle_A |p\rangle_B \langle y|_A \langle w|_B \,. \tag{5.38}$$

For the trace, we have

$$\mathrm{Tr}(\rho\,\hat{O}_A \otimes \hat{I}_B) = \sum_{u,v} \langle u|_A \langle v|_B \rho |u\rangle_A |v\rangle_B = \rho^{uv}{}_{xv} O^x{}_u \,, \tag{5.39}$$

where we used

$$\begin{aligned} \langle u|_A \langle v|_B |s\rangle_A |p\rangle_B &= \delta^u{}_s \delta^v{}_p \,, \\ \langle y|_A \langle w|_B |u\rangle_A |v\rangle_B &= \delta^y{}_u \delta^w{}_v \,. \end{aligned} \tag{5.40}$$

We have obtained (5.39) by writing the basis components explicitly. The same result can be obtained just considering the tensors.

Specifically, the density matrix is a rank-4 tensor with `shape=(2,2,2,2)`, i.e.,

$$\rho = \rho^{sp}{}_{qr} \tag{5.41}$$

where s, q refer to the first qubit A and p, r to the qubit B. The outer product of $\hat{O}_A$ and $\hat{I}_B$ has components

$$\hat{O}_A \otimes \hat{I}_B = O^x{}_y \delta^w{}_z \tag{5.42}$$

where x, y refer to the first qubit A and w, z to the qubit B. In general, in a tensor expression, the first upper and the first lower index correspond to the first qubit, the second upper and lower to the second qubit, and so on.

The two operators act on the two-qubit Hilbert space, and this is also the case for their product

$$\rho\,\hat{O}_A \otimes \hat{I}_B = \rho^{sp}{}_{xw} O^x{}_y \delta^w{}_z = (\rho\,\hat{O}_A \otimes \hat{I}_B)^{sp}{}_{yz} \,, \tag{5.43}$$

which we obtain by contracting the inner indices (q, x) for qubit A [see Eq. (5.41)] and the inner indices (r, w) for the qubit B. In the resulting indices s, y in Eq. (5.43), refer to qubit A and p, z to qubit B.

By contracting w, z because of $\delta^w{}_z$, we have

$$(\rho\, \hat{O}_A \otimes \hat{I}_B)^{sp}{}_{yz} = \rho^{sp}{}_{xz} O^x{}_y \,, \tag{5.44}$$

which is also a rank-4 tensor, where the free indices for qubit A are (s, y) and for qubit B are (p, z). The trace is obtained by contracting the indices for A and B.

$$\mathrm{Tr}(\rho\, \hat{O}_A \otimes \hat{I}_B) = (\rho\, \hat{O}_A \otimes \hat{I}_B)^{uv}{}_{uv} = \rho^{uv}{}_{xv} O^x{}_u \,, \tag{5.45}$$

which is a scalar as in (5.39), with a change in the name of repeated indices.

Next we introduce the partial trace of qubit A, $\rho_A = \mathrm{Tr}_B(\rho)$, by observing that (5.39) is equivalent to the following:

$$\mathrm{Tr}\,(\rho \hat{O}_A \otimes \hat{I}_B) = \mathrm{Tr}_A\,(\rho_A \hat{O}_A) = \mathrm{Tr}_A\left[\mathrm{Tr}_B(\rho)\hat{O}_A\right] . \tag{5.46}$$

The partial trace ρ_A, i.e., the density matrix for local operations on the qubit A, is defined as follows:

$$\rho_A = \mathrm{Tr}_B(\rho) = \sum_{x,y} \rho^{xu}{}_{yu} |x\rangle_A \langle y|_A \,, \tag{5.47}$$

i.e., the tensor obtained from ρ by contracting the indices for qubit B.

In other words, if the density matrix has components

$$\rho = \rho^{sp}{}_{qr} \,, \tag{5.48}$$

the partial trace for qubit A is

$$\rho_A = (\rho_A)^s{}_p = \rho^{su}{}_{pu} \,. \tag{5.49}$$

The partial trace for qubit B is

$$\rho_B = (\rho_B)^q{}_r = \rho^{sq}{}_{sr} \,. \tag{5.50}$$

Note the free indices (s, p) and (q, r) in ρ_A and ρ_B, respectively.

The partial trace transforms a tensor ρ in the two-qubit Hilbert space to a tensor in the one-qubit Hilbert space; it is obtained for the first qubit from the tensor ρ by contracting the indices referring to the second qubit.

If we have an operator $\hat{I}_A \otimes \hat{Q}_B$, which acts only on the second qubit, i.e., a local operation on B, we have

$$\mathrm{Tr}(\rho \hat{I}_A \otimes \hat{Q}_B) = \mathrm{Tr}_B\,(\rho_B \hat{Q}_B) = \mathrm{Tr}_B\left[\mathrm{Tr}_A(\rho)\hat{Q}_B\right] . \tag{5.51}$$

Fig. 5.1 Indices in the ρ tensor: the reduced density matrices and the mean value of operators for two qubits. Repeated indices are colorized

Two-qubit density matrix in tensor notation

$$\rho^{sp}{}_{qr}$$

Reduced density matrix

$$\rho_A = \mathrm{Tr}_B(\rho) = \rho^{su}{}_{qu} = (\rho_A)^p{}_q$$

$$\rho_B = \mathrm{Tr}_A(\rho) = \rho^{up}{}_{ur} = (\rho_B)^p{}_r$$

Mean value

$$\langle \hat{O}_A \rangle = \mathrm{Tr}\left(\rho_A \hat{O}_A\right) = \rho^{su}{}_{qu}(O_A)^q{}_s$$

$$\langle \hat{O}_B \rangle = \mathrm{Tr}\left(\rho_B \hat{O}_B\right) = \rho^{up}{}_{ur}(O_B)^r{}_p$$

The reduced density matrix in the space of the qubit B is

$$\rho_B = \mathrm{Tr}_A(\rho) = \sum_{z,w} \rho^{xz}{}_{xw} |z\rangle_B \langle w|_B \ . \tag{5.52}$$

Starting from the matrix elements of ρ_A and ρ_B, and given two operators $\hat{O}_A$ and $\hat{Q}_B$, we obtain *locally* (i.e., considering only a single subsystem) the mean values by contracting

$$\langle \hat{O}_A \rangle = \mathrm{Tr}_A(\rho_A \hat{O}_A) = (\rho_A)^x{}_y \hat{O}^y_x = \rho^{xu}{}_{yu} O^y{}_x \ , \tag{5.53}$$

and

$$\langle \hat{Q}_B \rangle = \mathrm{Tr}_B(\rho_B \hat{Q}_B) = (\rho_B)^x{}_y \hat{Q}^y_x = \rho^{ux}{}_{uy} Q^y{}_x \ , \tag{5.54}$$

where $Q^y{}_x$ are the matrix elements of $\hat{Q}_B$.

The previous notation is ready to be implemented in `TensorFlow`, which is the next step before considering how to quantify the entanglement (Fig. 5.1).

5.6 One-Qubit Density Matrix Using Tensors

We start reviewing the density matrix in tensor notation for one qubit and its implementation in `TensorFlow`. We consider one qubit

$$|\psi\rangle = \sum_{z\in\{0,1\}} \psi^z |z\rangle \ . \tag{5.55}$$

In our notation, it corresponds to a vector with contravariant index ψ^z, which is the component of the *ket* $|\psi\rangle$. Its covector, i.e., the components of the *bra*, is the tensor with complex conjugated components

$$|\psi\rangle = \sum_{z\in\{0,1\}} (\psi^z)^* \langle z| \,. \tag{5.56}$$

The components of the covector are denoted by

$$\psi_z \equiv (\psi^z)^* \,, \tag{5.57}$$

such that we can write (5.56) as

$$|\psi\rangle = \sum_{z\in\{0,1\}} (\psi^z)^* \langle z| = \sum_{z\in\{0,1\}} \psi_z \langle z| \,. \tag{5.58}$$

The scalar product in (3.19) reads

$$\langle\psi|\psi\rangle = \sum_{z\in\{0,1\}} (\psi^z)^* \psi^z = \psi_z \psi^z = \psi^z \psi_z \,, \tag{5.59}$$

where we omit the sum symbol in the last equation, as we have repeated lower and upper index. We adopt the Einstein notation. ψ_z is the conjugate of ψ^z, as used in the definition of the `TensorFlow` function `scalar` in Chap. 3.

The density matrix reads

$$\rho = |\psi\rangle\langle\psi| = \sum_{z,w\in\{0,1\}} \psi^z (\psi^w)^* |z\rangle\langle w| = \sum_{z,w\in\{0,1\}^2} \psi^z \psi_w |z\rangle\langle w| \,, \tag{5.60}$$

where we first write the contravariant components ψ^z following the order of a *ket* and a *bra* in the density matrix definition $\rho = |\psi\rangle\langle\psi|$.

The elements of the density matrix are those of the outer product of ψ^z and ψ_w. Namely,

$$\langle x|\rho|y\rangle = (\rho)^x{}_y = \psi^x \psi_y = \psi^x (\psi^y)^* \,, \tag{5.61}$$

where we used

$$\langle x|z\rangle = \delta^x{}_z \,, \tag{5.62}$$

and

$$\langle w|y\rangle = \delta^w{}_y \,. \tag{5.63}$$

The trace of the one-qubit density matrix is

$$\mathrm{Tr}(\rho) = \sum_{x\in\{0,1\}} \langle x|\rho|x\rangle$$

$$= \sum_{x\in\{0,1\}} \psi^z\psi_w\langle x|z\rangle\langle w|x\rangle$$

$$= \psi^z\psi_w\delta^x{}_z\delta^w{}_x = \psi^x\psi_x = 1\ , \tag{5.64}$$

being

$$\delta^x{}_z = \langle x|z\rangle\ , \tag{5.65}$$

and

$$\delta^w{}_x = \langle w|x\rangle\ . \tag{5.66}$$

5.7 Coding the One-Qubit Density Matrix

We can use `TensorFlow` for computing the density matrix and the covector of states for single qubits. For example,[3] let us consider the `ket` base `qubit0` and `qubit1`.

```
qubit0 = tf.constant([1,0],dtype=mytype)
print(qubit0)
# returns
tf.Tensor([1.+0.j 0.+0.j], shape=(2,), dtype=complex64)
qubit1 = tf.constant([0,1],dtype=mytype)
print(qubit1)
#returns
tf.Tensor([0.+0.j 1.+0.j], shape=(2,), dtype=complex64)
```

[3] Details in `jupyternotebooks/quantumfeaturemap/QubitsDensityMatrix.ipynb`.

The corresponding covectors (i.e., the `bras`) are

```
coqubit0=tf.transpose(qubit0,conjugate=True);
print(coqubit0)
#returns
tf.Tensor([1.-0.j 0.-0.j], shape=(2,), dtype=complex64)
coqubit1=tf.transpose(qubit1,conjugate=True);
print(coqubit1)
#returns
tf.Tensor([0.-0.j 1.-0.j], shape=(2,), dtype=complex64)
```

Note that for a `tensor` with `shape=(2,)`, the transposition still return a tensor with `shape=(2,)`, hence for a single qubit transposition and conjugation returns the same output of simple conjugation. We can equivalently define the covectors by a simple conjugation, as follows.

```
coqubit0=tf.math.conj(qubit0);
print(coqubit0)
#returns
tf.Tensor([1.-0.j 0.-0.j], shape=(2,), dtype=complex64)
coqubit1=tf.math.conj(qubit1);
print(coqubit1)
#returns
tf.Tensor([0.-0.j 1.-0.j], shape=(2,), dtype=complex64)
```

As the basis vectors have real components, vectors and covectors correspond to the same tensors. For a generic qubit, with complex components, this is not the case. We can use the covector of the qubit to compute the scalar product by tensor contraction.

```
# use tensor contraction for computing the scalar products <0|0>
↪   and <0|1>
# we use axis=1 as the tensors have only one index
#<0|0>
tf.tensordot(coqubit0,qubit0,axes=1)
#returns
# <tf.Tensor: shape=(), dtype=complex64, numpy=(1+0j)>
#<0|1>
tf.tensordot(coqubit0,qubit1,axes=1)
#returns
#<tf.Tensor: shape=(), dtype=complex64, numpy=0j>
# the two states are orthogonal as expected
```

By using the tensor outer product, we can build the density matrices. The density matrix for one qubit has `shape=(2,2)` being an operator on a single-qubit space.

```
# build |0><0|
# use axes=0 for the outer product
tf.tensordot(qubit0, coqubit0,axes=0)
# returns a (shape=(2,2)) tensor
#<tf.Tensor: shape=(2, 2), dtype=complex64, numpy=
#array([[1.+0.j, 0.+0.j],
#        [0.+0.j, 0.+0.j]], dtype=complex64)>
# build |1\rangle\langle 1|
tf.tensordot(qubit1, coqubit1,axes=0)
# returns
#<tf.Tensor: shape=(2, 2), dtype=complex64, numpy=
#array([[0.-0.j, 0.-0.j],
#        [0.-0.j, 1.-0.j]], dtype=complex64)>
```

For a qubit like

$$
\begin{aligned}
|\psi\rangle &= \frac{1}{\sqrt{2}}|0\rangle + \frac{\mathrm{i}}{\sqrt{2}}|1\rangle \\
\langle\psi| &= \frac{1}{\sqrt{2}}\langle 0| - \frac{\mathrm{i}}{\sqrt{2}}\langle 1| \,,
\end{aligned}
\tag{5.67}
$$

the density matrix reads

$$
\rho = |\psi\rangle\langle\psi| = \frac{1}{2}|0\rangle\langle 0| - \frac{\mathrm{i}}{2}|0\rangle\langle 1| + \frac{\mathrm{i}}{2}|1\rangle\langle 0| + \frac{1}{2}|1\rangle\langle 1| \,,
\tag{5.68}
$$

or in matrix form

$$
\rho = \begin{pmatrix} \frac{1}{2} & -\frac{\mathrm{i}}{2} \\ \frac{\mathrm{i}}{2} & \frac{1}{2} \end{pmatrix} .
\tag{5.69}
$$

In `TensorFlow` we have

```
# define the state
psi=qubit0/np.sqrt(2.0)+1j*qubit1/np.sqrt(2.0)
print(psi)
# returns
# tf.Tensor([0.70710677+0.j 0.+0.70710677j], shape=(2,),
↪ dtype=complex64)
# compute the dual
copsi=qubit0/np.sqrt(2.0)+1j*qubit1/np.sqrt(2.0)
print(copsi)
# returns
# tf.Tensor([0.70710677-0.j 0.-0.70710677j], shape=(2,),
↪ dtype=complex64)
# check if the state is normalized
tf.tensordot(copsi,psi, axes=1)
```

```
# returns
# <tf.Tensor: shape=(), dtype=complex64, numpy=(0.99999994+0j)>
# compute the density matrix
# note the order of psi and copsi in the outer product
rho=tf.tensordot(psi,copsi, axes=0)
print(rho)
# returns
# tf.Tensor(
# [[0.5+0.j  0. +0.5j]
# [0. -0.5j 0.5+0.j ]], shape=(2, 2), dtype=complex64)
# compute the trace of the density matrix
tf.linalg.trace(rho)
# returns
# <tf.Tensor: shape=(), dtype=complex64, numpy=(1+0j)>
```

Note that we also computed the trace of the density matrix by `tf.linalg.trace`; however this method only works for matrices. On the contrary, it is useful to compute trace by `tf.einsum`, which is a general expression for computing quantities out of tensor by the Einstein notation (i.e., omitting sum over repeated indices). For the trace of a rank-2 tensor $\rho^u{}_u$, with `shape=(2,2)`, we have

```
#compute the trace by tf.einsum
tf.einsum('uu',rho)
# returns
# <tf.Tensor: shape=(), dtype=complex64, numpy=(1.0000001+0j)>
# the string 'uu' represent the tensor operation
# in Einstein notation
# note that the letter 'u' for the index in the string
# is not significant
tf.einsum('ii',rho)
# returns
# <tf.Tensor: shape=(), dtype=complex64, numpy=(1.0000001+0j)>
```

5.8 Two-Qubit Density Matrix by Tensors

Let us consider a two-qubit state

$$|\psi\rangle = \sum_{z,w\in\{0,1\}^2} \psi^{zw}|zw\rangle \ . \tag{5.70}$$

Hereafter to simplify the notation, we will just write the variables in the summations, without their values {0, 1}, as all the summed variables are binaries, as follows:

$$|\psi\rangle = \sum_{z,w} \psi^{zw} |zw\rangle \,. \tag{5.71}$$

The corresponding *bra* is

$$\langle\psi| = \sum_{z,w} (\psi^{zw})^* \langle zw| = \sum_{z,w} \psi_{zw} \langle zw| \,, \tag{5.72}$$

where we defined the cotensor

$$\psi_{zw} = (\psi^{zw})^* \,. \tag{5.73}$$

The norm of the state is

$$\begin{aligned}\langle\psi|\psi\rangle &= \sum_{x,y,z,w} (\psi^{xy})^* \psi^{zw} \langle xy|zw\rangle \\ &= \sum_{x,y,z,w} (\psi^{xy})^* \psi^{zw} \delta_z^x \delta_z^y \\ &= \sum_{z,w} (\psi^{zw})^* \psi^{zw} = \psi_{zw} \psi^{zw} = 1 \,,\end{aligned} \tag{5.74}$$

omitting the sum over repeated upper and lower indices. The scalar product is the contraction of the outer product $\psi_{zw}\psi^{xy}$ with respect to the indices of the first qubit (x, z) and the indices of the second qubit (y, w).

For the density matrix, we have

$$\rho = |\psi\rangle\langle\psi| = \sum_{x,y,z,w} \psi^{xy} \psi_{zw} |xy\rangle\langle zw| \,. \tag{5.75}$$

The elements of the density matrix are

$$\langle xy|\rho|zw\rangle = \psi^{xy} \psi_{zw} \,. \tag{5.76}$$

The trace of the density matrix is

$$\mathrm{Tr}(\rho) = \sum_{x,y} \langle xy|\rho|xy\rangle = \psi^{xy} \psi_{xy} = 1 \,. \tag{5.77}$$

For a pure state, the trace of the density matrix corresponds to the unitary norm.

5.9 Coding the Two-Qubit Density Matrix

We obtain the density matrix in TensorFlow by considering a two-qubit state.[4] We have

$$|\psi\rangle = \frac{1}{4}|00\rangle + \frac{\mathrm{i}}{4}|01\rangle + \frac{1+3\mathrm{i}}{4}|10\rangle - \frac{1}{2}|11\rangle \tag{5.78}$$

as a combination of the basis tensors `q00`, `q01`, `q10`, and `q11`, also defined in `thqml.quantummap`, as follows.

```
# we repeat here the definition
# of the basis vector
q00 = tf.tensordot(qubit0,qubit0,axes=0);
q01 = tf.tensordot(qubit0,qubit1,axes=0);
q10 = tf.tensordot(qubit1,qubit0,axes=0);
q11 = tf.tensordot(qubit1,qubit1,axes=0);
# define the state
psi=0.25*q00+0.25*1j*q01+0.25*(1.0+3j)*q10-0.5*q11
print(psi)
# returns
tf.Tensor(
[[ 0.25+0.j     0.   +0.25j]
 [ 0.25+0.75j -0.5  +0.j   ]], shape=(2, 2), dtype=complex64)
# define the cotensor
copsi=tf.math.conj(psi)
print(copsi)
# returns
tf.Tensor(
[[ 0.25-0.j     0.   -0.25j]
 [ 0.25-0.75j -0.5  -0.j   ]], shape=(2, 2), dtype=complex64)
```

We check the normalization $\langle\psi|\psi\rangle = 1$, by contracting the product of `copsi` and `psi`. We contract the first index of `copsi` with the first index of `psi`, and seemingly for the second index, by specifying two lists in `axes=[[0,1],[0,1]]`.

```
tf.tensordot(copsi,psi,axes=[[0, 1],[0,1]])
# returns
<tf.Tensor: shape=(), dtype=complex64, numpy=(1+0j)>
```

We compute the density matrix as the outer product of `psi` and `copsi`.

---

[4] Details in `jupyternotebooks/quantumfeaturemap/QubitsDensityMatrix.ipynb`.

```
rho=tf.tensordot(psi,copsi,axes=0);
print(rho)
# returns
tf.Tensor(
[[[[ 0.0625+0.j       0.     -0.0625j]
   [ 0.0625-0.1875j -0.125 +0.j     ]]

  [[ 0.     +0.0625j  0.0625+0.j     ]
   [ 0.1875+0.0625j  0.     -0.125j ]]]


 [[[ 0.0625+0.1875j  0.1875-0.0625j]
   [ 0.625 +0.j     -0.125 -0.375j ]]

  [[-0.125 +0.j       0.     +0.125j ]
   [-0.125 +0.375j   0.25  +0.j     ]]]], shape=(2, 2, 2, 2),
 ↪  dtype=complex64)
```

`rho` is a rank-4 tensor, with four indices and `shape=(2,2,2,2)`.

We access the elements of ρ, as

$$\rho^{00}{}_{00} = \frac{1}{16} \tag{5.79}$$

and

$$\rho^{10}{}_{01} = \frac{3-\mathrm{i}}{16} \tag{5.80}$$

as follows.

```
print(rho[0,0,0,0])
# returns
tf.Tensor((0.0625+0j), shape=(), dtype=complex64)
print(rho[1,0,0,1])
# returns
tf.Tensor((0.1875-0.0625j), shape=(), dtype=complex64)
```

In `TensorFlow` each element of the tensor is also a tensor with empty `shape=()`, i.e., a rank-0 scalar.

The trace $\mathrm{Tr}(\rho) = \rho^{uv}{}_{uv} = 1$ is found by `tf.einsum` as follows.

```
# we compute the trace by tf.einsum
# with a string corresponding to the contractions
# of the first qubit indices,
```

```
# and the second qubit indices
tf.einsum('uvuv',rho)
#returns
<tf.Tensor: shape=(), dtype=complex64, numpy=(1+0j)>
```

The trace (a scalar tensor) is unitary as we are dealing with a two-qubit pure state.

5.10 Partial Trace with `tensors`

The partial trace is expected to be a rank-2 tensor with `shape=(2,2)` corresponding to an operator in a single-qubit Hilbert space.

The reduced density matrix for the qubit A is

$$\rho_A = \rho^{xu}{}_{yw} , \tag{5.81}$$

being the trace with respect to the qubit B, i.e., a contraction of the B indices.

```
# we compute the partial trace
# by the proper strin in tf.einsum
rhoA=tf.einsum('xuyu',rho);
print(rhoA)
# returns
tf.Tensor(
[[0.125 +0.j      0.0625-0.3125j]
 [0.0625+0.3125j 0.875 +0.j     ]], shape=(2, 2), dtype=complex64)
```

We can also check the trace of the reduced density matrix, which is unitary as it corresponds to $\mathrm{Tr}(\rho)$.

```
# trace of rhoA by einsum
tf.einsum('uu',rhoA)
# returns
<tf.Tensor: shape=(), dtype=complex64, numpy=(1+0j)>
# As rhoA is a matrix
# that is a tensor
# with shape=(2,2)
# we can also use linalg.trace
tf.linalg.trace(rhoA)
<tf.Tensor: shape=(), dtype=complex64, numpy=(1+0j)>
```

We also compute the reduced density matrix $\rho_B = \rho^{iu}{}_{iv}$ for B and the corresponding trace (see Table 5.1).

Table 5.1 Density matrix, reduced density matrices, and the corresponding `TensorFlow` shape

Density matrix	ρ	${\rho^{sp}}_{qr}$	`shape=(2,2,2,2)`
Reduced density matrix A	$\rho_A = \text{Tr}_B(\rho)$	${\rho^{su}}_{qu}$	`shape=(2,2)`
Reduced density matrix B	$\rho_B = \text{Tr}_A(\rho)$	${\rho^{ip}}_{ir}$	`shape=(2,2)`

```
rhoB=tf.einsum('iuiv',rho);
# returns
tf.Tensor(
[[ 0.6875+0.j      -0.125 -0.4375j]
 [-0.125 +0.4375j  0.3125+0.j    ]], shape=(2, 2),
 ↪  dtype=complex64)
# let us check trace
tf.einsum('vv',rhoB)
# returns
<tf.Tensor: shape=(), dtype=complex64, numpy=(1+0j)>
```

5.11 Entropy of Entanglement

If qubit A is entangled with qubit B, the reduced density matrix corresponds to a mixed state, which we write as

$$\rho_A = \sum_{r=0}^{d-1} p^r |\psi_r\rangle\langle\psi_r| \,, \tag{5.82}$$

with

$$\sum_{r=0}^{d-1} p^r = 1 \,. \tag{5.83}$$

Multiple kets $|\psi_r\rangle$, with $r = 0, 1, \ldots, d-1$, contribute to ρ_A. This circumstance is described by stating that we have a nonvanishing *entropy of entanglement*.

As we are dealing with qubits, the reduced Hilbert space of qubit A has dimension $N = 2$, and the number of states in (5.82) is $d \leq N$. Hence either $d = 1$, corresponding to the absence of entropy, or $d = 2$, which is the case with entanglement and nonvanishing entropy.

Also, the different kets in (5.82) do not provide the same amount of information on the system. Following the Shannon theory of information, if a state has high probability of being observed, it provides a limited amount of information, or little *surprisal*. Metaphorically, if we always look at the same side of the Moon, we do not get much additional information by repeating observations.

The Shannon information is determined by the p^r as follows:

$$I_r = \log_2\left(\frac{1}{p^r}\right) = -\log_2(p^r)\,, \tag{5.84}$$

with $r = 0, 1, \ldots, d-1$.

If we have a pure state $\rho_A = |\psi_0\rangle\langle\psi_0|$, such that $p^0 = 1$ and $p^1 = 0$, the surprisal in observing $|\psi_0\rangle$ is zero, while it is diverging for $|\psi_1\rangle$, as we never observe $|\psi_1\rangle$.

The mean information $\langle I\rangle$ in ρ_A is the Shannon entropy of entanglement, which is a measure of the average surprisal in our observations. This is also a measure of the jumps from one state to another. Maximal surprisal corresponds to maximal fluctuation. The mean information is

$$\mathcal{E}(\rho) = \langle I\rangle = \sum_r p^r I^r = -\sum_r p^r \log_2(p^r)\,. \tag{5.85}$$

For a pure state, $p^0 = 1 - p^1 = 1$, we have $\mathcal{E} = 0$. On the contrary, if we have two states with equal probability $p^0 = p^1 = \frac{1}{2}$, we have $\mathcal{E} = 1$, i.e., maximum entropy.

5.12 Schmidt Decomposition

The entropy is a simple measure of entanglement, for which we need the reduced density matrix ρ_A, or ρ_B. The Schmidt decomposition is a way to determine the reduced density matrix and the mean information $\mathcal{E}$.

Let us a consider a bipartite system with Hilbert space $\mathcal{H} = \mathcal{H}_A \otimes \mathcal{H}_B$ with $\mathcal{H}_A$ having dimension M and $\mathcal{H}_B$ dimension N. The case of two simple qubits corresponds to $M = N = 2$.

The general pure state in $\mathcal{H}$ is written as

$$|\psi\rangle = \sum_{i,j} \psi^{ij}|i\rangle \otimes |j\rangle \tag{5.86}$$

where $i = 0, 1, \ldots, M-1$ and $j = 0, 1, \ldots, N-1$. The density matrix is

$$\rho = |\psi\rangle\langle\psi| = \sum_{i,j,k,l} \psi^{ij}\psi_{kl}|ij\rangle\langle kl|\,, \tag{5.87}$$

being the cotensor $\psi_{kl} = (\psi^{ij})^*$. To compute the partial trace ρ_A, we cannot readily use (5.87), as contracting with the states in B does not immediately furnish a diagonal expression in A.

We first consider the matrix Ψ with elements ψ^{jk}. Ψ is a rectangular matrix with dimension $M \times N$ with complex elements. Using the *singular value decomposition* (SVD), we write Ψ as

$$\Psi = \mathrm{UDV}^\dagger \tag{5.88}$$

D being a real rectangular diagonal matrix with dimension $M \times N$, U a complex $M \times M$ unitary matrix, and V a complex $N \times N$ unitary matrix.[5]

In tensor notation, Eq. (5.88) reads

$$\psi^{ij} = \mathrm{U}^i{}_m \mathrm{D}^{mq} (\mathrm{V}^\dagger)_q{}^j \tag{5.89}$$

with

$$\mathrm{D}^{mq} = d^m \delta^{mq} \ . \tag{5.90}$$

Note that in (5.90) no sum is implicit, as m appear only as an upper index.

For example, when $M = 2$ and $N = 4$, (5.90) reads in matrix form

$$\mathrm{D} = \begin{pmatrix} d^0 & 0 & 0 & 0 \\ 0 & d^1 & 0 & 0 \end{pmatrix} . \tag{5.91}$$

The state in (5.86) is

$$|\psi\rangle = \sum_{ij} \mathrm{U}^i{}_m \mathrm{D}^{mq} (\mathrm{V}^\dagger)_q{}^j |i\rangle|j\rangle = \sum_{mq} D^{mq} |u_m\rangle|v_q\rangle = \sum_m d^m |u_m\rangle|v_m\rangle \tag{5.92}$$

letting

$$|u_m\rangle_A = \sum_i \mathrm{U}^i{}_m |i\rangle_A \ , \tag{5.93}$$

and

$$|v_q\rangle_B = \sum_j (\mathrm{V}^\dagger)_q{}^j |j\rangle_B \ , \tag{5.94}$$

where we specify the index A or B for denoting the corresponding Hilbert space. Equation (5.92) shows that a two-qubit state can be expressed as the sum of product states, while $|u_m\rangle_A$ and $|v_n\rangle_B$ form new bases as U and V are unitary matrices.

[5] Some authors use V instead of $\mathrm{V}^\dagger$, but we use this notation as `tf.linalg.svd` returns V in the notation of (5.88).

The principal values d^m are real and can be zero. In particular, if only one d^m is different from zero (e.g., $d^0 = 1$ and $d^m = 0$ for $m \geq 1$), the state is a product state [see Eq. (5.92)].

Counting the nonzero d^m gives an indication of the degree of entanglement. The number of nonzero d^m is the *rank* $R \leq \{M, N\}$ of the Schmidt decomposition.

The density matrix from Eq. (5.92) is

$$\rho = \sum_{nm} d^m d^n |u_m v_m\rangle\langle u_n v_n| \tag{5.95}$$

If we want the reduced density matrix for A, we trace out the system B. As the trace does not depend on the basis, we use $|v_n\rangle_B$:

$$\begin{aligned}\rho_A = \mathrm{Tr}_B(\rho) &= \sum_r \langle v_r|_B \rho |v_r\rangle_B = \\ &= \sum_{rnm} d^m d^n \delta^r{}_m \delta^n{}_r |u_m\rangle_A \langle u_n|_A \\ &= \sum_r (d^r)^2 |u_r\rangle_A \langle u_r|_A \end{aligned} \tag{5.96}$$

Equation (5.96) is a mixture as in Eq. (5.82) with $p^r = (d^r)^2$.

We can also trace out the A degrees of freedom and obtain the reduced density matrix for B; we get

$$\rho_B = \sum_r (d^r)^2 |v_r\rangle_B \langle v_r|_B \ . \tag{5.97}$$

Also the reduced density matrix for B appears as a mixture with $p^r = (d^r)^2$.

The outcome is that the entropy of entanglement is the same for the two subsystems A and B. We have

$$\mathcal{E}(\rho_A) = \mathcal{E}(\rho_B) = -\sum_r (d^r)^2 \log_2[(d^r)^2] = -\sum_r p^r \log_2 p^r \ , \tag{5.98}$$

with the sum involving the nonzero d^r.

Summarizing, to determine the entropy of entanglement, we need to perform the SVD of ψ^{ij}. In the particular case of two qubits with $N = M = 2$, this is equivalent to computing eigenvalues and eigenvectors of a 2×2 matrix and counting the nonzero eigenvalues. As an alternative, we can diagonalize the reduced density matrices ρ_A and ρ_B, finding (5.96) and (5.97), with eigenvalues p^0 and p^1.

We make examples in `TensorFlow` in the following.

5.13 Entropy of Entanglement with tensors

We consider the state in (5.78) and obtain the singular value decomposition by `tf.linalg.svd`. First, we use the singular values d^r to retrieve the entropy of entanglement, as follows.

```
# We compute the singular values.
# The flag compute_uv=False returns only
# the principal values
d=tf.linalg.svd(psi,compute_uv=False)
# returns
tf.Tensor([0.996055   0.08873843],
shape=(2,), dtype=float32)
# note this is a real tensor
# in this case the Schmidt rank = 2
# and the state is entangled
```

We have a state with Schmidt rank $R = 2$, as we obtain two nonvanishing real principal values d^r. The state is entangled.

Following (5.82), we have the probabilities $p^0 = (d^0)^2$ and $p^1 = (d^1)^2$ and the entropy of entanglement.

```
# compute the probabilities
p=tf.abs(d)**2
tf.print(p)
# returns
[0.992125571 0.00787450839]
# entropy of entanglement
-p[0]*np.log2(p[0])-p[1]*np.log2(p[1])
# returns
<tf.Tensor: shape=(), dtype=float32, numpy=0.0663473>
# note we used the numpy function <code>np.log2</code>
# and we still obtain a tf.tensor
```

We then consider the eigenvalues of the reduced density matrices ρ_A and ρ_B, which correspond to (5.96) and (5.97). We find the eigenvalues of `rhoA` and `rhoB` by `np.linalg.eig` as follows.

```
# np.linalg.eig returns eigenvalues
# and eigenvectors
eA=tf.linalg.eig(rhoA)
# the first elements are the eigenvectors
print(eA[0])
# returns
tf.Tensor([0.00787451-4.8248174e-09j 0.9921254 +4.8248174e-09j],
```

```
shape=(2,), dtype=complex64)
# these are positive and real
# and equal to the probabilities
# p[0] and p[1]
# If we consider the eigenvalues of rhoB
# we obtaine the same result
eB=tf.linalg.eig(rhoB);
print(eB[0])
# returns
tf.Tensor([0.00787451+1.5521238e-08j 0.9921255 -2.2971818e-08j],
shape=(2,), dtype=complex64)
# note that linalg.eig returns complex eigenvalues
# with numerical precision
# while linalg.svd returns real singular values
```

The eigenvalues of the reduced density matrices of ρ_A and ρ_B are equal to the probabilities p^0 and p^1, and they furnish the same value of the entanglement entropy.

5.14 Schmidt Basis with `tensors`

Let us compute the orthonormal bases $|u_r\rangle_A$ and $|v_r\rangle_B$. One approach is using the output of `tf.linalg.svd`, which returns U and V. Note that in (5.94), one uses $V^\dagger$; thus we have to use the adjoint of V.

```
# d, U, V
d, U, V=tf.linalg.svd(psi)
print(U)
#returns
tf.Tensor(
[[-0.05729683+0.34017158j  0.43057674+0.8340288j ]
 [-0.9381789 +0.02864844j  0.26953763-0.21528839j]],
shape=(2, 2), dtype=complex64)
print(V)
#returns
tf.Tensor(
[[-0.22828318+0.7989912j  0.15283479-0.5349218j]
 [ 0.5563271 +0.j         0.83096343+0.j       ]],
shape=(2, 2), dtype=complex64)
# we need the adjoint of V
VH=tf.math.conj(tf.transpose(V))
print(VH)
#returns
tf.Tensor(
[[-0.22828318-0.7989912j  0.5563271 -0.j       ]
 [ 0.15283479+0.5349218j  0.83096343-0.j       ]],
shape=(2, 2), dtype=complex64)
```

From (5.93), we see that the columns of U, i.e., the elements `U[:,0]` and `U[:,1]`, are the components of $|u_0\rangle_A$ and $|u_1\rangle_A$, respectively. From (5.94), the components of $|v_0\rangle_B$ and $|v_B\rangle$ are the rows of ($\mathrm{V}^\dagger$), namely, `VH[0,:]` and `VH[1,:]`. We have

```
# compute the components of the vector u
u0=U[:,0]
u1=U[:,1]
print(u0)
print(u1)
# returns
tf.Tensor([-0.05729683+0.34017158j -0.9381789 +0.02864844j],
shape=(2,), dtype=complex64)
tf.Tensor([0.43057674+0.8340288j  0.26953763-0.21528839j],
shape=(2,), dtype=complex64)
# compute the components of the vector v
tf.Tensor([-0.22828318-0.7989912j  0.5563271 -0.j        ],
shape=(2,), dtype=complex64)
tf.Tensor([0.15283479+0.5349218j 0.83096343-0.j        ],
shape=(2,), dtype=complex64)
```

It is also instructing to compute the vectors $|u_r\rangle$ and $|v_r\rangle$ by implementing Eqs. (5.93) and (5.94) in TensorFlow. We first build a list with the basis vectors as follows.

```
basis=[qubit0,qubit1]
# print basis
# returns
[<tf.Tensor: shape=(2,), dtype=complex64,
numpy=array([1.+0.j, 0.+0.j], dtype=complex64)>,
<tf.Tensor: shape=(2,), dtype=complex64,
numpy=array([0.+0.j, 1.+0.j], dtype=complex64)>]
# such that
# qubit0=basis[0]
# qubit1=basis[1]
# and the elemnts of qubit0 are
# basis[0][0] and basis[0][1]
# and the elemnts of qubit1 are
# basis[1][0] and basis[1][1]
```

Then we contract as in (5.93) with respect to the first index of `U` and the first of `basis` and as in (5.94) the second index of `VH` and the first of `basis`. `basis` is the same for A and B.

```
# we contract the first index of U and the first index of basis
ucomponents=tf.tensordot(U,basis,[[0],[0]])
# we contract the second index of V and the first index of basis
```

```
vcomponents=tf.tensordot(VH,basis,[[1],[0]])
# print u0 vector
ucomponents[0]
#returns
<tf.Tensor: shape=(2,), dtype=complex64,
numpy=array([-0.05729683+0.34017158j, -0.9381789 +0.02864844j],
dtype=complex64)>
# print u1 vector
ucomponents[1]
#returns
<tf.Tensor: shape=(2,), dtype=complex64,
numpy=array([0.43057674+0.8340288j, 0.26953763-0.21528839j],
dtype=complex64)>
# print v0 vector
vcomponents[0]
#returns
<tf.Tensor: shape=(2,), dtype=complex64,
numpy=array([-0.22828318-0.7989912j,  0.5563271 +0.j        ],
dtype=complex64)>
# print v1 vector
vcomponents[1]
# returns
<tf.Tensor: shape=(2,), dtype=complex64,
numpy=array([0.15283479+0.5349218j, 0.83096343+0.j        ],
dtype=complex64
```

We find the components of the reduced density matrix $|u_r\rangle\langle u_r|$ and $|v_r\rangle\langle v_r|$.

```
# |u0><u0|
print(tf.tensordot(U[:,0],tf.math.conj(U[:,0]),axes=0))
tf.Tensor(
[[0.11899963+0.j          0.06350006-0.31750035j]
 [0.06350006+0.31750035j 0.8810004 +0.j          ]],
shape=(2, 2), dtype=complex64)
# |u1><u1|
tf.Tensor(
[[ 0.88100034+0.j         -0.06350008+0.3175003j]
 [-0.06350008-0.3175003j  0.11899962+0.j         ]],
shape=(2, 2), dtype=complex64)
# |v0><v0|
tf.Tensor(
[[ 0.6905002 +0.j -0.46228746+0.j]
 [-0.46228746+0.j  0.30949983+0.j]],
shape=(2, 2), dtype=complex64
# |v1><v1|
tf.Tensor(
[[0.30949986+0.j 0.4622875 +0.j]
 [0.4622875 +0.j 0.6905002 +0.j]],
shape=(2, 2), dtype=complex64)
```

We have found the Schmidt basis by using the SVD decomposition. We obtain the same results by diagonalizing ρ_A and ρ_B, i.e., the reduced density matrices. First, we consider ρ_A and we find the eigenvalues and eigenvectors by `tf.linalg.eig`.

```
# compute eigenvalues and eigenvectors
# of rhoA
eA=tf.linalg.eig(rhoA);
# eA is a list of two elements
# eA[0] are the eigenvalues
# eA[1] is a matrix
# with eigenvectors as columns
print(eA[0])
tf.Tensor([0.00787451-4.8248174e-09j 0.9921254 +4.8248174e-09j],
shape=(2,), dtype=complex64)
print(eA[1])
# returns
tf.Tensor(
[[-0.9386162 +0.0000000e+00j  0.34496322+1.2747664e-08j]
 [ 0.06765284+3.3826429e-01j  0.18407774+9.2038894e-01j]],
shape=(2, 2), dtype=complex64)
```

Note that the eigenvectors of ρ_A differ from $|u_r\rangle$ within a phase factor. However, we can factor out the phase factor by considering the matrix elements.

For example, we compare the matrix elements $|u_0\rangle\langle u_0|$ which correspond to `U[:,0]` and two `eA[1][:,1]` and we have identical matrices.

```
print(tf.tensordot(eA[1][:,0],tf.math.conj(eA[1][:,0]),axes=0))
print(tf.tensordot(U[:,1],tf.math.conj(U[:,1]),axes=0))
tf.Tensor(
[[0.11899962+0.j          0.06350006-0.31750032j]
 [0.06350006+0.31750032j 0.8810004 +0.j         ]],
shape=(2, 2), dtype=complex64)
tf.Tensor(
[[0.11899963+0.j          0.06350006-0.31750035j]
 [0.06350006+0.31750035j 0.8810004 +0.j         ]],
shape=(2, 2), dtype=complex64)
```

Seemingly, we can consider the eigenvalues and eigenvectors of ρ_B.

```
# compute eigenvalues and eigenvectors
# of rhoB
eB=tf.linalg.eig(rhoB);
# eB is a list of two elements
# eB[0] are the eigenvalues
# eB[1] is a matrix
# with eigenvectors as columns
```

```
print(eB[0])
tf.Tensor([0.00787451+1.5521238e-08j 0.9921255 -2.2971818e-08j],
shape=(2,), dtype=complex64)
print(eB[1])
# returns
tf.Tensor(
[[-0.9386162 +0.0000000e+00j  0.34496322+1.2747664e-08j]
 [ 0.06765284+3.3826429e-01j  0.18407774+9.2038894e-01j]],
shape=(2, 2), dtype=complex64)
```

We compare the matrix elements of the corresponding eigenvectors.

For example, for $|v_1\rangle\langle v_1|$ we have

```
print(tf.tensordot(eB[1][:,0],tf.math.conj(eB[1][:,0]),axes=0))
print(tf.tensordot(VH[1,:],tf.math.conj(VH[1,:]),axes=0))
# returns
tf.Tensor(
[[0.30949992+0.j         0.12700014+0.44450054j]
 [0.12700014-0.44450054j 0.6905002 +0.j        ]],
shape=(2, 2), dtype=complex64)
tf.Tensor(
[[0.30949983+0.j         0.12700012+0.44450048j]
 [0.12700012-0.44450048j 0.6905002 +0.j        ]],
shape=(2, 2), dtype=complex64)
```

5.15 Product States and Maximally Entangled States with `tensors`

Let us consider the entanglement entropy of product states and maximally entangled states. First, we define a `tf.function` that returns the entropy.

```
@tf.function
def VonNeumannEntropy2(psi):
    """ Return the entanglement entropy for a two-qubit
    with tensor (2,2) psi
    Return the principal values and pr
    """
    d=tf.linalg.svd(psi,compute_uv=False)
    d2=tf.abs(d)**2
    logd2=tf.math.log(d2+1e-12)/tf.math.log(2.0)
    # note we use log and we change bases
    # we also add 1e-12 in the log
    # to avoid NaN when log(0)
    return tf.reduce_sum(-d2*logd2),d2
```

We test the function on $|\psi\rangle$ in Eq. (5.78).

```
# the functions returns the entropy and
# the probabilities
tf.print(VonNeumannEntropy2(psi))
# returns
(0.0663472936, [0.992125571 0.00787450839])
# the first value is the entropy
# the second is the list of
# the two probabilities
```

We then consider a product state $|\phi\rangle = |\phi\rangle_A|\phi\rangle_B$ with

$$
\begin{aligned}
|\phi\rangle_A &= \frac{1}{\sqrt{2}}|0\rangle + \frac{\mathrm{i}}{\sqrt{2}}|1\rangle \\
|\phi\rangle_B &= |1\rangle \\
|\phi\rangle &= \frac{1}{\sqrt{2}}|01\rangle + \frac{\mathrm{i}}{\sqrt{2}}|11\rangle \,.
\end{aligned}
$$

We define the state in TensorFlow and compute the entropy.

```
psiA=(qubit0+1j*qubit1)/np.sqrt(2.0)
psiB=qubit1
# compute the outer product
# for the two qubit states
psi=tf.tensordot(psiA,psiB,axes=0)
print(psi)
#returns
tf.Tensor(
[[0.        +0.j         0.70710677+0.j        ]
 [0.        +0.j         0.        +0.70710677j]], shape=(2, 2),
dtype=complex64)
# compute the entropy
tf.print(VonNeumannEntropy2(phi))
# returns
(1.71982634e-07, [0.999999881 0])
```

For a product state, the entropy is zero, and the Schmidt rank $R = 1$, as we have only one nonvanishing p^r, within numerical precision.

Finally, we consider *maximally entangled states*, as the Bell states

$$
|\Phi^+\rangle = \frac{1}{\sqrt{2}}|00\rangle + \frac{1}{\sqrt{2}}|11\rangle \,, \tag{5.99}
$$

$$|\Psi^+\rangle = \frac{1}{\sqrt{2}}|01\rangle + \frac{1}{\sqrt{2}}|10\rangle\ . \tag{5.100}$$

We define the Bell states in `TensorFlow` and compute the entropy.

```
PHIplus=(q00+q11)/np.sqrt(2.0)
tf.print(VonNeumannEntropy2(PHIplus))
# returns
(1, [0.49999997 0.49999997])
PSIplus=(q01+q10)/np.sqrt(2.0)
tf.print(VonNeumannEntropy2(PSIplus))
# returns
(1, [0.49999997 0.49999997])
```

The two Bell states have unitary (i.e., maximum) entropy, denoting the maximum variability or surprise when we make an observation. Indeed, the probabilities are equally distributed in the two composite states.

By using SVD, we verify that the density matrix of Ψ_+ for the qubit A is

$$\rho_A = \text{Tr}_B(|\Psi_+\rangle\langle\Psi_+|) = \frac{1}{2}|0\rangle\langle 0| + \frac{1}{2}|1\rangle\langle 1| \tag{5.101}$$

that is, a mixture between $|0\rangle$ and $|1\rangle$. The same holds for ρ_B. This conclusion holds true also for the other Bell states[6]

$$|\Phi^-\rangle = \frac{1}{\sqrt{2}}|00\rangle - \frac{1}{\sqrt{2}}|11\rangle$$

$$|\Psi^-\rangle = \frac{1}{\sqrt{2}}|01\rangle - \frac{1}{\sqrt{2}}|10\rangle\ .$$

5.16 Entanglement in the Two-Qubit TIM

We now compute the entanglement of the ground state of the TIM. We first consider the simple case $J > 0$, and $h_0 = h_1 = h > 0$ as in Eq. (5.9). For $h = 0$, the ground state can be a product state or a maximally entangled state because of the degeneracy.

Considering the eigenvalues in (5.13), the ground state has energy

$$E_0 = -\sqrt{4d^2 + J^2}\ . \tag{5.102}$$

[6] Details in `jupyternotebooks/quantumfeaturemap/QubitsDensityMatrix.ipynb`.

The corresponding eigenvector is[7]

$$|\psi_0\rangle = \frac{h}{4h^2 + J^2 - J\sqrt{4h^2+J^2}} \begin{pmatrix} 1 \\ -\frac{J-\sqrt{4h^2+J^2}}{2h} \\ -\frac{J-\sqrt{4h^2+J^2}}{2h} \\ 1\,. \end{pmatrix} . \tag{5.103}$$

By writing $|\psi_0\rangle$ in tensor notation as

$$|\psi_0\rangle = \sum_{i,j\in\{0,1\}^2} C^{ij}\,|ij\rangle \,, \tag{5.104}$$

the tensor C^{ij} is obtained by reordering the components in Eq. (5.103).

$$C^{ij} = \begin{pmatrix} \frac{h}{\sqrt{4h^2+J^2-J\sqrt{4h^2+J^2}}} & \frac{1}{2}\sqrt{1-\frac{J}{\sqrt{4h^2+J^2}}} \\ \frac{1}{2}\sqrt{1-\frac{J}{\sqrt{4h^2+J^2}}} & \frac{h}{\sqrt{4h^2+J^2-J\sqrt{4h^2+J^2}}} \end{pmatrix} . \tag{5.105}$$

As C^{ij} is a 2×2 matrix, the singular values of C^{ij} correspond to its eigenvalues, denoted as d^r_{TIM}, which are given by

$$d^0_{\mathrm{TIM}} = \frac{4h\sqrt{4h^2+J^2}-\sqrt{64h^4+48h^2J^2+8J^4-32h^2J\sqrt{4h^2+J^2}-8J^3\sqrt{4h^2+J^2}}}{4\sqrt{4h^2+J^2}\sqrt{4h^2+J^2-J\sqrt{4h^2+J^2}}} \tag{5.106}$$

and

$$d^1_{\mathrm{TIM}} = \frac{4h\sqrt{4h^2+J^2}+\sqrt{64h^4+48h^2J^2+8J^4-32h^2J\sqrt{4h^2+J^2}-8J^3\sqrt{4h^2+J^2}}}{4\sqrt{4h^2+J^2}\sqrt{4h^2+J^2-J\sqrt{4h^2+J^2}}} \tag{5.107}$$

The corresponding probabilities $p^r = (d^r_{\mathrm{TIM}})^2$ give the entropy by Eq. (5.98).

$$\mathcal{E} =$$

$$-\frac{1}{\log(2)}\left[\frac{4h\sqrt{4h^2+2J\left(J-\sqrt{4h^2+J}\right)}\coth^{-1}\left(\frac{\sqrt{2}\mathrm{h}}{\sqrt{2\mathrm{h}^2+\mathrm{J}^2-\mathrm{J}\sqrt{4\mathrm{h}^2+\mathrm{J}^2}}}\right)}{4h^2+J\left(J-\sqrt{4h^2+J^2}\right)} + \log\left(\frac{J}{2\sqrt{4h^2+J^2}}\right)\right] . \tag{5.108}$$

[7] Details in the `Mathematica` file `TwoQubitTransverseIsing.nb`.

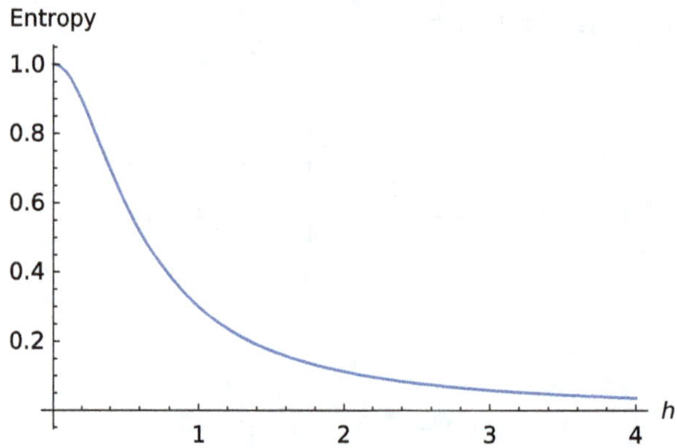

Fig. 5.2 Entropy of entanglement for the two-qubit transverse-field Ising Hamiltonian, when $J = 1.0$ and $h_0 = h_1 = h$

Figure 5.2 shows the entanglement entropy $\mathcal{E}$ in Eq. (5.108) as a function of h. We see that for $h \to 0$, one finds maximal $\mathcal{E}$, showing that even a vanishing small external field introduces entanglement and breaks the degeneracy by making the Bell state in Eq. (5.8) the leading ground state. Also, the entanglement entropy goes to zero when increasing the field, as the state tends to a product state.

5.16.1 $h_0 = h$ and $h_1 = 0$

One can argue about what happens when $h_0 = h$ and $h_1 = 0$, i.e., if the external field only acts on one single qubit. The Hamiltonian matrix form reads

$$\hat{\mathcal{H}} = \begin{pmatrix} -J & 0 & -h & 0 \\ 0 & J & 0 & -h \\ -h & 0 & J & 0 \\ 0 & -h & 0 & -J \end{pmatrix}, \tag{5.109}$$

with degenerate eigenvalues

$$\begin{aligned} E_{0,1} &= -\sqrt{h^2 + J^2}\,, \\ E_{2,3} &= \sqrt{h^2 + J^2}\,. \end{aligned} \tag{5.110}$$

For the lowest energy eigenvectors (within a normalization constant), we have

$$|\psi_0\rangle = \begin{pmatrix} 0 \\ -J + \sqrt{h^2 + J^2} \\ 0 \\ h \end{pmatrix} \tag{5.111}$$

and

$$|\psi_1\rangle = \begin{pmatrix} J + \sqrt{h^2 + J^2} \\ 0 \\ h \\ 0 \end{pmatrix} . \tag{5.112}$$

Let us consider the first eigenvector $|\psi_0\rangle$ in matrix form as in Eq. (5.104); we have

$$C^{ij} = \frac{h}{2h^2 + 2J^2 + 2J\sqrt{h^2 + J^2}} \begin{pmatrix} 0 & 1 \\ 0 & 1 \end{pmatrix} . \tag{5.113}$$

From Eq. (5.113) we realize that the singular values d^0 and d^1 are not both different from zero. We have $d^0 = 0$ and

$$d^1 = \frac{h}{\sqrt{2h^2 + 2J^2 - 2J\sqrt{h^2 + J^2}}} . \tag{5.114}$$

Correspondingly, the ground state with energy $-\sqrt{h^2 + J^2}$ is not entangled and has zero entropy. We have the same outcome for the eigenstate $|\psi_1\rangle$. The conclusion is that when $h_0 = h$ and $h_1 = 0$, the entropy of entanglement is zero for any h. Hence, to entangle the ground state we need a field acting on both qubits.

As one can realize, analytically handling the TIM is cumbersome, also in the simple case of two qubits. In the next chapter, we will use `TensorFlow` and a variational quantum ansatz to compute the ground state.

5.17 Further Reading

- Introduction to entanglement theory. A first introduction to entanglement and Schmidt decomposition can be found in the book II of Gardiner and Zoller [2].
- The ground state of the transverse-field Ising model. The ground state of the TIM has many interesting features and is a paradigm for the optimization of quantum models. The reader may consider [3].

References

1. V. Havlíček, A.D. Córcoles, K. Temme, A.W. Harrow, A. Kandala, J.M. Chow, J.M. Gambetta, **567**, 209 (2019). https://doi.org/10.1038/s41586-019-0980-2
2. C. Gardiner, P. Zoller, *The Quantum World of Ultra-Cold Atoms and Light Book II: The Physics of Quantum-Optical Devices* (Imperial College Press, London, 2015). https://doi.org/10.1142/p983
3. S.F.E. Oliviero, L. Leone, A. Hamma. ArXiv:2205.02247 (2022)

Chapter 6
Variational Algorithms, Quantum Approximate Optimization Algorithm, and Neural Network Quantum States with Two Qubits

Machine learning is a colossal fit!

Abstract We describe different variational ansatzes and study entanglement in the ground state of the two-qubit transverse-field Ising model.

6.1 Introduction

The goal is to introduce variational methods for the ground state of the two-qubit transverse-field Ising Hamiltonian. These methods apply to different problems and vary in terms of complexity. We do not have a specific recipe to choose one approach or another, but we learn by comparing them. It is also significant to try different implementations in terms of code design.

We write tensor representations and use them to find the ground state by training the resulting model.

6.2 Training the Two-Qubit Transverse-Field Ising Model

We create a model that returns the expected value of the TIM Hamiltonian $\hat{\mathcal{H}}$ for a given state $|\psi\rangle$, i.e.,

$$\langle\hat{\mathcal{H}}\rangle = \langle\phi|\hat{\mathcal{H}}|\phi\rangle \ . \tag{6.1}$$

As detailed in Chap. 5, the Hamiltonian is

$$\hat{\mathcal{H}} = -J\,\hat{\mathrm{Z}}_0 \otimes \hat{\mathrm{Z}}_1 - h_0\hat{\mathrm{X}}_0 - h_1\hat{\mathrm{X}}_1 \ . \tag{6.2}$$

C. Conti, *Quantum Machine Learning*, Quantum Science and Technology,
https://doi.org/10.1007/978-3-031-44226-1_6

We want to study different variational ansatzes, so we pass to the model different functions. For example, we start considering a simple two-qubit state, which is just a rotation.

```
@tf.function
def FeatureMapU1(theta, psi):
    """ Feature Map for the
    Variational Quantum Algorithm
    with no entanglement

    Params
    ------
    theta, shape=(6,) real
    psi,   two-qubit state

    Returns
    -------
    A two-qubit state
    """
    thetaX0=theta[0]
    thetaX1=theta[1]
    thetaY0=theta[1]
    thetaY1=theta[2]
    thetaZ0=theta[3]
    thetaZ1=theta[4]
    thetaZZ=theta[5]
    phi=quantummap.Gate(quantummap.EXI(thetaX0),psi)
    phi=quantummap.Gate(quantummap.EIX(thetaX1),phi)
    phi=quantummap.Gate(quantummap.EYI(thetaY0),phi)
    phi=quantummap.Gate(quantummap.EIY(thetaY1),phi)
    phi=quantummap.Gate(quantummap.EZI(thetaZ0),phi)
    phi=quantummap.Gate(quantummap.EIZ(thetaZ1),phi)
    phi=quantummap.Gate(quantummap.EZZ(thetaZZ),phi)
    return phi
```

We build the ansatz as a feature map, which takes a vector of parameters and a *bias* state $|\psi\rangle$ and returns a new state. The bias ket is an initial guess for the minimum, which we can freely choose. In the simplest case, we use vacuum $|\psi\rangle = |00\rangle$.

The `FeatureMapU1` uses the `Gate` methods and the rotation gates `EXI`, `EIX`, `etc.` defined in `thqml.quantummap`.[1]

Figure 6.1 gives a sketch of the links in the NN model definition. Specifically, `FeatureMapU1` corresponds (see Fig. 6.2) to

$$|\phi\rangle = e^{i\theta_0\hat{X}_0}e^{i\theta_1\hat{X}_1}e^{i\theta_2\hat{Y}_0}e^{i\theta_3\hat{Y}_1}e^{i\theta_4\hat{Z}_0}e^{i\theta_5\hat{Z}_1}|\psi\rangle \ . \tag{6.3}$$

[1] Details in `jupyternotebooks/quantumfeaturemap/TwoQubitTransverseField Ising.ipynb`.

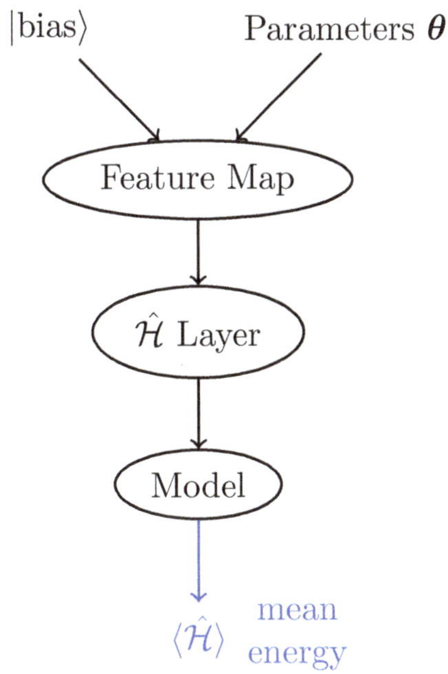

Fig. 6.1 The bias state and the parameter $\boldsymbol{\theta}$ are inputs to the feature map function, which is used by the layer for the Hamiltonian. The latter is used in the model that returns the expected value of the energy adopted for training

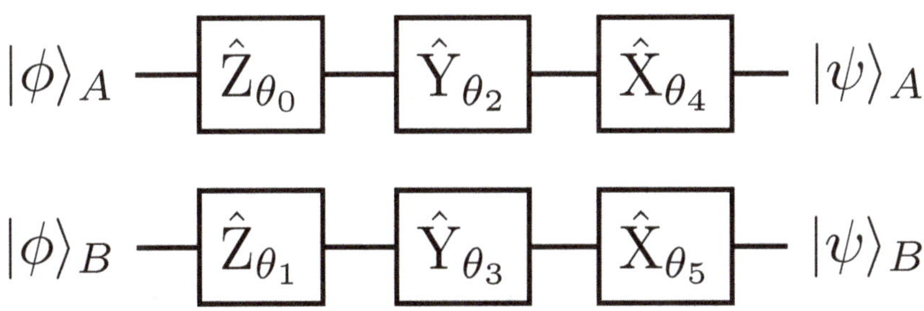

Fig. 6.2 The simplest non-entangled two-qubit feature map for the variational ansatz. The resulting state is $|\phi\rangle = |\phi\rangle_A \otimes |\phi\rangle_B$

`FeatureMapU1` does not introduce any entanglement on the two qubits. We create the model by passing `FeatureMapU1` to the layer definition, as follows.

```
class HamiltonianLayer(tf.keras.layers.Layer):
    # Define a Layer returning the mean value of the Hamiltonian
    # on the variational ansatz
    #
    # The layer has a trainable parameters, i.e.,
    # the  parameters theta of the trainable feature map
    #
    # The Layer use the functions
    # Hamiltonian
```

```
    # FeatureMapU1 or other feature map
    #
    # The constructor use parameters
    #
    #  Params
    #  ------
    #  J : coupling coefficient
    #      (default J=1)
    #  h : positive field shape=(2,)
    #      (default h=[0.0,0.0])
    #  map: a function containing the feature map
    #      (default is FeatureMapU1)
    #  nparams: the maximum number of trainable parameters
    #      (default is 10)
    #  bias: the bias state
    #      (default is ground state quantummap.q00)
    #
    #  Example
    #  -------
    #  HL=HL(J=.5, h=[1.2,-1.0], map=MyMap)
    #
    #  Returns
    #  -------
    #  The call return the real part of
    #  <\psi|H|\psi>
    #
    def __init__(self, J=1.0, h=[0.0,0.0],
                 FeatureMap=FeatureMapU1,
                 nparams=10,
                 bias=quantummap.q00,
                 **kwargs):
        super(HamiltonianLayer, self).__init__(**kwargs)
        # trainable parameter of the model
        # initially set as random (real variables)
        self.theta=tf.Variable(np.random.random(nparams),
                                 dtype=tf.float32,
                                name='theta')
        # gap coefficient
        self.J=J
        # field coefficient
        self.h=h
        # Hamiltonian
        self.H=Hamiltonian(J=self.J,h=self.h)
        # Bias state
        self.bias=bias
        # Feature map for the ground state
        self.FeatureMap=FeatureMap
    def ground_state(self):
        """ Return the current ground state """
        phi=self.FeatureMap(self.theta,self.bias)
        return phi
    def call(self,dummy):
        """ Return Re<H>
        Remark:  this layer has a dummy variable as input
```

```
        (required for compatibility with tensorflow call)
        """
        phi=self.ground_state()
        Hphi=quantummap.Gate2(self.H,phi)
        meanH=quantummap.Scalar(phi, Hphi)
        return tf.math.real(meanH)
```

The layer has a vector of weights `theta` with dimension `nparams`; the weights are initially set as random numbers. The function `Gate2` applies a two-qubit operator to a two-qubit state by tensor contraction, as described in the previous chapters.

```
# apply gate function for two qubits
@tf.function
def Gate2(A, psi):
    return tf.tensordot(A, psi, axes=[[1, 3], [0, 1]])
```

The layer uses the following function for the Hamiltonian `tensor`.

```
# import the gates from thqml.quantummap
from thqml import quantummap as quantummap
# define the Hamiltonian
@tf.function
def Hamiltonian(J=1, h=[0,0]):
    """ Two-qubit tranverse Ising model Hamiltonian

        H=-J Z0 Z1 -h0 X0 -h1 X0

        Params
        ------
        J = coupling shape=(1,)
        h = field shape=(2,)
    """
    out = -J*quantummap.ZZ-h[0]*quantummap.XI-h[1]*quantummap.IX

    return out
```

For example, we create an instance of the layer.

```
HL=HamiltonianLayer(FeatureMap=FeatureMapU1,nparams=6)
```

`FeatureMap` points to the function for the variational ansatz. `nparams` is the number of parameters to train; it must be equal to or greater than the parameters in the `FeatureMap` function. If smaller, we will get an error as `FeatureMap` will access elements not existent in the tensor `theta`. If greater, we will have non-trained dummy parameters. As bias state $|\psi\rangle$, we choose $|00\rangle$ as default.

The `HamiltonianLayer` returns $\langle\hat{\mathcal{H}}\rangle$; we need to complete the model by an input layer, an output, and a loss function.

```
# hyperparameter for the Hamiltonian
J=1.0
h=[0.0,0.0]
# Input layer (one dummy input)
xin1 = tf.keras.layers.Input(1, name='DummyInput1');
# Hamiltonian layer
HL1=HamiltonianLayer(J=J,h=h, nparams=6,FeatureMap=FeatureMapU1,
↪   name='H1')
# output
meanH1=HL1(xin1)
# trainable model returning meanH
Ising1 = tf.keras.Model(inputs = xin1, outputs=meanH1,
↪   name='model1')
# add loss function
Ising1.add_loss(meanH1)
```

The input layer is a dummy layer, which takes as input a real value not affecting the value of the Hamiltonian. We have this dummy input because we are working on a variational quantum algorithm with parameters inside the layer, and not with a quantum machine learning model that realizes an input/output function.

We run as

```
# run the layer with dummy input 1.1
tf.print(Ising1(1.1))
# returns the mean energy on
# the random ground states
[[0.00916039944]]
```

The model summary is the following.

```
Ising1.summary()
# returns
Model: "model1"
_________________________________________________________________
 Layer (type)                Output Shape              Param #
=================================================================
 input_4 (InputLayer)        [(None, 1)]               0

 H1 (HamiltonianLayer)       (1, 1)                    6

 add_loss (AddLoss)          (1, 1)                    0
```

```
==================================================================
Total params: 6
Trainable params: 6
Non-trainable params: 0
__________________________________________________________________
```

Note that the number of parameters is `nparams`. Figure 6.3 shows the plot of the model.

We use the feature map to minimize the Hamiltonian by compiling and training, as follows.

```
# we compile the model
# guessing a typical value for the learning rate
# and using the Adam optimizer
# as the typical first choice
Ising1.compile(optimizer=tf.keras.optimizers.Adam
(learning_rate=0.01))
# we train the model
history = Ising1.fit(x=np.zeros(10,),epochs=1000, verbose=0)
```

Figure 6.4 shows the resulting training history. The value of the mean energy $\langle\hat{\mathcal{H}}\rangle$ reaches the expected minimum $\langle\hat{\mathcal{H}}\rangle = -1$, which corresponds to the ground state energy as detailed in Chap. 5. If we inspect the ground state, we find that it corresponds to $|00\rangle$, within a phase factor.

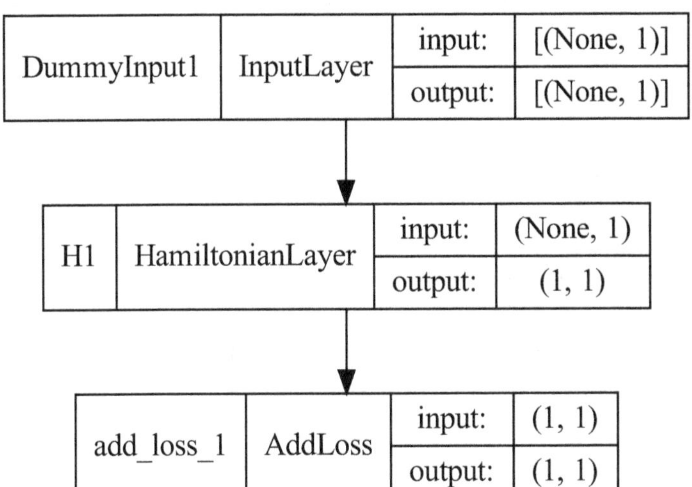

Fig. 6.3 The model `Ising1` for minimizing the two-qubit transverse-field Ising Hamiltonian

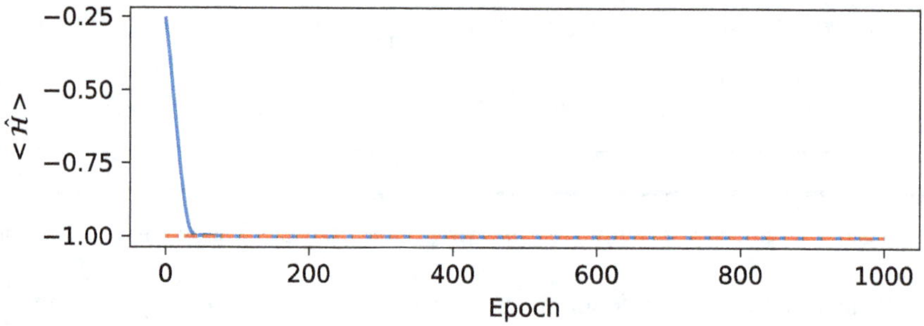

Fig. 6.4 Training the model `Ising1` with no external field. The final value of the loss $\langle\hat{\mathcal{H}}\rangle$ corresponds to the lowest eigenvalue $E_0 = -1.0$ of $\hat{\mathcal{H}}$ (dashed line). The ground state has zero entropy of entanglement

```
print(HL1.ground_state())
# returns
tf.Tensor(
[[-9.81e-02-9.95e-01j -2.81e-08-8.26e-08j]
 [-7.12e-08+5.04e-08j -3.84e-15-6.57e-15j]], shape=(2, 2),
 ↪  dtype=complex64)
```

We measure the entanglement entropy of the resulting ground state by using the `quantummap.VonNeumannEntropy2` defined in Chap. 5.

```
# compute the entanglement entropy
quantummap.VonNeumannEntropy2(HL1.ground_state())
# returns
(<tf.Tensor: shape=(), dtype=float32, numpy=0.0>,
 <tf.Tensor: shape=(2,), dtype=float32, numpy=array([1., 0.],
 ↪  dtype=float32)>)
# we have 0.0 for the entropy
# and [1.0 0.0] for the probabilities p0 and p1
```

The entanglement entropy is zero, as expected according to Chap. 5.

We now consider a model `Ising2` with a magnetic field $h_0 = 1$ and $h_1 = 0$.

```
# hyperparameter for the Hamiltonian
J=1.0
h=[1.0,0.0]
# Input layer (one dummy input)
xin2 = tf.keras.layers.Input(1);
# Hamiltonian layer
HL2=HamiltonianLayer(J=J,h=h, nparams=6,FeatureMap=FeatureMapU1,
↪   name='H2')
# output
meanH2=HL2(xin2)
# trainable model returning meanH
Ising2 = tf.keras.Model(inputs = xin2, outputs=meanH2,
↪   name='model2')
# add loss function
Ising2.add_loss(meanH2)\vspace*{12pt}
```

After compiling and training, as for the model `Ising1`, we obtain the history in Fig. 6.5. We find that the model reaches the lowest eigenvalue. The entropy of entanglement for the found ground state is also zero. This is not surprising because we are using a non-entangled variational ansatz. This ansatz can match the ground state, as we know from Chap. 5 that it is not entangled.

Finally, we consider the case $h_0 = h_1 = h > 0$. We know (Chap. 5) that the ground state has a nonvanishing entropy of entanglement. We expect that the variational ansatz `FeatureMapU1` will not be able to catch the minimum value of energy. We choose $h = 1.0$ and proceed as follows.

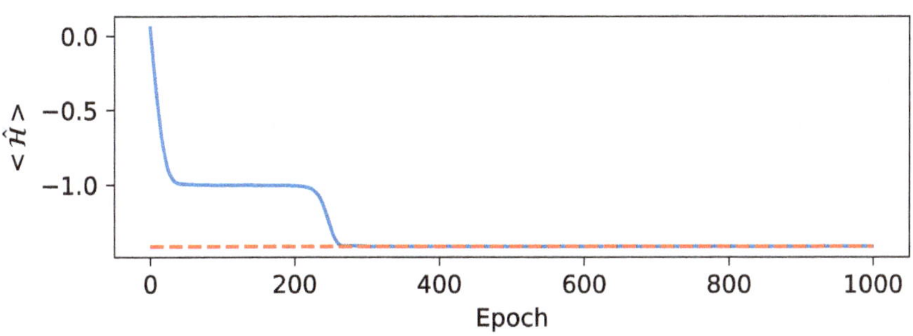

Fig. 6.5 Training the model `Ising2` with $h_0 = 1.0$ and $h_1 = 0.0$. The final value of the loss $\langle\hat{\mathcal{H}}\rangle$ corresponds to the lowest eigenvalue $E_0 = -\sqrt{2.0} \simeq -1.4$ of $\hat{\mathcal{H}}$ (red line). As in Fig. 6.4, the ground state has zero entropy of entanglement

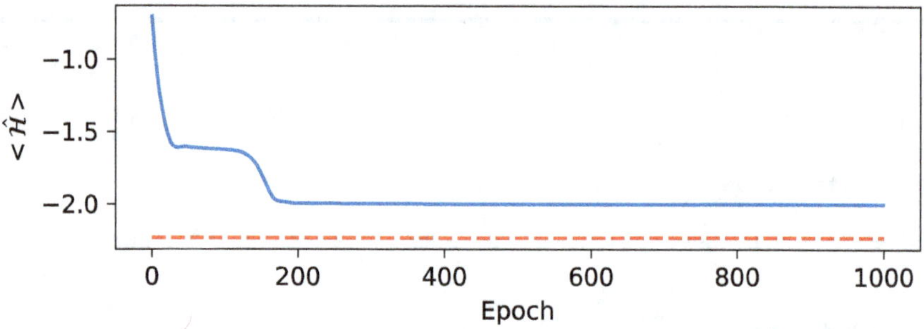

Fig. 6.6 Training the model `Ising3` with $h_0 = 1.0$ and $h_1 = 1.0$. The final value of the loss $\langle\hat{\mathcal{H}}\rangle$ does not correspond to the lowest eigenvalue $E_0 \simeq -2.24$ of $\hat{\mathcal{H}}$ (red line). As in Fig. 6.5, the ground state has zero entropy of entanglement, but the actual eigenstate has a nonzero entanglement (see Chap. 5), but the used variational ansatz cannot retrieve it, as the variational ansatz has zero entanglement

```
# hyperparameter for the Hamiltonian
J=1.0
h=[1.0,1.0]
# Input layer (one dummy input)
xin3 = tf.keras.layers.Input(1);
# Hamiltonian layer
HL3=HamiltonianLayer(J=J,h=h, nparams=6,FeatureMap=FeatureMapU1,
↪   name='H3')
# output
meanH3=HL3(xin3)
# trainable model returning meanH
Ising3 = tf.keras.Model(inputs = xin3, outputs=meanH3,
↪   name='model3')
# add loss function
Ising3.add_loss(meanH3)
```

Figure 6.6 shows the resulting training history. We have evidence that the final value for the loss $\langle\hat{\mathcal{H}}\rangle = -2.0$ is higher than the lowest eigenvalue. Hence, our variational ansatz is not able to find the lowest energy eigenstate of the system. The reason is that the target eigenstate is entangled, and we are using a product state as a variational guess. The ansatz is not able to explore the entire Hilbert space, and we miss the lowest energy.

This circumstance also shows that there is a link between energy and entropy. The ground state has a nonvanishing entropy.

6.3 Training with Entanglement

We need to develop a more general variational ansatz. First, we observe that we have a two-qubit state, which lies in a Hilbert space with dimension 4. Correspondingly, the state is determined by *four* complex parameters or, equivalently, the state is

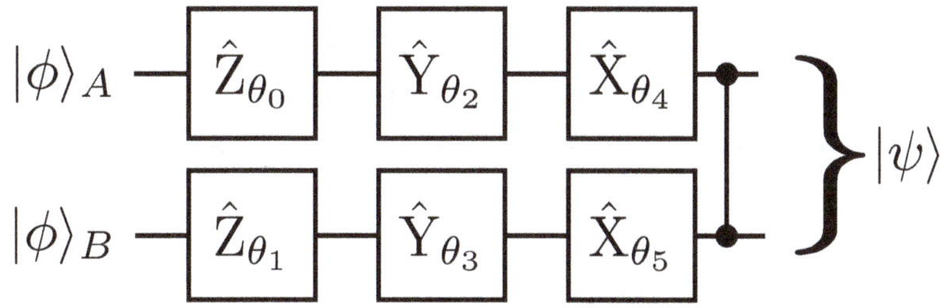

Fig. 6.7 A feature map including an entangling controlled-Z gate to the feature map in Fig. 6.2. The resulting state $|\phi\rangle$ is a variational ansatz for the TIM, useful when we expect an entangled ground state

determined by *eight* real weights. However, the normalization constraint and the arbitrary absolute phase reduce the minimal number of parameters to six. This is the optimal number for a variational ansatz, i.e., the minimum number of weights to be trained. However, it happens that one uses a much larger number of weights in applications, as this helps to avoid suboptimal local minima by starting from randomly initialized values.

For the moment, let us consider only six parameters as in Fig. 6.2. We know from Sect. 6.2 that we need entanglement. The first approach is introducing an entangling gate, as the controlled-Z gate $\widehat{\text{CZ}}$, discussed in Chap. 3.

Figure 6.7 shows the new feature map defined as follows.

```
@tf.function
def FeatureMapU2(theta, psi):
    """ Feature Map for the
    Variational Quantum Algorithm
    with entanglement
    Obtained by FeatureMapU1
    with an additional CZ gate

    Params
    ------
    theta, shape=(6,) real
    psi,   two-qubit state

    Returns
    -------
    A two-qubit state
    """
    phi=FeatureMapU1(theta,psi)
    # add an CZ gate that generates entanglement
    phi=quantummap.Gate(quantummap.CZ,phi)
    return phi
```

In our strategy, we generate the variational ansatz by applying the feature map on some bias state. We can test the feature map and also measure the corresponding entropy of entanglement. We are using $|00\rangle$ as bias state.

```
# we test the FeatureMapU2 for an entangled variational ansatz
# by using |00> as a bias
# generate random weights
theta_test=np.random.random(6)
# generate a state with bias q00
phi_test=FeatureMapU2(theta_test,quantummap.q00)
print(phi_test)
# returns
tf.Tensor(
[[-0.01+0.54j -0.34+0.54j]
 [-0.35+0.08j  0.16+0.4j ]], shape=(2, 2), dtype=complex64)
# We compute the VonNeumannEntropy
print(quantummap.VonNeumannEntropy2(phi_test)[0])
# returns
tf.Tensor(0.34071487, shape=(), dtype=float32)
```

As expected, the new variational ansatz has nonvanishing entropy.

We build a model for minimizing the two-qubit TIM by using the entangled feature map. We consider the case $h_0 = h_1 = 1.0$.

```
# hyperparameter for the Hamiltonian
J=1.0
h=[1.0,1.0]
# Input layer (one dummy input)
xin4 = tf.keras.layers.Input(1);
# Hamiltonian layer
HL4=HamiltonianLayer(J=J,h=h, nparams=6,FeatureMap=FeatureMapU2,
↪   name='H4')
# output
meanH4=HL4(xin4)
# trainable model returning meanH
Ising4 = tf.keras.Model(inputs = xin4, outputs=meanH4,
↪   name='model4')
# add loss function
Ising4.add_loss(meanH4)
```

Notably enough, also our entangled feature map is not able to match the ground state. The CZ gate in Fig. 6.7 does not represent all the states, in the sense that one qubit controls the other, and not vice versa. Our ansatz cannot describe all the possible entangled states (Fig. 6.8).

As a heuristic, a simple way to improve the variational ansatz is iterating the feature map as done in Sect. 3.13.2. By iterating, we improve the representation power of the feature map.

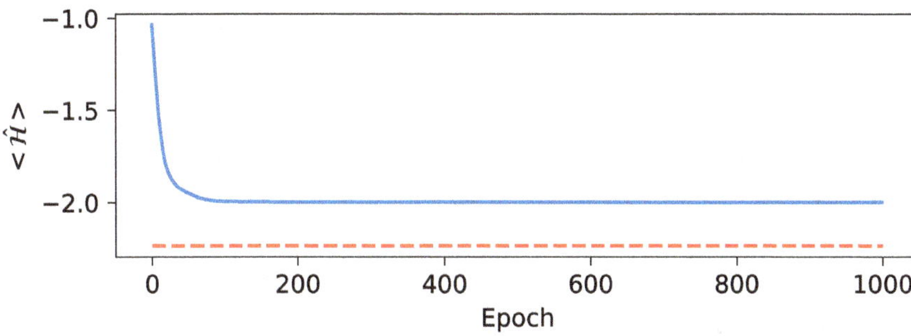

Fig. 6.8 Training the model `Ising4` with $h_0 = h_1 = 1.0$ with an entangled feature map. The final value of the loss $\langle\hat{\mathcal{H}}\rangle$ does not yet correspond to the lowest eigenvalue $E_0 \simeq -2.24$ of $\hat{\mathcal{H}}$ (red line). Even though the variational ansatz is entangled, it cannot match the ground state

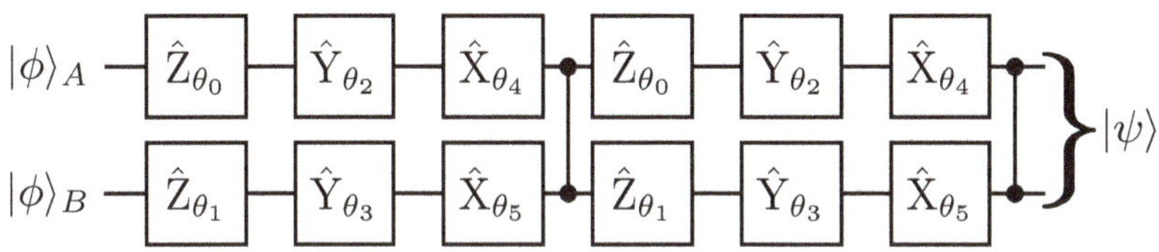

Fig. 6.9 An entangled feature map obtained by iterating the map in Fig. 6.7

```
@tf.function
def FeatureMapU3(theta, psi):
    # Define a new feature map
    # with entanglement
    # by iterating the FeatureMapU2
    psi = FeatureMapU2(theta, psi)
    psi = FeatureMapU2(theta, psi)
    return psi
```

We test the iterated map (see Fig. 6.9) by a new model as follows.

```
# hyperparameter for the Hamiltonian
J=1.0
h=[1.0,1.0]
# Input layer (one dummy input)
xin5 = tf.keras.layers.Input(1);
# Hamiltonian layer
HL5=HamiltonianLayer(J=J,h=h, nparams=6,FeatureMap=FeatureMapU3,
↪   name='H5')
# output
meanH5=HL5(xin5)
# trainable model returning meanH
```

```
Ising5 = tf.keras.Model(inputs = xin5, outputs=meanH5,
↪    name='model5')
# add loss function
Ising5.add_loss(meanH5)
```

The model `Ising5` is trained as in Fig. 6.8, and—remarkably—the loss function reaches the ground state energy E_0. We also compute the entanglement entropy of the resulting ground state and find that it is not vanishing (Fig. 6.10).

```
# print the entropy of the ground state
# and the probabilities
# of the model Ising5
quantummap.VonNeumannEntropy2(HL5.ground_state())
(<tf.Tensor: shape=(), dtype=float32, numpy=0.29811704>,
 <tf.Tensor: shape=(2,), dtype=float32,
numpy=array([0.95, 0.05], dtype=float32)>)
```

6.4 The Quantum Approximate Optimization Algorithm

As a further example of feature maps, we consider a widely used approach in combinatorial optimization: the quantum approximate optimization algorithm (QAOA). Introduced in [1], it is studied in different physical platforms.

The idea follows the feature maps in the previous sections with rotations and entangling gates. The novelty is in considering as a feature map the cost function $\hat{\mathrm{C}}$ itself. In our case, the cost function is the Hamiltonian $\hat{\mathrm{C}} = \hat{\mathcal{H}}$.

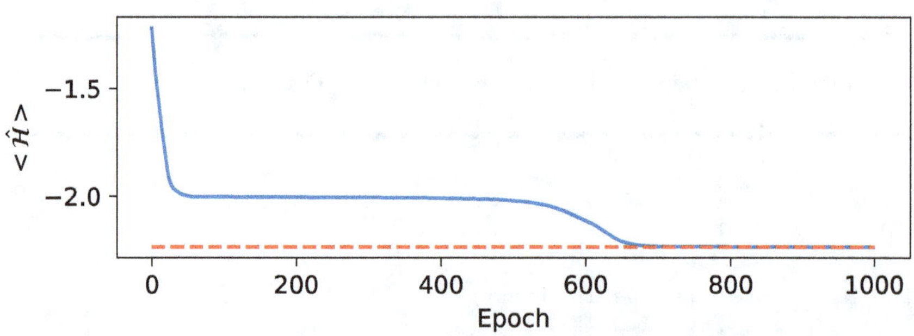

Fig. 6.10 Training the model `Ising5` with $h_0 = h_1 = 1.0$ with the iterated entangled feature map in Fig. 6.9. At variance with Fig. 6.8, the final loss value $\langle\hat{\mathcal{H}}\rangle$ does correspond to the lowest eigenvalue $E_0 \simeq -2.24$ of $\hat{\mathcal{H}}$ (red line). The iterated feature map can match the ground state

The reason for using the Hamiltonian is to conceive an algorithm that has the same locality of the cost functions, which means that it only involves the qubits coupled in $\hat{\mathcal{H}}$.

The first component of the feature map is the unitary operator

$$\hat{U}(\hat{C}, \gamma) = e^{-i\gamma\hat{C}} . \tag{6.4}$$

Let us consider, for example, (4.18), with $\hat{C} = \hat{\mathcal{H}}$; we have

$$\hat{U}(\hat{C}, \gamma) = e^{-i\gamma(-J\hat{Z}_0\hat{Z}_1 - h_0\hat{X}_0 - h_1\hat{X}_1)} . \tag{6.5}$$

The rationale beyond (6.4) is as follows: if the state is an eigenstate of the Hamiltonian, the effect of $\hat{U}$ is only a phase rotation.

To simplify the computation, Eq. (6.5) is replaced by a multiplication of exponents

$$\hat{U}(\hat{C}, \gamma) \simeq e^{-i\gamma(-J\hat{Z}_0\hat{Z}_1)} e^{-i\gamma\left(-h_0\hat{X}_0 - h_1\hat{X}_1\right)} . \tag{6.6}$$

The equivalence of Eqs. (6.5) and (6.6) is exact in absence of the field h, as typically the case in applications. Otherwise, one is neglecting further exponential terms in (6.6) arising from the commutator between the Z terms and the X terms in the Hamiltonian. The approximated expression (6.6) works in many cases and is readily implemented in real quantum processors.

In addition, one also needs a mixing operator or mixing Hamiltonian $\hat{B}$, which is commonly a rotation. Conventionally, the Ising couplings are in the Z-direction, as in Eq. (4.18), and one chooses the mixing in the X rotation as follows:

$$\hat{B} = \hat{X}_0 + \hat{X}_1 + \ldots = \sum_{l=0}^{N-1} \hat{X}_l , \tag{6.7}$$

with N the number of qubits. The corresponding unitary operator is (for $N = 2$)

$$\hat{U}(\hat{B}, \beta) = e^{-i\beta\hat{B}} = e^{-i\beta\hat{X}_0} e^{-i\beta\hat{X}_1} . \tag{6.8}$$

The QAOA feature map is

$$|\phi\rangle = \hat{U}(\hat{B}, \beta)\hat{U}(\hat{C}, \gamma)|\text{bias}\rangle \tag{6.9}$$

which has two parameters β and γ. As a bias state, one considers a balanced superposition of the basis vectors, which is obtained by the Hadamard gate $\hat{H}$

$$|\text{bias}\rangle = |+\rangle = \hat{H}^{\otimes N}|\mathbf{0}\rangle . \tag{6.10}$$

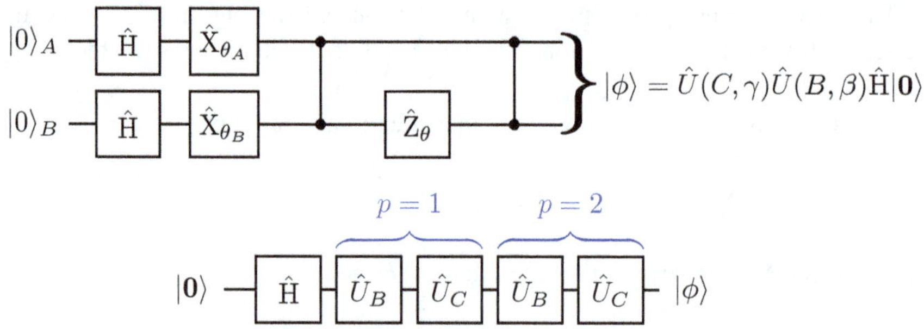

Fig. 6.11 A graphical representation of the QAOA ansatz. In the top panel, we show the Ising coupling gate $e^{\mathrm{i}J\hat{Z}_0\hat{Z}_1}$ as a combination of CNOT and R_Z gates

For two qubits, Eq. (6.10) reads

$$|+\rangle = \hat{H} \otimes \hat{H}|00\rangle = \frac{1}{2}\left(|00\rangle + |01\rangle + |10\rangle + |11\rangle\right) . \tag{6.11}$$

A further aspect in QAOA is iterating the feature map p times. The effectiveness of approximation increases with p [1].

Equation (6.9) corresponds to $p = 1$. For $p = 2$, we have

$$|\phi\rangle = \hat{\mathcal{U}}_2|+\rangle = \hat{U}(\hat{B}, \beta_1)\hat{U}(\hat{C}, \gamma_1)\hat{U}(\hat{B}, \beta_0)\hat{U}(\hat{C}, \gamma_0)|+\rangle . \tag{6.12}$$

Notice that, for each iteration, we use different parameters γ and β.

QAOA with p iterations has $2p$ parameters, as in the following:

$$\begin{aligned} |\phi\rangle &= \hat{\mathcal{U}}_p|+\rangle \\ &= \hat{U}(\hat{B}, \beta_0)\hat{U}(\hat{C}, \gamma_0)\hat{U}(\hat{B}, \beta_1)\hat{U}(\hat{C}, \gamma_1)\ldots\hat{U}(\hat{B}, \beta_{p-1})\hat{U}(\hat{C}, \gamma_{p-1})|+\rangle . \end{aligned} \tag{6.13}$$

Equation (6.12) can be represented as in Fig. 6.11.

The Ising coupling gate $e^{\mathrm{i}\theta\hat{Z}_0\otimes Z_1}$ is a sequence of CNOTs and rotations.

To understand this representation, we observe that, if the first qubit is $|0\rangle$, the gate is equivalent to $R_z(\theta)$. On the contrary, if the first qubit is $|1\rangle$, the gate corresponds to swapping $|0\rangle$ and $|1\rangle$, applying a rotation θ, and swapping again $|0\rangle$ and $|1\rangle$ the second qubit.

We implement the QAOA[2] by first coding a `FeatureMap` function for (6.13).

[2] Details in `jupyternotebooks/quantumfeaturemap/QAOATwoQubitTransverseFieldIsing.ipynb`.

```
@tf.function
def QAOA(theta, phi, layers=1, J=1, h=[0.0,0.0]):
    """ Feature Map for the
    Quantum Approximate Optimization Algorithm

    Params
    ------
    theta, shape=(2*layers,) real parameters
    psi,   two-qubit state
    layers=1, number of iterations

    Returns
    -------
    A two-qubit state
    """
    h0=h[0]
    h1=h[1]

    # extract parameters of the layers
    gamma=theta[:layers]
    beta=theta[layers:]

    # init the state as H|00>
    phi=quantummap.Gate(quantummap.HH,phi)
    # function for a step
    def UC(phi, gamma, beta):
        # e^{i gamma C}
        C=quantummap.EZZ(-J*gamma)
        phi=quantummap.Gate(C,phi)
        C=quantummap.EXI(-h0*gamma)
        phi=quantummap.Gate(C,phi)
        C=quantummap.EXI(-h1*gamma)
        phi=quantummap.Gate(C,phi)
        # e^{i beta B}
        phi=quantummap.Gate(quantummap.EXI(beta),phi)
        phi=quantummap.Gate(quantummap.EIX(beta),phi)
        return phi

    # iterate the layers
    for j in range(layers):
        phi=UC(phi, gamma[j], beta[j])

    return phi
```

The feature map uses the function `UC` that realizes a single iteration with parameters `gamma` and `beta`. In `UC` we build the iteration by first considering the exponent of the cost function as in (6.6) and then the mixer operator as in (6.9). The variable `layers` is the number of iterations p.

Note that at each iteration, we call `UC` with different parameters. All the parameters are in a single array `theta` in input, such that the first `layers` elements `gamma=theta[:layers]` correspond to $\gamma_0, \gamma_1, \ldots, \gamma_{p-1}$ and the second portion of `layers`, `beta=theta[layers:]`, consists of $\beta_0, \beta_1, \ldots, \beta_{p-1}$.

Next, we define a Hamiltonian layer with QAOA for the ground state.

```
class HamiltonianLayerQAOA(tf.keras.layers.Layer):
    # Define a Layer returning the mean value of the Hamiltonian
    # with
    # Quantum Approximate Optimization Annealing
    #
    # The layer has trainable parameters, i.e.,
    # the  parameters theta of the trainable feature map
    #
    # The Layer use the functions
    # Hamiltonian
    # QAOA
    #
    # The constructor use parameters
    #
    #  Params
    #  ------
    #  J : coupling coefficient
    #      (default J=1)
    #  h : positive field shape=(2,)
    #      (default h=[0.0,0.0])
    #  map: a function containing the feature map
    #      (default is QAOA)
    #  pQAOA: iterations of the QAOA feature map
    #      (default is 1)
    #  bias: the bias state
    #      (default is ground state quantummap.q00)
    #
    #  Example
    #  -------
    #  HL=HL(J=.5, h=[1.2,-1.0],pQAOA=3)
    #
    #  Returns
    #  -------
    #  The call return the real part of
    #  <\psi|H|\psi>
    #
    def __init__(self, J=1.0, h=[0.0,0.0],
                 FeatureMap=QAOA,
                 pQAOA=1,
                 q00=quantummap.q00,
                 **kwargs):
        super(HamiltonianLayerQAOA, self).__init__(**kwargs)
        # trainable parameter of the model
        # initially set as random (real variables)
        self.theta=tf.Variable(np.random.random(2*pQAOA),
                               dtype=tf.float32,
                               name='theta')
        # gap coefficient
```

```
        self.J=J
        # field coefficient
        self.h=h
        # Hamiltonian
        self.H=Hamiltonian(J=self.J,h=self.h)
        # Bias state
        self.q00=q00
        # Feature map for the ground state
        self.FeatureMap=FeatureMap
        # Number of layers of QAOA
        self.pQAOA=pQAOA
    def ground_state(self):
        """ Return the current ground state """
        phi=self.FeatureMap(self.theta,self.q00,self.pQAOA,
        ↪  self.J,self.h)
        return phi
    def call(self,dummy):
        """ Return Re<H>
        Remark:  this layer has a dummy variable as input
        (required for compatibility with tensorflow call)
        """
        phi=self.ground_state()
        Hphi=quantummap.Gate2(self.H,phi)
        meanH=quantummap.Scalar(phi, Hphi)
        return tf.math.real(meanH)
```

The layer `HamiltonianLayerQAOA` uses the QAOA in the `ground_state` and includes the parameters γ_j and β_j, with $j = 0, 1, \ldots, p-1$ in the weight array `theta` with `shape=(2*pQAOA,)`, with `pQAOA` the number of iterations. The weights `theta` are randomly initialized when the `HamiltonianLayerQAOA` is created.

We test and train the QAOA model as in the previous examples. Figure 6.12 shows the training for the case $J = 1$, $h_0 = h_1 = 1$, and $p = 3$. The training provides good performance; indeed the QAOA is one of the mostly studied variational ansatzes.

6.5 Neural Network Quantum States

In the previous examples, we built the variational quantum algorithm by using a quantum feature map with gates, which are experimentally realizable. The resulting ansatz is a function that takes as input a bias state (typically vacuum) and depends on some variational parameters `theta`.

If we are not interested at an ansatz that can be implemented in the real world, i.e., corresponding to unitary operations, we can use other formulations. This is helpful if we want to find the ground state for some theoretical analysis, for example,

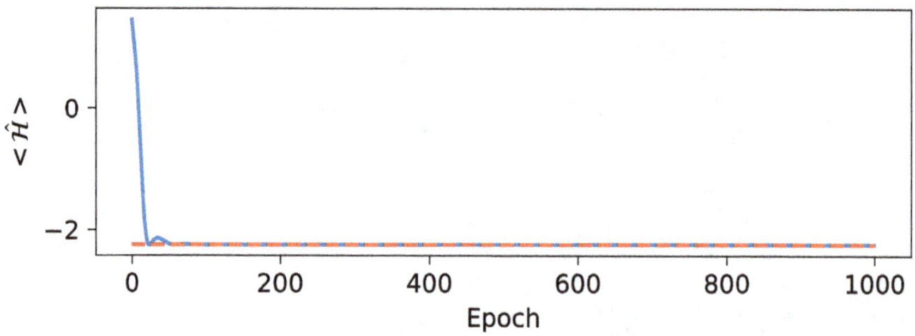

Fig. 6.12 Training the two-qubit transverse-field Ising model by using the Quantum Approximate Optimization Algorithm for $J = 1.0$, $h_0 = h_1 = 1.0$, and $p = 3$ iterations of the QAOA. The dashed line corresponds to the lowest eigenvalues of $\hat{\mathcal{H}}$

in quantum chemistry. A successful approach is representing the state as a neural network, which is indicated as a *neural network quantum state* (NNQS) [2, 3].

The neural network quantum states take as input some vector representation of the bias state $|\psi\rangle$, say, its components $s^i = \langle i|\psi\rangle$ in a basis $|i\rangle$, and return the components $\langle i|\phi\rangle$ of the variational ansatz $|\phi\rangle$, such that

$$\langle i|\phi, \boldsymbol{\theta}\rangle = F_i(s^0, s^1, \ldots, s^{N-1}, \boldsymbol{\theta}) . \tag{6.14}$$

In Eq. (6.14) $\boldsymbol{\theta}$ is the vector of parameters that define the ansatz. s^j with $j = 0, 1, \ldots, N-1$ are the components of the input bias state (see Fig. 6.13), and $\langle i|\phi, \boldsymbol{\theta}\rangle$ are the components of the target state ($N = 2$ for the two-qubit state).

The specific implementation varies from case to case, but the main idea is to use a neural network as a variational ansatz. As we do not need to have a unitary gate representing the NNQS, we use nonlinear activation functions, from one layer to another. In particular, we use deep neural models, i.e., dense neural networks built as a sequence of linear layers and "relu" activation function.

We give here an example for the two-qubit TIM. We start representing (6.14) as in Fig. 6.13, i.e., a two-branch deep model. An important aspect is that s_j and $\langle i|\phi, \boldsymbol{\theta}\rangle$ are complex coefficients. Hence the weights of the networks are also complex coefficients. However, training with complex coefficients does not guarantee convergence.

The simplest solution is to *decomplexify* the NNQS, which is separating real and imaginary parts of the weights. We consider two NNQS, one that returns the tensor `rephi` for real parts $\Re(\langle i|\phi\rangle)$ and the other returning `imphi` for the imaginary parts $\Im(\langle i|\phi\rangle)$. The advantage is that the two resulting tensors (with `shape=(2,2)` for two qubits), which we denote as `rephi` and `imphi`, are easily combined to obtain the final tensor state—without concatenation or reshaping—as `phi=tf.complex(rephi,imphi)`, that is,[3]

[3] Details in `jupyternotebooks/quantumfeaturemap/QuantumNeuralStateTwoQubitTransverseFieldIsing.ipynb`.

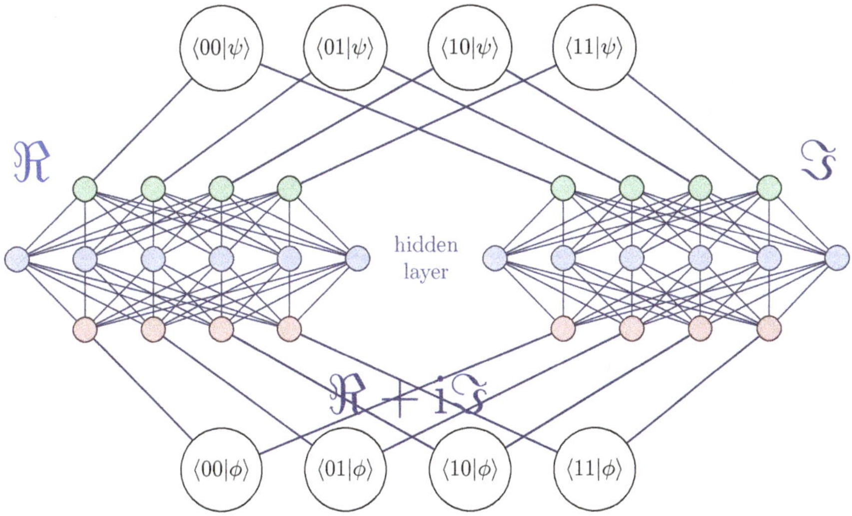

Fig. 6.13 A neural network quantum state is a neural network that returns the components of the variational ansatz and encodes parameters to be trained with a cost function, for example, to minimize the expected value of the Hamiltonian. We use two subnetworks for the real and imaginary parts of the components of the state vector

```
# complexify the output
# of two NNQS
phi=tf.complex(rephi,imphi)
```

The two NNQS in the branches in Fig. 6.13 have the complex s^j as inputs.

In order to use conventional NN methods, it is useful to also decomplexify the input.

One simple strategy is to concatenate the real and imaginary parts as follows.

```
# state is the complex bias state
# compute the real part with shape=(2,2)
rebias=tf.math.real(bias)
# compute the imaginary part shape=(2,2)
imbias=tf.math.imag(bias)
# concatenate real and imaginary part
# as a single tensor with shape=(4,2)
bias1=tf.concat([rebias,imbias],0)
# flatten the bias as a tensor
# with shape (8,)
bias2=tf.reshape(bias1,[-1])
# tf.reshape with [-1]
# corresponds to flattening
```

Once we have a single flattened real tensor as `bias2`, we use it as input to a dense neural network.

Let us first consider the real part. We consider two dense layers with weights `AR` and `HR` and `relu` activation function.

We define the weights `AR` as `tf.Variable` with `shape=(nhidden,8)`, `nhidden` being the number of nodes in the hidden layers, a hyperparameter that we tune for optimal learning.

```
# Define weights for the first layer
# and random initialize
AR=tf.Variable(tf.random.normal(shape=(nhidden,8)))
```

The hidden state `hideR` with `shape=(nhidden,)` for the real part NN, is the tensor computed as follows.

```
hideR=tf.nn.relu(tf.tensordot(AR,bias2,axes=1))
```

We first multiply the weights `AR` with the flattened bias `bias2` and then apply the activation function `tf.nn.relu`. Then, we define the second layer with weights `HR` with `shape=(4,hidden)` as they have as input the tensor `hideR` and compute the output corresponding to the real part of the state `rephi`, with `shape=(4,)` for real part of $\langle i|\phi\rangle$.

```
# define and initialize the weights for the second layer
HR=tf.Variable(tf.random.normal(shape=(4,nhidden)));
# compute the output state of the second layer
rephi=tf.nn.relu(tf.tensordot(HR,hideR,axes=1))
```

Seemingly, we define weights `AI` and `HI`, internal state `hideI`, and output `imphi` for the NN representing the imaginary part.

```
AI=tf.Variable(tf.random.normal(shape=(nhidden,8)));
HI=tf.Variable(tf.random.normal(shape=(4,nhidden)));
hideI=tf.nn.relu(tf.tensordot(self.AI,bias2,axes=1))
imphi=tf.nn.relu(tf.tensordot(self.HI,hideI,axes=1))
```

Note that we use different weights for the layers and the two NNs, as a higher number of weights is more favorable to training.

We combine the outputs of the two networks to return a complex tensor with `shape=(2,2)`. We also rescale the state to have a normalized vector.

```
# first combine real and imaginary part
cphi0=tf.complex(rephi,imphi)
# reshape as (2,2)
cphi=tf.reshape(cphi0,[2,2])
# compute the norm
norm0=quantum.Scalar(cphi,cphi)
# compute the squrt
sqrtnorm=tf.math.sqrt(norm0)
# return the normalized state
phi=cphi/sqrtnorm
```

The previous operations and weights are put in a new `HamiltonianLayer` for using the NNQS for the optimization. The number of hidden parameters `nhidden` is a hyperparameter of the layer.

```
class HamiltonianLayerNNQS(tf.keras.layers.Layer):
    # Define a Layer returning the mean value of the Hamiltonian
    # on the variational ansatz
    #
    # The layer uses quantum neural networks
    # formed by two dense layers with
    # relu activation
    #
    # The Layer use the function
    # Hamiltonian
    #
    # The constructor use parameters
    #
    #  Params
    #  ------
    #  J : coupling coefficient
    #      (default J=1)
    #  h : positive field shape=(2,)
    #      (default h=[0.0,0.0])
    #  nhidden: the number of internal layers
    #      (default is 16)
    #  bias: the bias state
    #      (default is ground state quantummap.q00)
    #
    #  Example
    #  -------
    #  HL=HL(J=.5, h=[1.2,-1.0], nhidden=100)
    #
    #  Returns
    #  -------
    #  The call return the real part of
    #  <\psi|H|\psi>
    #
    def __init__(self, J=1.0, h=[0.0,0.0],
```

```
                 nhidden=10,
                 bias=quantummap.q00,
                 **kwargs):
        super(HamiltonianLayerNNQS, self).__init__(**kwargs)
        # trainable parameter of the model
        # gap coefficient
        self.J=J
        # field coefficient
        self.h=h
        # Hamiltonian
        self.H=Hamiltonian(J=self.J,h=self.h)
        # Bias state
        self.bias=bias
        # Weights of the NNQS
        self.AR=tf.Variable(tf.random.normal(shape=(nhidden,8)));
        self.AI=tf.Variable(tf.random.normal(shape=(nhidden,8)));
        self.HR=tf.Variable(tf.random.normal(shape=(4,nhidden)));
        self.HI=tf.Variable(tf.random.normal(shape=(4,nhidden)));

    @tf.function
    def realNNQS(self,bias):
        """ Function returning
        the real part NN

        Input
        -----
        bias state shape=(2,2)

        Returns
        -------
        real phi shape=(4,)
        """
        rebias=tf.math.real(bias)
        imbias=tf.math.imag(bias)
        bias1=tf.concat([rebias,imbias],0)
        bias2=tf.reshape(bias1,[-1])
        hideR=tf.nn.relu(tf.tensordot(self.AR,bias2,axes=1))
        rephi=tf.nn.relu(tf.tensordot(self.HR,hideR,axes=1))
        return rephi

    @tf.function
    def imagNNQS(self,bias):
        """ Function returning
        the imaginary part NN

        Input
        -----
        bias state shape=(2,2)

        Returns
        -------
        imag phi shape=(4,)
        """
        rebias=tf.math.real(bias)
```

```
        imbias=tf.math.imag(bias)
        bias1=tf.concat([rebias,imbias],0)
        bias2=tf.reshape(bias1,[-1])
        hideI=tf.nn.relu(tf.tensordot(self.AI,bias2,axes=1))
        imphi=tf.nn.relu(tf.tensordot(self.HI,hideI,axes=1))
        return imphi

    @tf.function
    def complexify(self,rephi,imphi):
        """ Function combining
        real and imaginary part
        and normalizing
        """
        cphi0=tf.complex(rephi,imphi)
        cphi=tf.reshape(cphi0,[2,2])
        sqrtnorm=tf.math.sqrt(quantummap.Scalar(cphi,cphi))
        nphi=cphi/sqrtnorm
        return nphi

    @tf.function
    def ground_state(self):
        """ Return the current ground state"""
        rph=self.realNNQS(self.bias)
        iph=self.imagNNQS(self.bias)
        return self.complexify(rph,iph)

    def call(self,dummy):
        """ Return Re<H>
        Remark:  this layer has a dummy variable as input
        (required for compatibility with tensorflow call)
        """
        phi=self.ground_state()
        Hphi=quantummap.Gate2(self.H,phi)
        meanH=tf.math.real(quantummap.Scalar(phi, Hphi))
        return meanH
```

In the `HamiltonianLayerNNQS` we define methods for building the ground state by the NNQS.

Specifically, `realNNQS` and `imagNNQS` take as input the bias and return the real and imaginary parts of the components of $|\phi\rangle$.

`complexify` takes the two parts and returns the normalized variational state $|\phi\rangle$ components.

These functions are used in the `ground_state` method to return the current state with the actual values of the weights `AR`, `AI`, `HR`, and `HI`, stored as internal variables of the `HamiltonianLayerNNQS`.

We train the model as in previous sections. For example, we create an instance using `nhidden` $=$ `10` and having as bias $\hat{\mathrm{H}} \otimes \hat{\mathrm{H}}|00\rangle$ as follows.

```
# bias
Hbias=quantummap.Gate2(quantummap.HH,quantummap.q00)
# hyperparameter for the Hamiltonian
J=1.0
h=[1.0,1.0]
# Input layer (one dummy input)
xin10 = tf.keras.layers.Input(1,name='DummyInput10');
# Hamiltonianl layer
HL10=HamiltonianLayerNNQS(J=J,h=h, bias=Hbias,nhidden=100,
↪    name='H10')
# output
meanH10=HL10(xin10)
# trainable model returning meanH
IsingNNQS10 = tf.keras.Model(inputs = xin10, outputs=meanH10,
↪    name='model10')
# add loss function for the mean energy
IsingNNQS10.add_loss(meanH10)
```

The number of weights depends on the number of hidden nodes `nhidden`. For `nhidden=10`, we have 2400 weights, as in the following summary.

```
IsingNNQS.summary()
#returns
Model: "modelNNQS"
_________________________________________________________________
 Layer (type)                Output Shape              Param #
=================================================================
 DummyInput (InputLayer)     [(None, 1)]               0

 HL (HamiltonianLayerNNQS)   (1, 1)                    2400

 add_loss (AddLoss)          (1, 1)                    0

=================================================================
Total params: 2,400
Trainable params: 2,400
Non-trainable params: 0
```

We find that the system converges to the ground state in a way dependent on the number of hidden layers and the initial random weights. For example, in Fig. 6.14 we show the training history in the case $J = 1$ and $h_0 = h_1 = 1.0$ for `nhidden=10` and using as bias $|00\rangle$. One can notice that the lowest value of the loss $\langle \hat{H} \rangle$ is higher than the ground state energy $E_0 \simeq -2.23$. On the contrary if we use another bias and `nhidden=10` as in Fig. 6.15, we find the lowest energy level.

In general, we do not know the real energy minimum and have to tweak the hyperparameters and bias to have convergence. In addition, we have to run the model many times and select the lowest energy among all the runs.

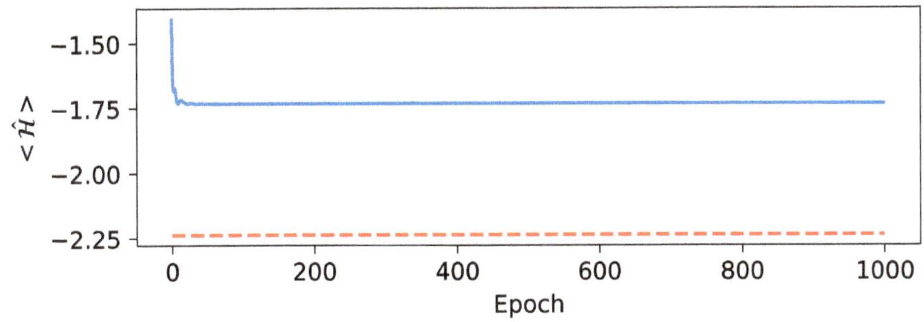

Fig. 6.14 Training QNSS with `nhidden=10` and $|00\rangle$ as a bias state. The model fails to converge to the ground state with energy $E_0 \simeq -2.233$

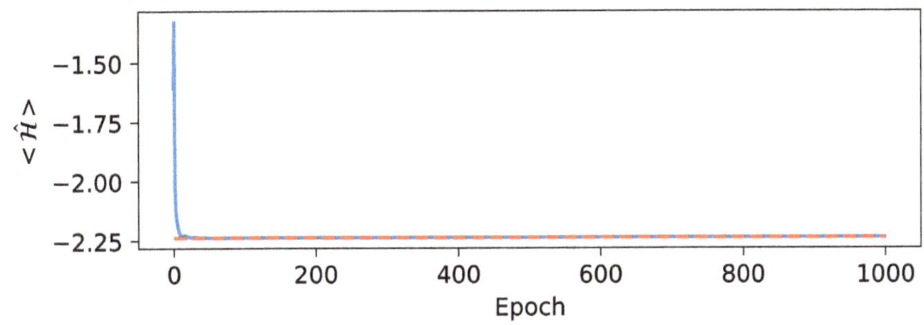

Fig. 6.15 Training QNSS with `nhidden=100` and $|+\rangle$ as a bias state. The model converges to the ground state with energy $E_0 \simeq -2.233$

6.6 Further Reading

- Quantum Approximate Optimization Algorithm: The QAOA [1] has been the subject of many investigations. An interesting aspect is the role of entanglement in QAOA [4–6].
- Neural network quantum state: There are different formulations; the reader may consider the review paper [2].

References

1. E. Farhi, J. Goldstone, S. Gutmann, arXiv:1411.4028 (2014)
2. G. Carleo, I. Cirac, K. Cranmer, L. Daudet, M. Schuld, N. Tishby, L. Vogt-Maranto, L. Zdeborová, Rev. Mod. Phys. **91**, 045002 (2019). https://doi.org/10.1103/RevModPhys.91.045002. https://link.aps.org/doi/10.1103/RevModPhys.91.045002
3. G. Carleo, M. Troyer, Science **355**, 602 (2016). https://doi.org/10.1126/science.aag2302
4. M. Dupont, N. Didier, M.J. Hodson, J.E. Moore, M.J. Reagor, Phys. Rev. A **106**, 022423 (2022). https://doi.org/10.1103/PhysRevA.106.022423
5. Y. Chen, L. Zhu, C. Liu, N.J. Mayhall, E. Barnes, S.E. Economou, arXiv:2205.12283 (2022)
6. R. Sreedhar, P. Vikstål, M. Svensson, A. Ask, G. Johansson, L. García-Álvarez, arXiv:2207.03404 (2022)

Chapter 7
Phase Space Representation

"Only the density matrix has meaning." – Steven Weinberg

Abstract Here we introduce the phase space representation of many-body quantum systems. We consider specifically Gaussian states and their characteristic function. We show how to compute expected values by derivation and define the covariance matrix in terms of different variables. We introduce a general method for linear transformations and gate as building blocks for quantum circuits and processors. First examples are gates with random coefficients.

7.1 Introduction

It is convenient to formulate the quantum evolution in the phase space to deal with the most general class of quantum states. This approach enables to work with a c-number representation of states (instead of abstract kets and operators in a Hilbert space) and to study phenomena like squeezing and entanglement.

In the phase space, we represent a state by a function of several variables encoded in a real vector $\boldsymbol{x}$. For a n-body state $\boldsymbol{x}$ is a vector of $N = 2n$ real variables.

Among the many representations, we consider the characteristic function $\chi(\boldsymbol{x})$. Gaussian states have the simplest characteristic function.

For Gaussian states [1], in symmetric ordering [2],

$$\chi(\boldsymbol{x}) = \exp\left(-\frac{1}{4}\boldsymbol{x} \cdot \boldsymbol{g} \cdot \boldsymbol{x} + \mathrm{i}\,\boldsymbol{x} \cdot \boldsymbol{d}\right), \tag{7.1}$$

with $\boldsymbol{g}$ the $N \times N$ covariance matrix and $\boldsymbol{d}$ the displacement vector. Gaussian states include vacuum, thermal states, coherent states, and squeezed vacuum.

$\chi(\boldsymbol{x})$ is a complex function with real and imaginary part, i.e.,

$$\chi(\boldsymbol{x}) = \chi_R(\boldsymbol{x}) + \mathrm{i}\chi_I(\boldsymbol{x}) \,. \tag{7.2}$$

C. Conti, *Quantum Machine Learning*, Quantum Science and Technology,
https://doi.org/10.1007/978-3-031-44226-1_7

As neural networks prominently deal with real-valued quantities, we will represent the characteristic functions by models with two real outputs corresponding to χ_R and χ_I. Hence any many-body state is a couple of real functions $\chi_{R,I}$ of N real variables. A noteworthy link between phase space functions and neural networks is that neural networks can represent arbitrary functions; therefore they can model any characteristic function.

7.2 The Characteristic Function and Operator Ordering

We introduce the complex vector z with components

$$z = (z_0, z_0, \dots, z_{n-1}, z_0^*, z_1^*, \dots, z_{n-1}^*), \tag{7.3}$$

which include the complex z_j components and their conjugate z_j^*. For a many-body system with density matrix ρ, the characteristic function is [2]

$$\chi(z, \mathcal{P}) = \mathrm{Tr}\left\{\rho \exp\left[\sum_k \left(z_k \hat{a}_k^\dagger - z_k^* \hat{a}_k\right)\right]\right\} \exp\left(\sum_k \frac{\mathcal{P}}{2} z_k z_k^*\right) \tag{7.4}$$

In Eq. (7.4) the variable $\mathcal{P} = 0, 1, -1$ refers to the adopted operator ordering. The characteristic functions depend on z and the ordering index $\mathcal{P}$. $\mathcal{P}$ is important when determining the expected value of operators as derivatives of χ.

Given the density matrix ρ, the expectation values of an observable $\hat{O}$ are

$$\langle \hat{O} \rangle = \mathrm{Tr}\left(\rho \hat{O}\right). \tag{7.5}$$

The complex function $\chi(z, \mathcal{P})$ represents a quantum state or a mixture. We evaluate the expectation value of combinations of the annihilation and creation operators by derivatives, as follows:

$$\langle \left(\hat{a}_j^\dagger\right)^m (\hat{a}_k)^n \rangle_{\mathcal{P}} = \left(\frac{\partial}{\partial z_j}\right)^m \left(-\frac{\partial}{\partial z_k^*}\right)^n \chi(z, \mathcal{P})\bigg|_{z=0}. \tag{7.6}$$

One realizes that the exponential term in (7.4) weighted by $\mathcal{P}$ changes the derivatives, and we have different expectation values when varying $\mathcal{P}$.

The simplest one is the mean value of the field operator

$$\langle \hat{a}_k \rangle = -\frac{\partial \chi}{\partial z_k^*}, \tag{7.7}$$

which does not depend on the ordering index $\mathcal{P}$, such as

$$\langle \hat{a}_k^\dagger \rangle = \frac{\partial \chi}{\partial z_k}. \tag{7.8}$$

On the other hand, for the mean photon (or particle) number of the mode k,

$$\hat{n}_k = \hat{a}_k^\dagger \hat{a}_k \ , \tag{7.9}$$

we have

$$\langle \hat{a}_k^\dagger \hat{a}_k \rangle_{\mathcal{P}} = -\frac{\partial^2 \chi(z, \mathcal{P})}{\partial z_k \partial z_k^*}\bigg|_{z=0} \ . \tag{7.10}$$

For the symmetric ordering $\mathcal{P} = 0$, one has

$$\langle \hat{a}_k^\dagger \hat{a}_k \rangle_{\mathcal{P}=0} = \langle \hat{a}_k^\dagger \hat{a}_k + \hat{a}_k \hat{a}_k^\dagger \rangle; \tag{7.11}$$

for the normal ordering $\mathcal{P} = 1$, one has

$$\langle \hat{a}_k^\dagger \hat{a}_k \rangle_{\mathcal{P}=1} = \langle \hat{a}_k^\dagger \hat{a}_k \rangle; \tag{7.12}$$

finally, for antinormal ordering $\mathcal{P} = -1$,

$$\langle \hat{a}_k^\dagger \hat{a}_k \rangle_{\mathcal{P}=-1} = \langle \hat{a}_k \hat{a}_k^\dagger \rangle \ . \tag{7.13}$$

In general, we have

$$\langle \hat{a}_k^\dagger \hat{a}_k \rangle_{\mathcal{P}} = \langle \hat{a}_k^\dagger \hat{a}_k \rangle + \frac{1}{2}(1 - \mathcal{P}), \tag{7.14}$$

which will be adopted below to write the neural network layer that returns the mean particle number. When non-explicitly stated, we refer to symmetric ordering $\mathcal{P} = 0$.

7.3 The Characteristic Function in Terms of Real Variables

For use with the machine learning application programming interfaces (APIs), it is more convenient to use real variables, as NNs usually deal with real-valued vectors.

We consider the quadrature operators $\hat{x}_k$ and $\hat{p}_k$ in

$$\hat{a}_k = \frac{\hat{q}_k + \mathrm{i}\hat{p}_k^\dagger}{\sqrt{2}}. \tag{7.15}$$

We cast the quadrature operators in $n \times 1$ vectors

$$\hat{\boldsymbol{q}} = \begin{pmatrix} \hat{q}_0 \\ \hat{q}_1 \\ \vdots \\ \hat{q}_{n-1} \end{pmatrix}, \qquad \hat{\boldsymbol{p}} = \begin{pmatrix} \hat{p}_0 \\ \hat{p}_1 \\ \vdots \\ \hat{p}_{n-1} \end{pmatrix}. \tag{7.16}$$

In addition, we consider the real quantities

$$q_k = \frac{z_k - z_k^*}{\sqrt{2}\mathrm{i}} \qquad p_k = \frac{z_k + z_k^*}{\sqrt{2}}, \tag{7.17}$$

and we collect them in two real $1 \times n$ vectors

$$\begin{aligned} \boldsymbol{q} &= \begin{pmatrix} q_0 \ q_1 \ \cdots \ q_{n-1} \end{pmatrix}, \\ \boldsymbol{p} &= \begin{pmatrix} p_0 \ p_1 \ \cdots \ p_{n-1} \end{pmatrix}. \end{aligned} \tag{7.18}$$

We have

$$\chi = \chi(\boldsymbol{q}, \boldsymbol{p}, \mathcal{P} = 0) = \mathrm{Tr}[\rho \exp\left(\mathrm{i}\, \boldsymbol{q} \cdot \hat{\boldsymbol{q}} + \mathrm{i}\, \boldsymbol{p} \cdot \hat{\boldsymbol{p}}\right)]. \tag{7.19}$$

As $\hat{\boldsymbol{q}}$ and $\hat{\boldsymbol{p}}$ are column vectors, and $\boldsymbol{q}$ and $\boldsymbol{p}$ are row vectors, we can omit the dot product symbol and write

$$\chi(\boldsymbol{q}, \boldsymbol{p}) = \mathrm{Tr}[\rho \exp\left(\mathrm{i}\, \boldsymbol{q}\, \hat{\boldsymbol{q}} + \mathrm{i}\, \boldsymbol{p}\, \hat{\boldsymbol{p}}\right)]. \tag{7.20}$$

We simplify the notation by defining the $1 \times N$ real row vector

$$\boldsymbol{x} = \begin{pmatrix} q_0 \ p_0 \ q_1 \ p_2 \cdots q_n \ p_n \end{pmatrix}, \tag{7.21}$$

such that

$$\chi = \chi(\boldsymbol{x}) = \mathrm{Tr}\left[\rho \exp\left(\mathrm{i}\, \boldsymbol{x}\, \hat{\boldsymbol{R}}\right)\right] = \chi_R(\boldsymbol{x}) + \mathrm{i}\chi_I(\boldsymbol{x}), \tag{7.22}$$

being

$$\begin{aligned} \boldsymbol{x}\, \hat{\boldsymbol{R}} &= q_0\hat{q}_0 + p_0\hat{p}_0 + q_1\hat{q}_1 + \ldots + q_{n-1}\hat{q}_{n-1} + p_{n-1}\hat{p}_{n-1} \\ &= x_0\hat{R}_0 + x_1\hat{R}_1 + \ldots + x_{N-1}\hat{R}_{N-1}, \end{aligned} \tag{7.23}$$

after introducing the $N \times 1$ column vector

$$\hat{\boldsymbol{R}} = \begin{pmatrix} \hat{q}_0 \\ \hat{p}_0 \\ \hat{q}_1 \\ \hat{p}_1 \\ \vdots \\ \hat{q}_{n-1} \\ \hat{p}_{n-1} \end{pmatrix} . \tag{7.24}$$

The commutators of the quadratures in units with $\hbar = 1$ are

$$\begin{aligned} \left[\hat{q}_j, \hat{q}_k\right] &= 0 \\ \left[\hat{p}_j, \hat{p}_k\right] &= 0 \\ \left[\hat{q}_j, \hat{p}_k\right] &= \mathrm{i}\delta_{jk} \ , \end{aligned} \tag{7.25}$$

which result into

$$[\hat{R}_p, \hat{R}_q] = \mathrm{i}J_{pq}, \tag{7.26}$$

where we introduced the $N \times N$ symplectic matrix $\mathbf{J}$

$$\mathbf{J} = \bigoplus_j \mathbf{J}_1 = \begin{pmatrix} 0 & 1 & 0 & 0 & \cdots & 0 & 0 \\ -1 & 0 & 0 & 0 & \cdots & 0 & 0 \\ 0 & 0 & 0 & 1 & \cdots & 0 & 0 \\ 0 & 0 & -1 & 0 & \cdots & 0 & 0 \\ \vdots & \vdots & \vdots & \vdots & \ddots & \vdots & \vdots \\ 0 & 0 & 0 & 0 & \cdots & 0 & 1 \\ 0 & 0 & 0 & 0 & \cdots & -1 & 0 \end{pmatrix}, \tag{7.27}$$

being $\mathbf{J}_1 = \left(\begin{smallmatrix} 0 & 1 \\ -1 & 0 \end{smallmatrix}\right)$ [1]. In Eq. (7.27), the symbol $\bigoplus$ is the direct sum for block matrices with j running in $0, 1, \dots n-1$. The matrix $\mathbf{J}$ is such that $\mathbf{J}^{-1} = \mathbf{J}^\top$, $\mathbf{J}^2 = -\mathbf{1}_N$, and $\mathbf{J}^\top\mathbf{J} = \mathbf{1}_N$ is the identity $N \times N$ matrix.

The expectation value of $\hat{\boldsymbol{R}}$ is determined by the derivatives of the characteristic function. By using the definition of $\chi(\boldsymbol{x})$

$$\chi(\boldsymbol{x}) = \mathrm{Tr}\left(\rho e^{\mathrm{i}\,\boldsymbol{x}\,\hat{\boldsymbol{R}}}\right), \tag{7.28}$$

one has

$$\langle \hat{R}_j \rangle = \mathrm{Tr}[\rho \hat{R}_j] = -\mathrm{i} \left.\frac{\partial \chi}{\partial x_j}\right|_{\boldsymbol{x}=0} = \left.\frac{\partial \chi_I}{\partial x_j}\right|_{\boldsymbol{x}=0} \tag{7.29}$$

or

$$\langle \hat{\boldsymbol{R}} \rangle = \mathrm{Tr}[\rho \hat{\boldsymbol{R}}] = -\mathrm{i}\, \nabla \chi \,|_{\boldsymbol{x}=0} = \nabla \chi_I |_{\boldsymbol{x}=0} \;. \tag{7.30}$$

As the components of $\hat{\boldsymbol{R}}$ are self-adjoint, their quantum expectation is real, and it is the derivative of the imaginary part χ_I at $\boldsymbol{x} = 0$.

Seemingly, we have the mean value of the particle number in the mode j

$$\langle \hat{a}_j^\dagger \hat{a}_j \rangle_{\mathcal{P}=0} = -\frac{\partial^2}{\partial z_j \partial z_j^*} \chi \bigg|_{\boldsymbol{x}=0} \;. \tag{7.31}$$

By expressing χ in terms of the real components $\boldsymbol{q}$ and $\boldsymbol{p}$, we have

$$-\frac{\partial^2}{\partial z_j \partial z_j^*} \chi = \\ -\left(\frac{\partial}{\partial q_j}\frac{\partial q_j}{\partial z_j} + \frac{\partial}{\partial p_j}\frac{\partial p_j}{\partial z_j} \right) \left(\frac{\partial}{\partial q_j}\frac{\partial q_j}{\partial z_j^*} + \frac{\partial}{\partial p_j}\frac{\partial p_j}{\partial z_j^*} \right) \chi = -\frac{1}{2} \left(\frac{\partial^2}{\partial q_j^2} + \frac{\partial^2}{\partial p_j^2} \right) \chi \,, \tag{7.32}$$

after using (7.17).

We obtain the expected particle number with the second derivative of χ_R because its imaginary part is vanishing

$$\langle \hat{a}_j^\dagger \hat{a}_j \rangle_{\mathcal{P}=0} = -\frac{1}{2} \left(\frac{\partial^2}{\partial x_{2j}^2} + \frac{\partial^2}{\partial x_{2j+1}^2} \right) \chi_R \bigg|_{\boldsymbol{x}=0} \,, \tag{7.33}$$

with the r.h.s. evaluated at $\boldsymbol{x} = 0$ and $j = 0, 1, 2, \ldots, n-1$.

For the expected value of total number of particles $\mathcal{N}$, we have

$$\langle \hat{\mathcal{N}} \rangle_{\mathcal{P}=0} = -\frac{1}{2} \nabla^2 \chi_R \bigg|_{\boldsymbol{x}=0} \,, \tag{7.34}$$

with the Laplacian computed with the xs in $\boldsymbol{x}$.

Equation (7.34) refers to symmetric ordering with $\mathcal{P} = 0$.

For normal ordering $\mathcal{P} = 1$, we use Eq. (7.14)

$$\langle \hat{a}_k^\dagger \hat{a}_k \rangle = \langle \hat{a}_k^\dagger \hat{a}_k \rangle_{\mathcal{P}=1} = \langle \hat{a}_k^\dagger \hat{a}_k \rangle_{\mathcal{P}=0} - \frac{1}{2} \tag{7.35}$$

in (7.34), and we get

$$\begin{aligned} \langle \mathcal{N} \rangle = \langle \mathcal{N} \rangle_{\mathcal{P}=1} &= \langle \mathcal{N} \rangle_{\mathcal{P}=0} - \frac{n}{2} \\ &= -\frac{1}{2} \nabla^2 \chi_R \bigg|_{\boldsymbol{x}=0} - \frac{n}{2} = -\frac{1}{2} \left(\nabla^2 + n \right) \chi \Big|_{\boldsymbol{x}=0} \;. \end{aligned} \tag{7.36}$$

7.4 Gaussian States

Gaussian states are relevant for many applications; they include coherent and squeezed states. The characteristic function in symmetric ordering is

$$\chi(\boldsymbol{x}) = \exp\left(-\frac{1}{4}\boldsymbol{x}\,\boldsymbol{g}\,\boldsymbol{x}^{\top} + \mathrm{i}\,\boldsymbol{x}\,\boldsymbol{d}\right) \tag{7.37}$$

with $\boldsymbol{g}$ the $N \times N$ *covariance matrix* and $\boldsymbol{d}$ the $N \times 1$ *displacement column vector*. Correspondingly, recalling that $\boldsymbol{x}$ is a row vector,

$$\begin{aligned} \chi_R(\boldsymbol{x}) &= e^{-\frac{\boldsymbol{x}\boldsymbol{g}\boldsymbol{x}^{\top}}{4}}\cos(\boldsymbol{x}\boldsymbol{d}) \\ \chi_I(\boldsymbol{x}) &= e^{-\frac{\boldsymbol{x}\boldsymbol{g}\boldsymbol{x}^{\top}}{4}}\sin(\boldsymbol{x}\boldsymbol{d}) \end{aligned} \tag{7.38}$$

From Eq. (7.30), we have for the components $\boldsymbol{d}$, with $q = 0, 1, \ldots, N-1$,

$$\langle \hat{R}_q \rangle = \mathrm{Tr}[\hat{\rho}\hat{R}_q] = d_q = \left.\frac{\partial \chi_I}{\partial x_q}\right|_{\boldsymbol{x}=0} \tag{7.39}$$

or

$$\langle \hat{\boldsymbol{R}} \rangle = \boldsymbol{d} = \nabla \chi_I|_{\boldsymbol{x}=0}\ . \tag{7.40}$$

The displacement d_q is the first moment of the Gaussian χ.

In general, the derivatives of the characteristic functions can be expressed by the displacement and covariance matrix d_q and g_{pq}. A Gaussian state is determined by the N displacements and the $N(N-1)/2$ elements $g_{pq} = g_{qp}$.

The number of real parameters for Gaussian states is $\frac{N^2}{2} + \frac{N}{2}$, which is important to determine the trainable variables in the corresponding NN model.

We use this feature of Gaussian states in speeding up computing the derivatives of χ and the observable quantities. Indeed, for the particle number, we have

$$\begin{aligned} \langle \hat{a}_j^{\dagger}\hat{a}_j \rangle_{\mathcal{P}=0} &= -\frac{1}{2}\left(\frac{\partial^2}{\partial x_{2j}^2} + \frac{\partial^2}{\partial x_{2j+1}^2}\right)\chi_R\bigg|_{\boldsymbol{x}=0} \\ &= \frac{g_{2j,2j} + g_{2j+1,2j+1}}{4} + \frac{d_{2j}^2 + d_{2j+1}^2}{2} \end{aligned} \tag{7.41}$$

and

$$\begin{aligned} \langle \hat{a}_j^{\dagger}\hat{a}_j \rangle &= -\frac{1}{2}\left(\frac{\partial^2}{\partial x_{2j}^2} + \frac{\partial^2}{\partial x_{2j+1}^2}\right)\chi_R\bigg|_{\mathbf{x}=0} - \frac{1}{2} \\ &= \frac{g_{2j,2j} + g_{2j+1,2j+1}}{4} + \frac{d_{2j}^2 + d_{2j+1}^2}{2} - \frac{1}{2} \end{aligned} \tag{7.42}$$

while the second derivatives of χ_I at $\boldsymbol{x} = 0$ are vanishing, and we used (7.14). [1]

The covariance matrix $\boldsymbol{g}$ is obtained by the second moments

$$g_{pq} = \langle\{\hat{R}_p, \hat{R}_q\}\rangle - 2\langle\hat{R}_p\rangle\langle\hat{R}_q\rangle \tag{7.43}$$

with

$$\{\hat{R}_p, \hat{R}_q\} = \hat{R}_p\hat{R}_q + \hat{R}_q\hat{R}_p. \tag{7.44}$$

Equation (7.43) shows that $\boldsymbol{g}$ is a real symmetric matrix obtained by expectation values of self-adjoint operators. We also have after Eq. (7.26)

$$\begin{aligned} g_{pq} &= \langle\hat{R}_p\hat{R}_q\rangle + \langle\hat{R}_q\hat{R}_p\rangle - 2\langle R_p\rangle\langle R_q\rangle \\ &= \langle\hat{R}_p\hat{R}_q\rangle + \langle\hat{R}_q\hat{R}_p\rangle - 2\langle R_p\rangle\langle R_q\rangle \\ &= 2\langle\hat{R}_p\hat{R}_q\rangle - \mathrm{i}J_{pq} - 2\langle R_p\rangle\langle R_q\rangle = \\ &= 2\langle\left(\hat{R}_p - d_p\right)\left(\hat{R}_q - d_q\right)\rangle - \mathrm{i}J_{pq} \end{aligned} \tag{7.45}$$

7.4.1 Vacuum State

Vacuum states have $\boldsymbol{d} = \mathbf{0}_{N\times 1}$ and $\boldsymbol{g} = \mathbf{1}_N$ the $N \times N$ identity matrix. By using these expressions in (7.42), we get

$$\langle\hat{a}_j^\dagger\hat{a}_j\rangle_{\mathcal{P}=0} = \frac{1}{2}\,. \tag{7.46}$$

Correspondingly,

$$\langle\hat{a}_j^\dagger\hat{a}_j\rangle = \langle\hat{a}_j^\dagger\hat{a}_j\rangle_{\mathcal{P}=0} - \frac{1}{2} = 0. \tag{7.47}$$

We have $\langle\hat{\boldsymbol{R}}\rangle = 0$ and for the average photon number after Eq. (7.42)

$$\langle\mathcal{N}\rangle = \sum_{j=0}^{n-1}\langle\hat{a}_j^\dagger\hat{a}_j\rangle = 0 \tag{7.48}$$

[1] The detailed derivation of (7.42) is in Sect. 7.4.4.

7.4.2 Coherent State

A coherent state has a nonvanishing displacement $\boldsymbol{d} \neq \mathbf{0}$ and $\boldsymbol{g} = \mathbf{1}_N$. If we consider a single mode $|\alpha\rangle$, with $n = 1$ and $N = 2$,

$$d_0 = \sqrt{2}\Re\alpha = \frac{\alpha + \alpha^*}{\sqrt{2}}, \tag{7.49}$$

$$d_1 = \sqrt{2}\Im\alpha = \frac{\alpha - \alpha^*}{\mathrm{i}\sqrt{2}}. \tag{7.50}$$

From (7.42), we have ($j = 0$)

$$\langle \hat{a}_j^\dagger \hat{a}_j \rangle = \langle \hat{a}_j^\dagger \hat{a}_j \rangle_{\mathcal{P}=0} - \frac{1}{2} = |\alpha|^2 = \frac{d_0^2 + d_1^2}{2}. \tag{7.51}$$

For n distinct modes with $j = 0, 1, \ldots, n-1$, we have

$$\langle \hat{a}_j^\dagger \hat{a}_j \rangle = \frac{d_{2j}^2 + d_{2j+1}^2}{2}, \tag{7.52}$$

and

$$\langle \mathcal{N} \rangle = \sum_{j=0}^{n-1} \langle \hat{a}_j^\dagger \hat{a}_j \rangle = \sum_{q=0}^{N-1} \frac{d_q^2}{2}. \tag{7.53}$$

7.4.3 Thermal State

Single-mode thermal states are mixed number states $|n\rangle$ with a thermal distribution [2], such that (in our units $\hbar\omega = 1$)

$$\rho = (1 - e^{-\beta})e^{-\beta\hat{n}} = \sum_{n=0}^{\infty} P(n)|n\rangle\langle n|, \tag{7.54}$$

and

$$P(n) = (1 - e^{-\beta})e^{-\beta n}. \tag{7.55}$$

β is the *inverse temperature*. The mean particle number $\bar{n}$ is

$$\bar{n} = \langle \hat{n} \rangle = \frac{1}{e^\beta - 1}, \tag{7.56}$$

or equivalently

$$e^{-\beta} = \frac{\bar{n}}{1+\bar{n}} . \tag{7.57}$$

The characteristic function for a thermal state is [2]

$$\chi(z, \mathcal{P}) = e^{-\frac{1}{2}(2\bar{n}+1-\mathcal{P})|z|^2} . \tag{7.58}$$

For symmetric ordering $\mathcal{P} = 0$, we have

$$\chi(z) = \chi(z, 0) = e^{-\frac{1}{2}(2\bar{n}+1)|z|^2} . \tag{7.59}$$

Introducing the variables q and p for a single mode as in Eq. (7.17) (we omit the index k as $n = 1$), such that $|z|^2 = q^2 + p^2$ and $\boldsymbol{x} = (q\ p)$, we have

$$\chi(\boldsymbol{x}) = e^{-\frac{1}{4}\boldsymbol{x}\boldsymbol{g}\boldsymbol{x}^\top} , \tag{7.60}$$

with

$$\boldsymbol{g} = \begin{pmatrix} 2(2\bar{n}+1) & 0 \\ 0 & 2(2\bar{n}+1) \end{pmatrix} . \tag{7.61}$$

The previous arguments generalize to a multimode thermal state ($n > 1$) at inverse temperature β (the temperature is the same for all modes at equilibrium) and give a Gaussian mixed state with

$$\begin{aligned} \boldsymbol{d} &= \boldsymbol{0} , \\ \boldsymbol{g} &= 2(2\bar{n}+1)\mathbf{1}_N = 2\frac{e^{\beta}+1}{e^{\beta}-1}\mathbf{1}_N . \end{aligned} \tag{7.62}$$

7.4.4 Proof of Eq. (7.42)

Proof We start from the expression of χ with the Einstein summation convention, i.e., omitting the $\sum$ symbol over repeated symbols

$$\chi(\boldsymbol{x}) = \exp\left(-\frac{1}{4}g_{pq}x_p x_q + \mathrm{i}d_r x_r\right) . \tag{7.63}$$

We compute the derivatives of χ as follows:

$$\frac{\partial \chi}{\partial x_s} = \left(-\frac{1}{4} g_{ps} x_p - \frac{1}{4} g_{sq} x_q + \mathrm{i} d_s\right) \chi(\boldsymbol{x}) \tag{7.64}$$

being

$$\frac{\partial x_p}{\partial x_s} = \delta_{ps} . \tag{7.65}$$

Note that in the r.h.s. of Eq. (7.64), there are sums with respect to repeated indices. Evaluating (7.64) at $\boldsymbol{x} = \boldsymbol{0}$, being $\chi(\boldsymbol{0}) = 1$, we have

$$\left.\frac{\partial \chi}{\partial x_s}\right|_{\boldsymbol{x}=0} = \mathrm{i} d_s . \tag{7.66}$$

For the second derivatives, we have

$$\begin{aligned} \frac{\partial^2 \chi}{\partial x_s \partial x_u} =& \left(-\frac{1}{4} g_{us} - \frac{1}{4} g_{su}\right) \chi(\boldsymbol{x}) \\ &+ \left(-\frac{1}{4} g_{ps} x_p - \frac{1}{4} g_{sq} x_q + \mathrm{i} d_s\right) \\ &\times \left(-\frac{1}{4} g_{wu} x_w - \frac{1}{4} g_{uw} x_w + \mathrm{i} d_u\right) \chi(\boldsymbol{x}) , \end{aligned} \tag{7.67}$$

and

$$\left.\frac{\partial^2 \chi}{\partial x_s \partial x_u}\right|_{\boldsymbol{x}=0} = -\frac{1}{4} g_{us} - \frac{1}{4} g_{su} - d_s d_u = -\frac{g_{su}}{2} - d_s d_u , \tag{7.68}$$

where we used the fact that the covariance matrix is symmetric, i.e., $g_{su} = g_{us}$. Note that the imaginary part of l.h.s. in (7.68) is vanishing. Being

$$\langle \hat{a}_j^\dagger \hat{a}_j \rangle = -\frac{1}{2} \left(\frac{\partial^2}{\partial x_{2j}^2} + \frac{\partial^2}{\partial x_{2j+1}^2}\right) \chi \Bigg|_{\boldsymbol{x}=0} - \frac{1}{2} \tag{7.69}$$

we have the proof. □

7.5 Covariance Matrix in Terms of the Derivatives of χ

Given a generic χ, one obtains the covariance matrix and the displacement operator by derivation. The covariance matrix is defined by Eq. (7.43), here reported as

$$\begin{aligned} g_{pq} &= \langle\{\hat{R}_p, \hat{R}_q\}\rangle - 2\langle\hat{R}_p\rangle\langle\hat{R}_q\rangle \\ &= \langle\hat{R}_p\hat{R}_q\rangle + \langle\hat{R}_q\hat{R}_p\rangle - 2\langle R_p\rangle\langle R_q\rangle \\ &= 2\,\mathrm{Tr}[\rho(\hat{R}_p - d_p)(\hat{R}_q - d_q)] - \mathrm{i}J_{pq}\,, \end{aligned} \tag{7.70}$$

and the displacement vector is

$$d_p = \langle\hat{R}_p\rangle\,. \tag{7.71}$$

$\boldsymbol{d}$ is determined by the first derivatives, as in Eq. (7.29), here written as

$$d_q = \langle\hat{R}_q\rangle = \mathrm{Tr}[\hat{\rho}\hat{R}_q] = \left.\frac{\partial\chi_I}{\partial x_q}\right|_{\boldsymbol{x}=0}. \tag{7.72}$$

Seemingly, the covariance matrix is

$$g_{pq} = -2\left.\frac{\partial^2\chi_R}{\partial x_p\partial x_q}\right|_{\boldsymbol{x}=0} - 2d_pd_q = -2\frac{\partial^2\chi_R}{\partial x_p\partial x_q} - 2\left.\frac{\partial\chi_I}{\partial x_p}\frac{\partial\chi_I}{\partial x_q}\right|_{\boldsymbol{x}=0}. \tag{7.73}$$

The former expressions (proved in Sect. (7.5.1)) are not limited to Gaussian states. They are used to obtain $\boldsymbol{g}$ and $\boldsymbol{d}$ from a NN model. The particular case of a Gaussian state is in Sect. (7.5.2).

7.5.1 Proof of Eqs. (7.72) and (7.73) for General States*

Proof The definition of the characteristic function in Eq. (7.28) is reported here:

$$\chi(\boldsymbol{x}) = \mathrm{Tr}\left(\rho e^{\mathrm{i}\boldsymbol{x}\hat{\mathbf{R}}}\right), \tag{7.74}$$

We consider the first moment by deriving Eq. (7.74) w.r.t. x_q; we have

$$\frac{\partial\chi}{\partial x_q} = \mathrm{Tr}\left(\mathrm{i}\hat{R}_q\rho e^{\mathrm{i}\boldsymbol{x}\hat{\mathbf{R}}}\right), \tag{7.75}$$

which gives at $\boldsymbol{x} = 0$

$$\left.\frac{\partial\chi}{\partial x_q}\right|_{\boldsymbol{x}=0} = \mathrm{i}\,\mathrm{Tr}\left(\hat{R}_q\rho\right) = \mathrm{i}\langle\hat{R}_q\rangle, \tag{7.76}$$

as $\chi = \chi_R + \mathrm{i}\chi_I$ and $\langle\hat{R}_q\rangle$ is real because $\hat{R}_q$ is self-adjoint. We obtain the following:

$$
\begin{aligned}
\langle \hat{R}_q \rangle &= -\mathrm{i}\, \frac{\partial \chi}{\partial x_q}\bigg|_{\boldsymbol{x}=0} = \frac{\partial \chi_I}{\partial x_q}\bigg|_{\boldsymbol{x}=0} , \\
\frac{\partial \chi_R}{\partial x_q}\bigg|_{\boldsymbol{x}=0} &= 0
\end{aligned}
\tag{7.77}
$$

For the second derivatives, we write

$$
\begin{aligned}
\langle \hat{R}_p \hat{R}_q \rangle = \mathrm{Tr}\left(\rho \hat{R}_p \hat{R}_q\right) &= \mathrm{Tr}\left[\rho \frac{\partial e^{\mathrm{i}\, x_p \hat{R}_p}}{\partial(\mathrm{i} x_p)} \frac{\partial e^{\mathrm{i}\, x_q \hat{R}_q}}{\partial(\mathrm{i} x_q)}\right]\Bigg|_{\boldsymbol{x}=0} \\
&= -\frac{\partial^2}{\partial x_p \partial x_q} \mathrm{Tr}\left(\rho e^{\mathrm{i}\, x_p \hat{R}_p} e^{\mathrm{i}\, x_q \hat{R}_q}\right)\bigg|_{\boldsymbol{x}=0} .
\end{aligned}
\tag{7.78}
$$

We have

$$
e^{\mathrm{i}\, x_p \hat{R}_p} e^{\mathrm{i}\, x_q \hat{R}_q} = e^{\mathrm{i}\, x_q \hat{R}_q + \mathrm{i}\, x_q \hat{R}_q} e^{-\frac{1}{2}\left[\hat{R}_p, \hat{R}_q\right] x_p x_q} = e^{\mathrm{i}\, x_q \hat{R}_q + \mathrm{i}\, x_q \hat{R}_q} e^{-\frac{\mathrm{i}}{2} J_{pq} x_p x_q} ,
\tag{7.79}
$$

which is used in (7.78):

$$
\begin{aligned}
\langle \hat{R}_p \hat{R}_q \rangle = &- \frac{\partial^2}{\partial x_p \partial x_q} \mathrm{Tr}\left(\rho e^{\mathrm{i}\, x_p \hat{R}_p} e^{\mathrm{i}\, x_q \hat{R}_q}\right)\bigg|_{\boldsymbol{x}=0} = \\
&- \frac{\partial^2}{\partial x_p \partial x_q} \mathrm{Tr}\left(\rho e^{\mathrm{i}\, x_q \hat{R}_q + \mathrm{i}\, x_q \hat{R}_q} e^{-\frac{\mathrm{i}}{2} J_{pq} x_p x_q}\right)\bigg|_{\boldsymbol{x}=0} = \\
&- \frac{\partial^2}{\partial x_p \partial x_q} \mathrm{Tr}\left(\rho e^{\mathrm{i}\, \boldsymbol{x} \hat{\boldsymbol{R}}} e^{-\frac{\mathrm{i}}{2} J_{pq} x_p x_q}\right)\bigg|_{\boldsymbol{x}=0} = \\
&- \frac{\partial^2}{\partial x_p \partial x_q} \left[\chi(\boldsymbol{x}) e^{-\frac{\mathrm{i}}{2} J_{pq} x_p x_q}\right]\bigg|_{\boldsymbol{x}=0} = \\
&- \frac{\partial^2}{\partial x_p \partial x_q} \chi(\boldsymbol{x})\bigg|_{\boldsymbol{x}=0} + \frac{\mathrm{i}}{2} J_{pq} .
\end{aligned}
\tag{7.80}
$$

By using the previous result in [see Eq. (7.70)]

$$
g_{pq} = \langle \hat{R}_p \hat{R}_q \rangle + \langle \hat{R}_q \hat{R}_p \rangle - 2 \langle R_p \rangle \langle R_q \rangle ,
\tag{7.81}
$$

we have ($J_{pq} = -J_{qp}$)

$$
g_{pq} = -2 \frac{\partial^2}{\partial x_p \partial x_q} \chi(\boldsymbol{x})\bigg|_{\boldsymbol{x}=0} - 2 d_p d_q ,
\tag{7.82}
$$

as g_{pq} and d_p are real (we have the proof). □

7.5.2 Proof of Eqs. (7.72) and (7.73) for a Gaussian State*

Proof Equation (7.73) for a Gaussian characteristic model is derived by writing the Gaussian characteristic function in terms of the components of $\boldsymbol{x}$

$$\chi(\boldsymbol{x}) = \exp\left(-\frac{1}{4}\sum_{pq} g_{pq}x_p x_q + \mathrm{i}\sum_{p} x_p d_p\right), \tag{7.83}$$

with $p, q = 0, 1, ..., N-1$.

We have for the first derivative, by $g_{pq} = g_{qp}$,

$$\frac{\partial\chi}{\partial x_m} = \left(-\frac{1}{2}\sum_p g_{mp}x_p + \mathrm{i}d_m\right)\chi(\boldsymbol{x}), \tag{7.84}$$

with $m = 0, 1, ..., N-1$.

Evaluating Eq. (7.84) at $\boldsymbol{x} = 0$, we obtain, being $\chi(\mathbf{0}) = 1$,

$$\left.\frac{\partial\chi}{\partial x_m}\right|_{\boldsymbol{x}=0} = \mathrm{i}\left.\frac{\partial\chi_I}{\partial x_m}\right|_{\boldsymbol{x}=0} = \mathrm{i}d_m \, . \tag{7.85}$$

For the second derivative of Eq. (7.83), we have

$$\frac{\partial^2\chi}{\partial x_m\partial x_n} = -\frac{1}{2}g_{mn}\chi(\boldsymbol{x}) + \left(-\frac{1}{2}\sum_p g_{mp}x_p + \mathrm{i}d_m\right)\left(-\frac{1}{2}\sum_p g_{np}x_p + \mathrm{i}d_n\right)\chi(\boldsymbol{x}) \tag{7.86}$$

with $n = 0, 1, ..., N-1$. Evaluating Eq. (7.86) at $\boldsymbol{x} = 0$, we have

$$\left.\frac{\partial^2\chi_R}{\partial x_m\partial x_n}\right|_{\boldsymbol{x}=0} = -\frac{1}{2}g_{mn} - d_m d_n \, , \tag{7.87}$$

$$\left.\frac{\partial^2\chi_I}{\partial x_m\partial x_n}\right|_{\boldsymbol{x}=0} = 0 \, , \tag{7.88}$$

and using Eq. (7.85) in the first of Eq. (7.87), we have the proof. □

7.6 Covariance Matrices and Uncertainties

Throughout this book, we use the column vector with dimensions $1 \times N$

$$\hat{\boldsymbol{R}} = \begin{pmatrix} \hat{q}_0 \\ \hat{p}_0 \\ \vdots \\ \hat{q}_{n-1} \\ \hat{p}_{n-1} \end{pmatrix} . \tag{7.89}$$

In our notation at each column operator vector, we associate a row vector of real numbers. The expectation value is $\boldsymbol{d} = \langle \hat{\boldsymbol{R}} \rangle$ and corresponds to a real `tf.Variable` with shape (N, 1).

The c-number column vector conjugated to $\hat{\boldsymbol{R}}$ is $\boldsymbol{x}$ with shape `(1,N)`:

$$\begin{aligned} \boldsymbol{x} &= \begin{pmatrix} q_0, & p_0, & \ldots, & q_{n-1}, & p_{n-1} \end{pmatrix} \\ &= \begin{pmatrix} x_0, & x_1, & \ldots, & x_{N-2}, & x_{N-1} \end{pmatrix} . \end{aligned} \tag{7.90}$$

Given a vector of operators, one can define the covariance matrix with their second moments. The $\hat{\boldsymbol{R}}$ covariance matrix is denoted as $\boldsymbol{\sigma}^R$ and given by

$$\sigma^R_{pq} = \frac{1}{2}\langle\{\hat{R}_p \hat{R}_q\}\rangle - \langle \hat{R}_p \rangle \langle \hat{R}_q \rangle. \tag{7.91}$$

The covariance matrix $\boldsymbol{\sigma}^R$ is also written as

$$\begin{aligned} \sigma^R_{pq} &= \frac{1}{2}\langle \hat{R}_p \hat{R}_q + \hat{R}_q \hat{R}_p \rangle - d_p d_q \\ &= \langle \hat{R}_p \hat{R}_q \rangle + \mathrm{i} J_{pq} - d_p d_q \\ &= \langle (\hat{R}_p - d_p)(\hat{R}_q - d_q) \rangle + \mathrm{i} J_{pq} . \end{aligned} \tag{7.92}$$

The link with the covariance matrix $\boldsymbol{g}$ is

$$\boldsymbol{g} = 2\boldsymbol{\sigma}^R . \tag{7.93}$$

The `shape` of $\boldsymbol{g}$ and $\boldsymbol{\sigma}^R$ is `(N,N)`. $\boldsymbol{g} = \mathbf{1}_N$ for the vacuum state.

The elements of $\boldsymbol{g}$ are related, by definition, to the uncertainties relations. For example, given that

$$\begin{aligned} \Delta \hat{q}_j &= \hat{q}_j - \langle \hat{q}_j \rangle \\ \Delta \hat{p}_j &= \hat{p}_j - \langle \hat{p}_j \rangle \end{aligned} \tag{7.94}$$

and the uncertainty (with units $\hbar = 1$) principle holds

$$\langle \Delta \hat{q}_i \rangle^2 \langle \Delta \hat{p}_i \rangle^2 \geq \frac{1}{4} , \tag{7.95}$$

we have

$$\begin{aligned} g_{00} &= 2\langle\Delta\hat{q}_0\rangle\rangle^2 \\ g_{11} &= 2\langle\Delta\hat{p}_0\rangle^2 \\ g_{01} &= 2\langle\Delta\hat{q}_0\rangle\langle\Delta\hat{p}_0\rangle - \mathrm{i} \\ &\vdots \end{aligned} \tag{7.96}$$

One realizes that uncertainties relations pose bounds on $\boldsymbol{g}$. Specifically, one has [3]

$$\boldsymbol{g} + \mathrm{i}\mathbf{J} \geq 0. \tag{7.97}$$

Also, for a pure state, we must have

$$\det \boldsymbol{g} = 1 \,. \tag{7.98}$$

For $\boldsymbol{\sigma}^R$, the previous equations reads

$$\boldsymbol{\sigma}^R + \frac{1}{2}\mathbf{J} \geq 0 \,, \tag{7.99}$$

and for a pure state

$$\det \boldsymbol{\sigma}^R = \left(\frac{1}{2}\right)^N \,. \tag{7.100}$$

7.6.1 The Permuted Covariance Matrix

There are various representations for Gaussian states [1, 4]. One can use the vector of observable $\hat{\mathbf{S}}$ obtained by a permutation the elements of $\hat{\boldsymbol{R}}$

$$\hat{\mathbf{S}} = \begin{pmatrix} \hat{q}_0 \\ \hat{q}_1 \\ \vdots \\ \hat{q}_{n-1} \\ \hat{p}_0 \\ \hat{p}_1 \\ \vdots \\ \hat{p}_{n-1} \end{pmatrix} \,. \tag{7.101}$$

The corresponding c-number row vector with shape `(1,N)` is

$$\begin{aligned} \boldsymbol{y} &= \left(q_0,\ q_1,\ \ldots,\ p_{n-2},\ p_{n-1}\right) \\ &= \left(y_0,\ y_1,\ \ldots,\ y_{N-2},\ y_{N-1}\right) . \end{aligned} \tag{7.102}$$

$\hat{\boldsymbol{R}}$ and $\hat{\boldsymbol{S}}$ are related by a permutation matrix $\mathbf{P}$ as follows [4]:

$$\hat{\boldsymbol{S}} = \mathbf{P}\hat{\boldsymbol{R}} , \tag{7.103}$$

with the elements of $\mathbf{P}$ defined as

$$\begin{aligned} \mathrm{P}_{qr} &= \delta_{2q,r} && \text{for } q \leq n , \\ \mathrm{P}_{qr} &= \delta_{2(q-n)+1,2r} && \text{for } q \geq n . \end{aligned}$$

For the mean value, we have

$$\langle \hat{\boldsymbol{R}} \rangle = \boldsymbol{d} = \mathbf{P}\langle \hat{\boldsymbol{S}} \rangle . \tag{7.104}$$

Seemingly

$$\boldsymbol{y} = \boldsymbol{x}\mathbf{P}^T . \tag{7.105}$$

For the commutator, we have

$$\begin{aligned} \left[\hat{S}_p, \hat{S}_q\right] &= \hat{S}_p\hat{S}_q - \hat{S}_q\hat{S}_p \\ &= P_{pr}\hat{R}_r\hat{R}_s\mathrm{P}_{qs} - \mathrm{P}_{pr}\hat{R}_s\hat{R}_r\mathrm{P}_{qs} \\ &= \mathrm{P}_{pr}\mathrm{i}\mathrm{J}_{rs}\mathrm{P}_{qs} \\ &= \mathrm{i}\mathrm{J}^{\mathrm{P}}_{pq} , \end{aligned} \tag{7.106}$$

being

$$\mathbf{J}^{\mathrm{P}} = \mathbf{PJP}^{\top} = \begin{pmatrix} \mathbf{0}_n & \mathbf{1}_n \\ -\mathbf{1}_n & \mathbf{0}_n \end{pmatrix} \tag{7.107}$$

with $\mathbf{0}_n$ the $n \times n$ matrix of 0s and $\mathbf{1}_{n\times n}$ the $n \times n$ identity matrix.

The covariance matrix for the elements of the $\hat{\boldsymbol{S}}$ vector is

$$\sigma^S_{pq} = \frac{1}{2}\langle\{\hat{S}_p, \hat{S}_q\}\rangle - \langle\hat{S}_p\rangle\langle\hat{S}_q\rangle . \tag{7.108}$$

For the covariance matrices $\boldsymbol{\sigma}^R$ and $\boldsymbol{\sigma}^S$, we have

$$\boldsymbol{\sigma}^S = \mathbf{P}\boldsymbol{\sigma}^R\mathbf{P}^\top . \tag{7.109}$$

The covariance matrices have the same determinant as the permutation matrix is orthogonal, i.e., $\det\left(\mathbf{P}\mathbf{P}^\top\right) = 1$; hence

$$\det(\boldsymbol{\sigma}^S) = \det(\boldsymbol{\sigma}^R) .$$

Proof of Eq. (7.109)*

Proof First, we observe that

$$\sigma^S_{pq} = \frac{1}{2}\langle\{\hat{S}_p, \hat{S}_q\}\rangle - \langle\hat{S}_p\rangle\langle\hat{S}_q\rangle = \frac{1}{2}\langle\{\Delta\hat{S}_p, \Delta\hat{S}_q\}\rangle \tag{7.110}$$

being

$$\Delta\hat{S}_p = \hat{S}_p - \langle\hat{S}_p\rangle. \tag{7.111}$$

Seemingly

$$\sigma^R_{pq} = \frac{1}{2}\langle\{\Delta\hat{R}_p, \Delta\hat{R}_q\}\rangle. \tag{7.112}$$

Then we have

$$\Delta\hat{S}_q = \mathrm{P}_{qr}\hat{R}_r - \mathrm{P}_{qr}\langle\hat{R}_r\rangle = \mathrm{P}_{qr}\Delta\hat{R}_r, \tag{7.113}$$

$$\langle\hat{S}_q\rangle = \mathrm{P}_{qr}\langle\hat{R}_r\rangle, \tag{7.114}$$

which gives

$$\sigma^S_{pq} = \frac{1}{2}\mathrm{P}_{pr}\langle\{\Delta\hat{R}_r, \Delta\hat{R}_s\}\rangle\mathrm{P}_{qs}, \tag{7.115}$$

corresponding to

$$\boldsymbol{\sigma}^S = \mathbf{P}\boldsymbol{\sigma}^R\mathbf{P}^\top, \tag{7.116}$$

as we expected. □

7.6.2 Ladder Operators and Complex Covariance matrix

To the canonical operators $\hat{q}_j$ and $\hat{p}_j$, we associate the annihilation and creation operators

$$\hat{a}_j = \frac{1}{\sqrt{2}}(\hat{q}_j + \mathrm{i}\hat{p}_j) ,$$
$$\hat{a}_j^\dagger = \frac{1}{\sqrt{2}}(\hat{q}_j - \mathrm{i}\hat{p}_j) ,$$

which also form a vector with dimensions $2n \times 1$

$$\hat{\boldsymbol{A}} = \begin{pmatrix} \hat{a}_0 \\ \hat{a}_1 \\ \vdots \\ \hat{a}_{n-1} \\ \hat{a}_0^\dagger \\ \hat{a}_1^\dagger \\ \vdots \\ \hat{a}_{n-1}^\dagger \end{pmatrix} . \tag{7.117}$$

The conjugate row vector of complex variables with shape `(N,1)` is

$$\boldsymbol{\alpha} = \left(\alpha_0,\, \alpha_1, \ldots,\, \alpha_{n-1}, \ldots,\, \alpha_0^*,\, \alpha_1^*,\, \alpha_{n-1}^*\right) .$$

As $\left[\hat{a}_i, \hat{a}_j\right] = 0$ and $\left[\hat{a}_i, \hat{a}_j\right] = \mathrm{i}\delta_{ij}$, the commutator for the elements of $\boldsymbol{A}$ is

$$\left[\hat{A}_p, \hat{A}_q^\dagger\right] = (\boldsymbol{\Sigma}_3)_{pq} \tag{7.118}$$

being

$$\boldsymbol{\Sigma}_3 = \begin{pmatrix} \mathbf{1}_n & \mathbf{0}_n \\ \mathbf{0}_n & -\mathbf{1}_n \end{pmatrix} , \tag{7.119}$$

or, equivalently, as $\boldsymbol{A}$ contains both annihilation and creation operators,

$$\left[\hat{A}_p, \hat{A}_q\right] = \mathrm{J}^{\mathrm{P}}_{pq} . \tag{7.120}$$

The complex vectors are related to the real vectors by the matrix

$$\boldsymbol{\Omega} = \frac{1}{\sqrt{2}} \begin{pmatrix} \mathbf{1}_n & \mathrm{i}\mathbf{1}_n \\ \mathbf{1}_n & -\mathrm{i}\mathbf{1}_n \end{pmatrix}$$

such that

$$\hat{\boldsymbol{A}} = \boldsymbol{\Omega}\hat{\boldsymbol{S}} = \boldsymbol{\Omega}\mathbf{P}\hat{\boldsymbol{R}}. \tag{7.121}$$

The covariance matrix for the complex quantities is also complex,

$$\sigma_{ij}^{A} = \frac{1}{2}\langle\{\hat{\zeta}_i, \hat{\zeta}_j^\dagger\}\rangle - \langle\hat{\zeta}_i\rangle\langle\hat{\zeta}_j^\dagger\rangle \,, \tag{7.122}$$

with $\hat{\zeta}_i \in \{\hat{a}_j, \hat{a}_j^\dagger\}$.

$\boldsymbol{\sigma}^A$ is a $N \times N$ Hermitian matrix, and we have

$$\boldsymbol{\sigma}^A = \boldsymbol{\Omega}\boldsymbol{\sigma}^S\boldsymbol{\Omega}^\dagger = \boldsymbol{\Omega}\mathbf{P}\boldsymbol{\sigma}^R\mathbf{P}^\top\boldsymbol{\Omega}^\dagger \tag{7.123}$$

Proof of Eq. (7.123)*

Proof We have

$$\sigma_{pq}^{A} = \frac{1}{2}\langle\{\Delta\hat{A}_p, \Delta\hat{A}_q^\dagger\}\rangle \tag{7.124}$$

and

$$\Delta\hat{A}_p = \hat{A}_p - \langle\hat{A}_p\rangle = \Omega_{pr}\Delta S_r \,,$$
$$\Delta\hat{A}_q^\dagger = \hat{A}_q^\dagger - \langle\hat{A}_q^\dagger\rangle = \Omega_{qs}^*\Delta S_s^\dagger,$$

which used in (7.124) gives

$$\sigma_{pq}^{A} = \frac{1}{2}\Omega_{pr}\langle\{\Delta\hat{S}_r, \Delta\hat{S}_s^\dagger\}\rangle\Omega_{qs}^*, \tag{7.125}$$

that is,

$$\boldsymbol{\sigma}^A = \boldsymbol{\Omega}\boldsymbol{\sigma}^S\boldsymbol{\Omega}^\dagger. \tag{7.126}$$

□

7.7 Gaussian Characteristic Function

We have introduced the Gaussian characteristic function in Eq. (7.37) as

$$\chi(\boldsymbol{x}) = \exp\left(-\frac{1}{4}\boldsymbol{x}\,\boldsymbol{g}\,\boldsymbol{x}^\top + \mathrm{i}\,\boldsymbol{x}\,\boldsymbol{d}\right) , \tag{7.127}$$

which is equal to

$$\chi = \chi(\boldsymbol{x}) = \exp\left(-\frac{1}{2}\boldsymbol{x}\,\boldsymbol{\sigma}^R\,\boldsymbol{x}^\top + \mathrm{i}\,\boldsymbol{x}\,\langle\hat{\boldsymbol{R}}\rangle\right) . \tag{7.128}$$

Being $\langle\hat{\boldsymbol{R}}\rangle = \mathbf{P}^\top\langle\hat{\mathbf{S}}\rangle$ and $\boldsymbol{y} = \boldsymbol{x}\mathbf{P}^\top$, we have

$$\chi = \chi(\boldsymbol{y}) = \exp\left(-\frac{1}{2}\boldsymbol{y}\,\boldsymbol{\sigma}^S\,\boldsymbol{y}^\top + \mathrm{i}\,\boldsymbol{y}\,\langle\hat{\mathbf{S}}\rangle\right) \tag{7.129}$$

where we have used Eq. (7.109), which links the covariance matrices $\boldsymbol{\sigma}^S$ and $\boldsymbol{\sigma}^R$.

A Gaussian characteristic function looks the same in the different representations.

We express χ in terms of $\boldsymbol{\alpha}$, starting from Eq. (7.129), having

$$\langle\hat{\mathbf{S}}\rangle = \boldsymbol{\Omega}^\dagger\langle\hat{\mathbf{a}}\rangle ,$$

$$\chi = \chi(\boldsymbol{\alpha}, \boldsymbol{\alpha}^*) = \exp\left(-\frac{1}{2}\boldsymbol{\alpha}^*\,\boldsymbol{\sigma}^A\,\boldsymbol{\alpha}^\top + \mathrm{i}\,\boldsymbol{\alpha}^*\,\langle\hat{\mathbf{a}}\rangle\right) \tag{7.130}$$

with

$$\boldsymbol{\alpha}^* = \boldsymbol{y}\boldsymbol{\Omega}^\dagger ,$$

$$\boldsymbol{\sigma}^A = \boldsymbol{\Omega}\boldsymbol{\sigma}^S\boldsymbol{\Omega}^\dagger ,$$

and

$$\det(\boldsymbol{\sigma}) = \det(\boldsymbol{\sigma}^S) .$$

Equation (7.130) can be written in a different way that thus does not contain the conjugate of $\boldsymbol{\alpha}$. One has

$$\boldsymbol{\alpha}^* = \boldsymbol{\alpha}\mathbf{X} \tag{7.131}$$

where $\mathbf{X}$ is the permutation matrix

$$\mathbf{X} = \begin{pmatrix} \mathbf{0}_n & \mathbf{1}_n \\ \mathbf{1}_n & \mathbf{0}_n \end{pmatrix} , \tag{7.132}$$

and we have

$$\chi = \chi(\boldsymbol{\alpha}) = \exp\left(-\frac{1}{2}\boldsymbol{\alpha}\,\mathbf{X}\sigma^A\,\boldsymbol{\alpha}^\top + \mathrm{i}\,\boldsymbol{\alpha}\,\mathbf{X}\langle\hat{\mathbf{a}}\rangle\right) . \tag{7.133}$$

Having an expression for χ that does not include the conjugate of $\boldsymbol{\alpha}$ is convenient because complex conjugate variables are to be considered independent variables. For example, when dealing with the derivatives of χ, one has to retain both the derivatives with respect to $\boldsymbol{\alpha}$ and also $\boldsymbol{\alpha}^*$, if the latter appears explicitly as in Eq. (7.128). We have

$$\frac{\partial\chi(\boldsymbol{\alpha},\boldsymbol{\alpha}^*)}{\partial\alpha_j} = \frac{\partial\boldsymbol{\alpha}}{\partial\alpha_j}\cdot\nabla_{\boldsymbol{\alpha}}\chi + \frac{\partial\boldsymbol{\alpha}^*}{\partial\alpha_j}\cdot\nabla_{\boldsymbol{\alpha}^*}\chi \tag{7.134}$$

and

$$\frac{\partial\chi(\boldsymbol{\alpha},\boldsymbol{\alpha}^*)}{\partial\alpha_j^*} = \frac{\partial\boldsymbol{\alpha}}{\partial\alpha_j^*}\cdot\nabla_{\boldsymbol{\alpha}}\chi + \frac{\partial\boldsymbol{\alpha}^*}{\partial\alpha_j^*}\cdot\nabla_{\boldsymbol{\alpha}^*}\chi \tag{7.135}$$

By using Eq. (7.133), one has only to derive with respect $\boldsymbol{\alpha}$.

7.7.1 Remarks on the `shape` of a Vector

As we use vectors, vectors of operators and variables, we distinguish between their *dimensions* and their `shape`. A vector and a vector of operators have dimensions, but only variables that can be represented as `TensorFlow` variables have a `shape`.

For example, vectors of operators cannot be represented as `TensorFlow` variables, and $\hat{\boldsymbol{A}}$ has dimension $N \times 1$. On the contrary $\boldsymbol{y}$ has dimension $1 \times N$ but also `shape` = `(1, N)`.

Also, the array of operators as $\hat{\boldsymbol{R}}$ has dimensions $N \times 1$ but has not a `shape`, as it cannot be directly mapped to a `TensorFlow` variable, as it is not a c-number. On the contrary, its mean value $\boldsymbol{d} = \langle\hat{\boldsymbol{R}}\rangle$ has dimension $1 \times N$ but also a `shape` equal to `(1,N)` which is the shape of the corresponding `TensorFlow` tensor.

To each array of operators, we associate an array of variables by a conjugation operation. For example, when we introduce the characteristic function with the Fourier transform operator $e^{\mathrm{i}\boldsymbol{x}\hat{\boldsymbol{R}}}$ as

$$\chi(\boldsymbol{x}) = \mathrm{Tr}\left(\rho e^{\mathrm{i}\boldsymbol{x}\hat{\boldsymbol{R}}}\right) ,$$

we associate to q-number (i.e., operator) vector $\hat{\boldsymbol{R}}$, the conjugate c-number vector $\boldsymbol{x}$ with dimensions $1 \times N$ and shape `(1,N)`. We summarize in Table 7.1 the shape of vectors and matrices.

Table 7.1 Dimensions and `TensorFlow` `shape` of Gaussian characteristic vectors

Operator	*Dimension*	*Variable*	*Shape*	*Matrix* ($N \times N$)	*Properties*
$\hat{\boldsymbol{R}}$	$N \times 1$	$\boldsymbol{x}$	`(1,N)`	$\boldsymbol{\sigma}^R = 2\boldsymbol{g}$	Real symmetric
$\hat{\boldsymbol{S}}$	$N \times 1$	$\boldsymbol{y}$	`(1,N)`	$\boldsymbol{\sigma}^S = \mathbf{P}\boldsymbol{\sigma}^R\mathbf{P}^\top$	Real symmetric
$\hat{\boldsymbol{A}}$	$N \times 1$	$\boldsymbol{\alpha}$	`(1,N)`	$\boldsymbol{\sigma}^A = \boldsymbol{\Omega}\boldsymbol{\sigma}^S\boldsymbol{\Omega}^\dagger$	Hermitian

7.8 Linear Transformations and Symplectic Matrices

Linear transformations, or gates, are the building blocks for complex quantum circuits and processors, corresponding to unitary operations.

We start considering the gates with the output annihilation operators $\tilde{\hat{\boldsymbol{a}}}$ expressed as a linear combination of the input operators $\hat{\boldsymbol{a}}$. These are a particular case of the general transformation

$$\tilde{\hat{\boldsymbol{a}}} = \hat{U}^\dagger \hat{\boldsymbol{a}} \hat{U}. \tag{7.136}$$

$\hat{U}$ is the unitary transformation for the gate. In terms of the density matrix, Eq. (7.136) reads [2]

$$\tilde{\rho} = \hat{U} \rho \hat{U}^\dagger. \tag{7.137}$$

We consider the case with (7.136) written as

$$\tilde{\hat{\boldsymbol{a}}} = \mathbf{U}\hat{\boldsymbol{a}} , \tag{7.138}$$

with $\mathbf{U}$ a $n \times n$ complex matrix and $\hat{\boldsymbol{a}}$ and $\tilde{\hat{\boldsymbol{a}}}$ column $n \times 1$ vectors.

In a later section, we will consider the more general case with

$$\tilde{\hat{\boldsymbol{a}}} = \mathbf{U}\hat{\boldsymbol{a}} + \mathbf{W}\hat{\boldsymbol{a}}^\dagger. \tag{7.139}$$

For the moment, we set $\mathbf{W} = 0$.

In terms of the canonical variables, the transformation (7.136) reads [1]

$$\tilde{\hat{\boldsymbol{R}}} = \hat{U}^\dagger \hat{\boldsymbol{R}} \hat{U} = \mathbf{M}\,\hat{\boldsymbol{R}} + \boldsymbol{d}' . \tag{7.140}$$

Equation (7.140) transforms Gaussian states into Gaussian states [1].

A Gaussian state with a covariance matrix $\boldsymbol{g}$ and displacement vector $\boldsymbol{d}$ turns into a new Gaussian state with covariance matrix

$$\tilde{\boldsymbol{g}} = \mathbf{M}\,\boldsymbol{g}\,\mathbf{M}^\top \tag{7.141}$$

and displacement vector

$$\widetilde{\boldsymbol{d}} = \mathbf{M}\,\boldsymbol{d} + \boldsymbol{d}'. \tag{7.142}$$

If the transformation $\mathbf{U}$ is unitary, the matrix $\mathbf{M}$ is symplectic, that is, the following relation holds:

$$\mathbf{M}\mathbf{J}\mathbf{M}^{\top} = \mathbf{J} \tag{7.143}$$

with $\mathbf{J}$ given in Eq. (7.27). Equivalently, the inverse of $\mathbf{M}$ is

$$\mathbf{M}^{-1} = \mathbf{J}\mathbf{M}^{\top}\mathbf{J}^{\top}\,. \tag{7.144}$$

7.8.1 Proof of Eqs. (7.141) and (7.142)*

Proof We consider the definition of the characteristic function in the

$$\chi = \mathrm{Tr}\left(\rho e^{\mathrm{i}\boldsymbol{x}\hat{\boldsymbol{R}}}\right) \tag{7.145}$$

Under the action of a unitary operator $\hat{U}$, the density matrix transforms as

$$\tilde{\rho} = \hat{U}\rho\hat{U}^{\dagger}\,, \tag{7.146}$$

so that the corresponding characteristic function reads

$$\tilde{\chi} = \mathrm{Tr}\left(\tilde{\rho}e^{\mathrm{i}\boldsymbol{x}\hat{\boldsymbol{R}}}\right) = \mathrm{Tr}\left(\hat{U}\rho\hat{U}^{\dagger}\rho e^{\mathrm{i}\boldsymbol{x}\hat{\boldsymbol{R}}}\right) = \mathrm{Tr}\left(\rho\hat{U}^{\dagger}e^{\mathrm{i}\boldsymbol{x}\hat{\boldsymbol{R}}}\hat{U}\right) = \mathrm{Tr}\left(\rho e^{\mathrm{i}\hat{\boldsymbol{x}}U^{\dagger}\hat{\boldsymbol{R}}\hat{U}}\right)\,, \tag{7.147}$$

where we used the cyclic property of the trace and

$$\hat{U}^{\dagger}f(\hat{C})\hat{U} = f(\hat{U}^{\dagger}\hat{C}\hat{U}) \tag{7.148}$$

as $\hat{U}$ is unitary [2]. By Eq. (7.140)

$$\tilde{\chi} = \mathrm{Tr}\left(\tilde{\rho}e^{\mathrm{i}\boldsymbol{x}\mathbf{M}\hat{\boldsymbol{R}}+\mathrm{i}\boldsymbol{x}\boldsymbol{d}'}\right) = \chi(\boldsymbol{x}\mathbf{M})e^{\mathrm{i}\boldsymbol{x}\boldsymbol{d}'}\,. \tag{7.149}$$

For a Gaussian χ as in Eq. (7.37), the transformation reads

$$\tilde{\chi}(\boldsymbol{x}) = \exp\left[-\frac{1}{4}\boldsymbol{x}\mathbf{M}\,\boldsymbol{g}\,\mathbf{M}^{\top}\boldsymbol{x}^{\top} + \mathrm{i}\,\boldsymbol{x}\,(\mathbf{M}\boldsymbol{d} + \boldsymbol{d}')\right]\,, \tag{7.150}$$

which gives Eqs. (7.141) and (7.142). □

7.9 The U and M Matrices*

It is instructive to deepen the link of the matrix **U** with the matrix **M** for later use. We consider the $n \times 1$ vector $\hat{\boldsymbol{a}}$ of the annihilation operators

$$\hat{\boldsymbol{a}} = \begin{pmatrix} \hat{a}_0 \\ \hat{a}_1 \\ \vdots \\ \hat{a}_{n-1} \end{pmatrix}, \tag{7.151}$$

and the corresponding $n \times 1$ vectors of positions $\hat{\boldsymbol{q}}$ and momenta $\hat{\boldsymbol{p}}$,

$$\hat{\boldsymbol{q}} = \begin{pmatrix} \hat{q}_0 \\ \hat{q}_1 \\ \vdots \\ \hat{q}_{n-1} \end{pmatrix} \qquad \hat{\boldsymbol{p}} = \begin{pmatrix} \hat{p}_0 \, , \\ \hat{p}_1 \\ \vdots \\ \hat{p}_{n-1} \end{pmatrix}. \tag{7.152}$$

We build the $\hat{\boldsymbol{R}}$ vector in Eq. (7.24) by using auxiliary rectangular matrices $\mathbf{R}_\mathrm{q}$ and $\mathbf{R}_\mathrm{p}$, as follows:

$$\hat{\boldsymbol{R}} = \mathbf{R}_\mathrm{q}\hat{\boldsymbol{q}} + \mathbf{R}_\mathrm{p}\hat{\boldsymbol{p}}. \tag{7.153}$$

$\mathbf{R}_\mathrm{q}$ and $\mathbf{R}_\mathrm{p}$ are constant matrices with size $N \times n$.

For $N = 4$, we have

$$\mathbf{R}_\mathrm{q} = \begin{pmatrix} 1 & 0 \\ 0 & 0 \\ 0 & 1 \\ 0 & 0 \end{pmatrix}, \tag{7.154}$$

and

$$\mathbf{R}_\mathrm{p} = \begin{pmatrix} 0 & 0 \\ 1 & 0 \\ 0 & 0 \\ 0 & 1 \end{pmatrix}, \tag{7.155}$$

such that

$$\mathbf{R}_\mathrm{q}\hat{\boldsymbol{q}} = \begin{pmatrix} \hat{q}_0 \\ 0 \\ \hat{q}_1 \\ 0 \end{pmatrix}, \tag{7.156}$$

and

$$\mathbf{R}_{\mathrm{p}}\hat{\boldsymbol{p}} = \begin{pmatrix} 0 \\ \hat{p}_0 \\ 0 \\ \hat{p}_1 \end{pmatrix} . \tag{7.157}$$

Hence,

$$\hat{\boldsymbol{R}} = \mathbf{R}_{\mathrm{q}}\hat{\boldsymbol{q}} + \mathbf{R}_{\mathrm{p}}\hat{\boldsymbol{p}} = \begin{pmatrix} \hat{q}_0 \\ \hat{p}_0 \\ \hat{q}_1 \\ \hat{p}_1 \end{pmatrix} , \tag{7.158}$$

being

$$\hat{\boldsymbol{q}} = \begin{pmatrix} \hat{q}_0 \\ \hat{q}_1 \end{pmatrix} , \tag{7.159}$$

and

$$\hat{\boldsymbol{p}} = \begin{pmatrix} \hat{p}_0 \\ \hat{p}_1 \end{pmatrix} . \tag{7.160}$$

From the previous expressions, one has

$$\begin{aligned} \hat{\boldsymbol{q}} &= \mathbf{R}_{\mathrm{q}}^{\top}\hat{\boldsymbol{R}} \\ \hat{\boldsymbol{p}} &= \mathbf{R}_{\mathrm{p}}^{\top}\hat{\boldsymbol{R}} \end{aligned} \tag{7.161}$$

with the rectangular matrices $\mathbf{R}_{\mathrm{q}}$ and $\mathbf{R}_{\mathrm{p}}$ satisfying [2]

$$\begin{aligned} &\mathbf{R}_{\mathrm{q}}^{\top}\mathbf{R}_{\mathrm{q}} = \mathbf{1}_n \\ &\mathbf{R}_{\mathrm{q}}^{\top}\mathbf{R}_{\mathrm{p}} = \mathbf{0}_n \\ &\mathbf{R}_{\mathrm{p}}^{\top}\mathbf{R}_{\mathrm{p}} = \mathbf{1}_n \\ &\mathbf{R}_{\mathrm{p}}^{\top}\mathbf{R}_{\mathrm{q}} = \mathbf{0}_n \\ &\mathbf{R}_{\mathrm{q}}\mathbf{R}_{\mathrm{q}}^{\top} + \mathbf{R}_{\mathrm{p}}\mathbf{R}_{\mathrm{p}}^{\top} = \mathbf{1}_N \\ &\mathbf{J}\mathbf{R}_{\mathrm{q}} = -\mathbf{R}_{\mathrm{p}} \\ &\mathbf{J}\mathbf{R}_{\mathrm{p}} = \mathbf{R}_{\mathrm{q}} \\ &\mathbf{R}_{\mathrm{q}}\mathbf{R}_{\mathrm{p}}^{\top} - \mathbf{R}_{\mathrm{p}}\mathbf{R}_{\mathrm{q}}^{\top} = \mathbf{J} . \end{aligned} \tag{7.162}$$

[2] The identities in (7.162) are reported in the symbolic `MATLAB` file `matlabsymbolic/test_RqRp.m`.

We recall that, for $N = 4$ and $n = 2$, we have

$$\mathbf{J} = \bigoplus_{j=1}^{n} \mathbf{J}_1 = \begin{pmatrix} 0 & 1 & 0 & 0 \\ -1 & 0 & 0 & 0 \\ 0 & 0 & 0 & 1 \\ 0 & 0 & -1 & 0 \end{pmatrix} . \tag{7.163}$$

By the matrices $\mathbf{R}_\mathrm{q}$ and $\mathbf{R}_\mathrm{p}$, we write the $\hat{\boldsymbol{a}}$ vector as follows:

$$\hat{\boldsymbol{a}} = \frac{\hat{\boldsymbol{q}} + \mathrm{i}\hat{\boldsymbol{p}}}{\sqrt{2}} = \frac{\mathbf{R}_\mathrm{q}^\top + \mathrm{i}\mathbf{R}_\mathrm{p}^\top}{\sqrt{2}} \hat{\boldsymbol{R}}. \tag{7.164}$$

For a linear transformation $\hat{U}$ from the vector $\hat{\boldsymbol{a}}$ to the new $\hat{\tilde{\boldsymbol{a}}}$,

$$\begin{aligned} \hat{\tilde{\boldsymbol{a}}} &= \mathbf{U}\hat{\boldsymbol{a}} \\ \hat{\tilde{\boldsymbol{a}}}^\dagger &= \mathbf{U}^*\hat{\boldsymbol{a}}^\dagger \end{aligned} \tag{7.165}$$

with $\mathbf{U} = \mathbf{U}_R + \mathrm{i}\mathbf{U}_I$ a complex matrix with real part $\mathbf{U}_R$ and imaginary part $\mathbf{U}_I$ and $\mathbf{U}^* = \mathbf{U}_R - \mathrm{i}\mathbf{U}_I$. It is important to stress that $\hat{\tilde{\boldsymbol{a}}}^\dagger$ in Eq. (7.165) is the column vector with elements $\hat{a}_j^\dagger$, and not $\left(\hat{\tilde{\boldsymbol{a}}}\right)^\dagger$. Correspondingly,

$$\begin{aligned} \hat{\tilde{\boldsymbol{q}}} &= \mathbf{U}_R\hat{\boldsymbol{q}} - \mathbf{U}_I\hat{\boldsymbol{p}} \, , \\ \hat{\tilde{\boldsymbol{p}}} &= \mathbf{U}_I\hat{\boldsymbol{q}} + \mathbf{U}_R\hat{\boldsymbol{p}} \, . \end{aligned} \tag{7.166}$$

As $\hat{\boldsymbol{R}}$ transforms in $\hat{\tilde{\boldsymbol{R}}}$, we have

$$\hat{\tilde{\boldsymbol{R}}} = \mathbf{R}_\mathrm{q}\hat{\tilde{\boldsymbol{q}}} + \mathbf{R}_\mathrm{p}\hat{\tilde{\boldsymbol{p}}} = \mathbf{M}\hat{\boldsymbol{R}} = (\mathbf{M}_1 + \mathbf{M}_2)\,\hat{\boldsymbol{R}} \tag{7.167}$$

with $\mathbf{M}_1$, $\mathbf{M}_2$, and $\mathbf{M}$ real matrices such that

$$\begin{aligned} \mathbf{M} &= \mathbf{M}_1 + \mathbf{M}_2 \, , \\ \mathbf{M}_1 &= \mathbf{R}_\mathrm{q}\mathbf{U}_R\mathbf{R}_\mathrm{q}^\top + \mathbf{R}_\mathrm{p}\mathbf{U}_R\mathbf{R}_\mathrm{p}^\top \, , \\ \mathbf{M}_2 &= \mathbf{R}_\mathrm{p}\mathbf{U}_I\mathbf{R}_\mathrm{q}^\top - \mathbf{R}_\mathrm{q}\mathbf{U}_I\mathbf{R}_\mathrm{p}^\top \, . \end{aligned} \tag{7.168}$$

The matrix $\mathbf{M}$ is symplectic if $\mathbf{MJM}^\top = \mathbf{J}$, and the matrix $\mathbf{U}$ is unitary if $\mathbf{U}^\dagger\mathbf{U} = (\mathbf{U}^*)^\top\mathbf{U} = \mathbf{1}_n$. In general, for an arbitrary matrix $\mathbf{U}$, the matrix $\mathbf{M}$ is not a symplectic matrix; hence it does not represent a linear transformation.

We show in the following that if $\hat{U}$ is unitary, the matrix $\mathbf{M}$ is symplectic.

Theorem 7.1 *If $\hat{U}$ is a unitary operator, i.e., $\hat{U}^\dagger\hat{U} = 1$, the matrix* $\mathbf{U}$ *is unitary.*

Proof If $\hat{U}$ is a unitary transformation, the expected values must be invariant, and we have

$$\begin{aligned}\langle\hat{\tilde{a}}^{\dagger}\hat{\tilde{a}}\rangle &= \text{Tr}\left(\tilde{\rho}\hat{\tilde{a}}^{\dagger}\hat{\tilde{a}}\right) = \text{Tr}\left(\hat{U}\rho\hat{U}^{\dagger}\hat{\tilde{a}}^{\dagger}\hat{\tilde{a}}\right) \\ &= \text{Tr}\left(\rho\hat{U}^{\dagger}\hat{\tilde{a}}^{\dagger}\hat{\tilde{a}}\hat{U}\right) \\ &= \text{Tr}\left(\rho\hat{U}^{\dagger}\hat{\tilde{a}}^{\dagger}\hat{U}^{\dagger}\hat{U}\hat{\tilde{a}}\hat{U}\right) \\ &= \text{Tr}\left(\rho\boldsymbol{a}^{\dagger}\mathbf{U}^{\dagger}\mathbf{U}\boldsymbol{a}\right) ,\end{aligned} \tag{7.169}$$

where we used the cyclic property for the trace and the unitarity for $\hat{U}$.

Being

$$\langle\hat{\tilde{a}}^{\dagger}\hat{\tilde{a}}\rangle = \langle\hat{\boldsymbol{a}}^{\dagger}\hat{\boldsymbol{a}}\rangle = \text{Tr}\left(\rho\boldsymbol{a}^{\dagger}\boldsymbol{a}\right) , \tag{7.170}$$

we find

$$\mathbf{U}^{\dagger}\mathbf{U} = \mathbf{1}_n, \tag{7.171}$$

which is the proof. □

Theorem 7.2 *If* **U** *is unitary,* **M** *is symplectic.*[3]

Proof We have after Eq. (7.168)

$$\mathbf{M} = \mathbf{R}_{\rm q}\mathbf{U}_R\mathbf{R}_{\rm q}^{\top} + \mathbf{R}_{\rm p}\mathbf{U}_R\mathbf{R}_{\rm p}^{\top} + \mathbf{R}_{\rm p}\mathbf{U}_I\mathbf{R}_{\rm q}^{\top} - \mathbf{R}_{\rm q}\mathbf{U}_I\mathbf{R}_{\rm p}^{\top} \tag{7.172}$$

and using Eqs. (7.162)

$$\mathbf{M}^{\top} = \mathbf{R}_{\rm q}\mathbf{U}_R^{\top}\mathbf{R}_{\rm q}^{\top} + \mathbf{R}_{\rm p}\mathbf{U}_R^{\top}\mathbf{R}_{\rm p}^{\top} + \mathbf{R}_{\rm q}\mathbf{U}_I^{\top}\mathbf{R}_{\rm p}^{\top} - \mathbf{R}_{\rm p}\mathbf{U}_I^{\top}\mathbf{R}_{\rm q}^{\top} \tag{7.173}$$

$$\mathbf{J}\mathbf{M}^{\top} = -\mathbf{R}_{\rm p}\mathbf{U}_R^{\top}\mathbf{R}_{\rm q}^{\top} + \mathbf{R}_{\rm q}\mathbf{U}_R^{\top}\mathbf{R}_{\rm p}^{\top} - \mathbf{R}_{\rm p}\mathbf{U}_I^{\top}\mathbf{R}_{\rm p}^{\top} - \mathbf{R}_{\rm q}\mathbf{U}_I^{\top}\mathbf{R}_{\rm q}^{\top} , \tag{7.174}$$

which give, after some algebra,

$$\begin{aligned}\mathbf{M}\mathbf{J}\mathbf{M}^{\top} = &\ \mathbf{R}_{\rm q}\left(\mathbf{U}_R\mathbf{U}_R^{\top} + \mathbf{U}_I\mathbf{U}_I^{\top}\right)\mathbf{R}_{\rm p}^{\top} - \mathbf{R}_{\rm p}\left(\mathbf{U}_R\mathbf{U}_R^{\top} + \mathbf{U}_I\mathbf{U}_I^{\top}\right)\mathbf{R}_{\rm q}^{\top} \\ - &\ \mathbf{R}_{\rm q}\left(\mathbf{U}_R\mathbf{U}_I^{\top} - \mathbf{U}_I\mathbf{U}_R^{\top}\right)\mathbf{R}_{\rm q}^{\top} - \mathbf{R}_{\rm p}\left(\mathbf{U}_R\mathbf{U}_I^{\top} - \mathbf{U}_I\mathbf{U}_R^{\top}\right)\mathbf{R}_{\rm p}^{\top} .\end{aligned} \tag{7.175}$$

As **U** is a unitary matrix, one has

[3] The various equations are tested in the file `jupyternotebooks/phasespace/symplectic.ipynb`.

$$\mathbf{U}^{\dagger}\mathbf{U} = (\mathbf{U}_R - \mathrm{i}\mathbf{U}_I)^{\top} (\mathbf{U}_R + \mathrm{i}\mathbf{U}_I) = \mathbf{1}_n \,, \tag{7.176}$$

which is

$$\mathbf{U}_R\mathbf{U}_R^{\top} + \mathbf{U}_I\mathbf{U}_I^{\top} = \mathbf{1}_n \tag{7.177}$$
$$\mathbf{U}_R\mathbf{U}_I^{\top} - \mathbf{U}_I\mathbf{U}_R^{\top} = \mathbf{0}_n. \tag{7.178}$$

After Eq. (7.175), we have

$$\mathbf{MJM}^{\top} = \mathbf{R}_{\mathrm{q}}\mathbf{R}_{\mathrm{p}}^{\top} - \mathbf{R}_{\mathrm{p}}\mathbf{R}_{\mathrm{q}}^{\top} = \mathbf{J} \,, \tag{7.179}$$

where we used the last of Eqs. (7.162). □

7.9.1 *Coding the Matrices* $\mathbf{R}_{\mathrm{p}}$ *and* $\mathbf{R}_{\mathrm{q}}$

The former equations are helpful when we want to represent the linear transformations as part of gates of neural network models.

We report below the code to generate the projection matrices $\mathbf{R}_{\mathrm{p}}$ and $\mathbf{R}_{\mathrm{q}}$.[4]

```
def RQRP(N=10, dtype=np.float32):
    """
     Parameters
     ----------
     N : TYPE, dimension of R (2 times the number of bodies n)
         DESCRIPTION. The default is 10.

     The default precision for the output matrices is np.float32

     Returns
     -------
     The projection matrices RQ and RP, and the matrix J

    """
    assert n%2 ==0,\
         " Dimension must be even "
    n = np.floor_divide(N, 2)
    RQ = np.zeros((N,n), dtype = dtype)
    RP = np.zeros((N,n), dtype = dtype)
    c=-1
    d=-1
    for j in range(N):
        if j%2==0:
```

[4] The extended version of the function is in `phasespace.py`.

```
            c=c+1
            RQ[j,c]=1.0
        else:
            d=d+1
            RP[j,d]=1.0
    J = np.matmul(RQ, RP.transpose())-\
        np.matmul(RP, RQ.transpose())
    return RQ, RP, J
```

7.10 Generating a Symplectic Matrix for a Random Gate

As a first example, we apply Theorem 7.2 to model a random energy-conserving medium as a neural network, or—equivalently—a random unitary gate or interferometer.

For representing a random gate by a neural network layer, we need to generate a symplectic operator starting from real matrices.

The first goal is to generate a random unitary matrix $\mathbf{U} = \mathbf{U}_R + \mathrm{i}\mathbf{U}_I$. We use the fact that a unitary matrix is found by exponentiation of an Hermitian matrix $\mathbf{H} = \mathbf{H}^\dagger$

$$\mathbf{U} = \exp(\mathrm{i}\mathbf{H}) = \exp(\mathrm{i}\mathbf{H}_R - \mathbf{H}_I) \tag{7.180}$$

where $\mathbf{H} = \mathbf{H}_R + \mathrm{i}\mathbf{H}_I$ has a symmetric real part $\mathbf{H}_R = \mathbf{H}_R^\top$ and antisymmetric imaginary part $\mathbf{H}_I = -\mathbf{H}_I^\top$. We build $\mathbf{H}_R$ and $\mathbf{H}_I$ by two independent random matrices $\mathbf{W}_R$ and $\mathbf{W}_I$ and letting

$$\begin{aligned} \mathbf{H}_R &= \mathbf{W}_R + \mathbf{W}_R^\top \\ \mathbf{H}_I &= \mathbf{W}_I - \mathbf{W}_I^\top \end{aligned} \tag{7.181}$$

After exponentiation as in (7.180), taking the real and imaginary parts, we have the matrices $\mathbf{U}_R$ and $\mathbf{U}_I$, real and imaginary parts of a unitary matrix $\mathbf{U}$.

With $\mathbf{U}_R$ and $\mathbf{U}_I$, we build $\mathbf{M}$ as follows (Eq. 7.168):

$$\mathbf{M} = \mathbf{R}_\mathrm{q}\mathbf{U}_R\mathbf{R}_\mathrm{q}^\top + \mathbf{R}_\mathrm{p}\mathbf{U}_R\mathbf{R}_\mathrm{p}^\top + \mathbf{R}_\mathrm{p}\mathbf{U}_I\mathbf{R}_\mathrm{q}^\top - \mathbf{R}_\mathrm{q}\mathbf{U}_I\mathbf{R}_\mathrm{p}^\top \tag{7.182}$$

This may be realized by the following `python` function. [5]

```
def RandomSymplectic(N=10):
    """ Return a Random Square Symplectic Matrix_np M
```

[5] The extended version of the function is in `phasespace.py`.

```
    and its inverse MI with dimension N
    generated by a Random Complex Unitary U_np with dimension N/2

    here M is symplectic with respect to J

    param: N size of output
    """
    n = np.floor_divide(N, 2) #n = N/2
    wr = np.random.random((n, n))
    wi = np.random.random((n, n))
    # generate symmetric matrix
    HR = wr+wr.transpose()
    # generate an antisymmetric matrix
    HI = wi-wi.transpose()
    # exponentiate the Hermitian matrix
    # times the imaginary unit to have a unitary operator
    U = expm(-HI+1j*HR)
    # return the real and imaginary part
    UR = np.real(U)
    UI = np.imag(U)
    RQ, RP, J = RQRP(n)
    M = np.matmul(RQ, np.matmul(UR, RQ.transpose()))+ \
        np.matmul(RP, np.matmul(UR, RP.transpose()))- \
        np.matmul(RQ, np.matmul(UI, RP.transpose()))+ \
        np.matmul(RP, np.matmul(UI, RQ.transpose()))
    # also return the symplectic inverse
    MI = np.matmul(J.transpose(), np.matmul(M.transpose(), J))
    return M, MI
```

In different applications, we need to train unitary operators with unknown response functions. Starting with a random gate is a strategy. The matrices $\mathbf{W}_R$ and $\mathbf{W}_I$ are trainable weights, but they are redundant as we only use their symmetric and antisymmetric parts. A rigorous method to generate random unitary matrices can be found in [5].

7.11 Further Reading

- Phase space methods in quantum mechanics: The starting point is [2]. A detailed analysis with many formulas useful for Gaussian states is [6].
- Gaussian states: Gaussian states are considered in many review papers as in [1]. A detailed analysis of the different notations for Gaussian states is in [4].
- Linear transformation and phase space methods in quantum mechanics: High level introduction are in [2] and [6].
- Linear transformations and Gaussian states: Linear transformations of Gaussian states are considered in [1] and detailed in [4].

References

1. X. Wang, T. Hiroshima, A. Tomita, M. Hayashi, Phys. Rep. **448**, 1 (2007). https://doi.org/10.1016/j.physrep.2007.04.005
2. S.M. Barnett, P.M. Radmore, *Methods in Theoretical Quantum Optics* (Oxford University Press, New York, 1997)
3. R. Simon, N. Mukunda, B. Dutta, Phys. Rev. A **49**, 1567 (1994). https://doi.org/10.1103/PhysRevA.49.1567. https://link.aps.org/doi/10.1103/PhysRevA.49.1567
4. A. Ferraro, S. Olivares, M.G.A. Paris, arXiv:0503237 (2005)
5. K. Zyczkowski, M. Kus, J. Phys. A: Math. Gen. **27**, 4235 (1994)
6. C.W. Gardiner, P. Zoller, *Quantum Noise*, 3rd edn. (Springer, Berlin, 2004)

Chapter 8
States as a Neural Networks and Gates as Pullbacks

You're Lower Than Dirt. You're An Ant!
(A Bug's Life)

Abstract We represent the characteristic function of a state as a neural network. We detail the case of Gaussian states. We deepen linear transformations and show how to map cascaded gates in neural network models. We introduce the bug train representation and the concept of "pullback" for concatenating transformations. We present layered models in the phase space.

8.1 Introduction

Neural networks are powerful mathematical tools to represent unknown functions of many variables; they depend on weights that we fine-tune to fit a target model. Here, we use neural networks to represent the characteristic function $\chi(\boldsymbol{x})$ of a many-body quantum state. We consider the simplest case of Gaussian states.

A Gaussian density matrix is a "multi-head model" that is a NN with a computational backbone processing the inputs and multiple output layers returning different quantities. We start considering as "heads" (the NN outputs) the real and imaginary parts χ_R and χ_I of χ.

8.2 The Simplest Neural Network for a Gaussian State

Figure 8.1 shows the NN model for a Gaussian state with one input $\boldsymbol{x}$ and two outputs (the "heads") χ_R and χ_I. The input is the row vector $\boldsymbol{x}$ and seeds the input layer. The outputs χ_R and χ_I arise from the output layers.

The inner layer denoted "Gaussian state" computes the Gaussian characteristic function. The parameters of this layer, i.e., the covariance matrix $\boldsymbol{g}$ and the displacement $\boldsymbol{d}$, are indicated.

C. Conti, *Quantum Machine Learning*, Quantum Science and Technology,
https://doi.org/10.1007/978-3-031-44226-1_8

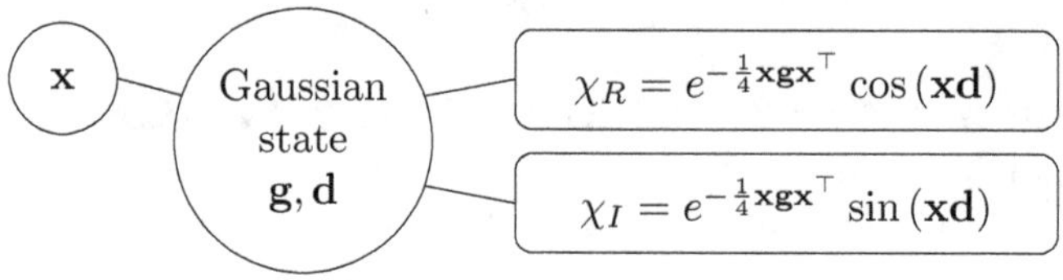

Fig. 8.1 Double-head model for a layer representing a Gaussian state

For a Gaussian state with covariance matrix **g** and displacement vector **d**, we have

$$\chi(\boldsymbol{x}) = \chi_R(\boldsymbol{x}) + \mathrm{i}\chi_I(\boldsymbol{x}) \tag{8.1}$$

with the real and imaginary part

$$\begin{aligned} \chi_R(\boldsymbol{x}) &= \exp\left(-\frac{1}{4}\boldsymbol{x}\boldsymbol{g}\boldsymbol{x}^\top\right)\cos\left(\boldsymbol{x}\boldsymbol{d}\right) , \\ \chi_I(\boldsymbol{x}) &= \exp\left(-\frac{1}{4}\boldsymbol{x}\boldsymbol{g}\boldsymbol{x}^\top\right)\sin\left(\boldsymbol{x}\boldsymbol{d}\right) . \end{aligned} \tag{8.2}$$

For reasons that will be clear in the following, it is convenient to consider a more general model that includes a further input vector $\boldsymbol{a}$.

8.3 Gaussian Neural Network with Bias Input

We generalize the Gaussian layer to include a further bias input **a** with the same dimensions of $\boldsymbol{d}$, in addition to $\boldsymbol{x}$, such that the output is

$$\chi(x) = e^{-\frac{1}{4}x\mathbf{g}x^T} e^{\mathrm{i}x(\boldsymbol{d}+\boldsymbol{a})}, \tag{8.3}$$

Here $\boldsymbol{a}$ is a bias in the displacement, which will be useful for cascading layers as detailed in the following chapters.

Equation (8.3) corresponds to the two heads of the model returning

$$\chi_R(\boldsymbol{x}) = \exp\left(-\frac{1}{4}\boldsymbol{x}\,\boldsymbol{g}\,\boldsymbol{x}^T\right)\cos\left[\boldsymbol{x}\,(\boldsymbol{d}+\boldsymbol{a})\right] , \tag{8.4}$$

$$\chi_I(\boldsymbol{x}) = \exp\left(-\frac{1}{4}\boldsymbol{x}\,\boldsymbol{g}\,\boldsymbol{x}^T\right)\sin\left[\boldsymbol{x}\,(\boldsymbol{d}+\boldsymbol{a})\right] . \tag{8.5}$$

Figure 8.2 shows a graphical representation of the generalized Gaussian NN with bias.

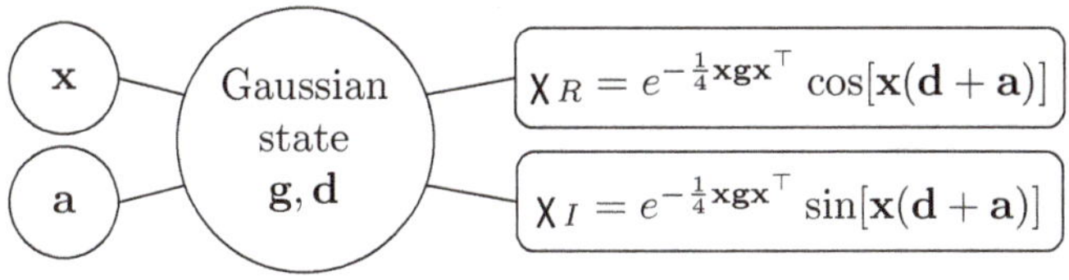

Fig. 8.2 (a) Double-head model for a layer representing a Gaussian with bias input

The code for the Gaussian state follows[1].

```
class GaussianLayer(layers.Layer):

    # define an init function
    def __init__(self, g_in, d_in, **kwargs):
            super(Gaussian, self).__init__(**kwargs)
            self.g = tf.Variable(g_in)
            self.d = tf.Variable(d_in)
            self.DotLayer = tf.keras.layers.Dot(axes=1)

    def call(self, x, ain=None):
        # [chir, chii] = self.call(x,a)
        # [chir, chii] = self.call(x)
        # param: x the input row vector (Nbatch, n)
        # param: ain the input bias column vector (n,1),
        # ain can be absent, if absent a is zero
        # output: chir, real part characteristic function
        # output: chii, imaginary part
        if ain is None:
            a = tf.constant(np.zeros((self.n, 1)),
                dtype=self.dtype)
        else:
            a = ain
        yi = tf.matmul(x, self.d+a)
        y = tf.matmul(x, self.g, transpose_b=True)
        # note the transpose above, such that  y = x g^T,
        # with y shape (Nbatch, n)
        # this transpose is for DotLayer below
        # The DotLayer below evaluates
        # x y^T= x g x^T for each input in the batch
        # (for understanding think to the case Nbatch=1)
        gDot = self.DotLayer([x, y])
        # As an alternative, one could do
        # y = tf.matmul(self.g, x)
        # and then
        # gDot = self.DotLayer([x, tf.transpose(g)])
        yr = tf.multiply(-0.25, gDot) # scale by -1/4
        return [tf.exp(yr)*tf.cos(yi), tf.exp(yr)*tf.sin(yi)]
```

[1] It is defined as class in the library phasespace.py.

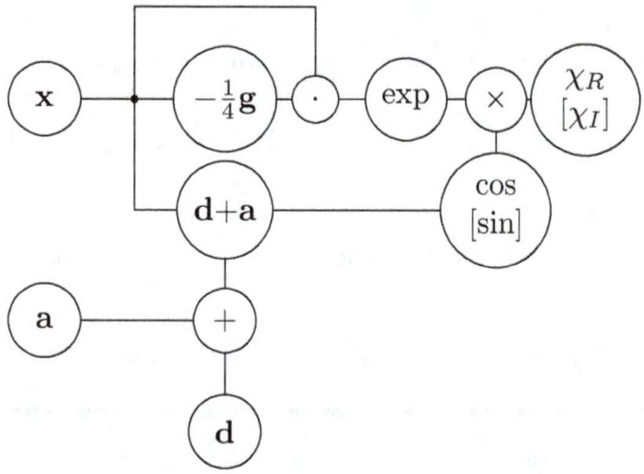

Fig. 8.3 Residual neural network model for the Gaussian layer with weights $\boldsymbol{g}$ and $\boldsymbol{d}$, with input $\boldsymbol{x}$ and bias $\boldsymbol{a}$. The cos (sin) function corresponds to χ_R (χ_I)

We define a new layer that takes as parameters the covariance matrix $\boldsymbol{g}$ and the displacement vector $\boldsymbol{d}$ stored in tensors.

The implementation is a sort of residual neural network: the input $\boldsymbol{x}$ is mixed in a scalar product with the vector obtained by the same $\boldsymbol{x}$ and the weights $\boldsymbol{g}$. Figure 8.3 shows the Gaussian layer call function.

In general, the input to the network has dimension $N_{\text{batch}} \times N$, corresponding to a shape (`Nbatch`, `N`) with N_{batch} simultaneous vectors $\boldsymbol{x}$ run in the NN. Thus, the NN must return N_{batch} output values at each head.

The important ingredient is the `DotLayer`, which takes as input a matrix $\boldsymbol{x}$ with shape $N_{\text{batch}} \times N$ and a vector $\boldsymbol{y}$ with shape $N_{\text{batch}} \times N$ and returns a vector with shape (N_{batch}, 1) containing the N_{batch} values of the scalar products of each column in $\boldsymbol{x}$ with each column in $\boldsymbol{y}$. This `DotLayer` is used in the `call` method of the `ResidualGaussianMultiHead` layer.

We test the layer by the following `Python` code. First, we create the layer.

```
import numpy as np
import phasespace as ps
N=20 # dimension
g=np.eye(np.eye(N)) # set a unitary covariance matrix
d=np.ones((N,1)) # set a unitary displacement vector
G=ps.GaussianLayer(g,d) #create the layer
```

Then we create a batch input as `Nbatch` random vectors of dimensions `N` stored in the two-dimensional array `xdata`, and run the layer, which returns `Nbatch` values.

```
Nbatch=5 # batch size
xdata=np.random.rand(Nbatch,N) # dataset
chir, chii=G(xdata) # run the layer, notice it has 2 outputs
#print the two output tensors
print(chir,chir)
# output :
 tf.Tensor(
  [[-0.24564974]
  [-0.22113787]
  [-0.21141261]
  [-0.08494049]
  [ 0.10695022]], shape=(5, 1), dtype=float32)
tf.Tensor(
  [[-0.00964258]
  [-0.02237024]
  [ 0.10472579]
  [ 0.3030984 ]
  [-0.05022743]], shape=(5, 1), dtype=float32)
```

8.4 The Vacuum Layer

As the vacuum state is a Gaussian state with unitary covariance matrix $\boldsymbol{g}$ and zero displacement $\boldsymbol{d} = \mathbf{0}$, we use the general `GaussianLayer` class to realize a function that returns the special class layer representing a many-body vacuum.

The code is as follows[2]

```
def VacuumLayer(N):
    """ Return a non-trainable GaussianLayer
        with unitary covariance matrix and no displacement
        corresponding to vacuum
    """
    return GaussianLayer(np.eye(N), np.zeros((N,1)),
                                trainable=False)
```

We test the `VacuumLayer` as follows[3].

```
N=20; # define the size
vacuum=VacuumLayer(N); # create the model
# run the layer, notice it has 2 outputs
chir_vacuum, chii_vacuum=vacuum(xdata)
```

[2] The complete code is in the package `phasespace.py`.

[3] See `jupyternotebooks/phasespace/testGaussianLayer.ipynb`.

```
#print the two output tensors
print(chir_vacuum,chir_vacuum)
#output :
tf.Tensor(
[[0.26409665]
 [0.09018651]
 [0.12887916]
 [0.17496228]
 [0.2342092 ]], shape=(5, 1), dtype=float32) tf.Tensor(
[[0.]
 [0.]
 [0.]
 [0.]
 [0.]], shape=(5, 1), dtype=float32)
```

Notice that the χ_I is zero, as expected as we have $\boldsymbol{d} = \boldsymbol{0}$.

8.5 Building a Model and the "Bug Train"

The previous examples use a single layer, and they do not use a `model`, which is an object formed by many layers in `TensorFlow` enabling training and other features.

We give an example of a simple model formed by an input and the vacuum. `model` We use the `VacuumLayer` to build more complex states by repeated transformations.

The code for creating a model representing vacuum is as follows.[4]

```
import tensoflow.keras as tfk
import phasespace as ps
N=20; # define the size
vacuum=VacuumLayer(N); # create the model
x = tfk.layers.Input(N); # this is the input layer
chir, chir = vacuum(x) # connect the layers
#create the model
vacuum_model = tfk.Model(inputs=x, outputs=[chir,chir])
# run the model and print output
print(vacuum_model(xdata))
[<tf.Tensor: shape=(5, 1), dtype=float32, numpy=
array([[0.26409665],
       [0.09018651],
       [0.12887916],
       [0.17496228],
       [0.2342092 ]], dtype=float32)>,
```

[4] See `jupyternotebooks/phasespace/testGaussianLayer.ipynb`.

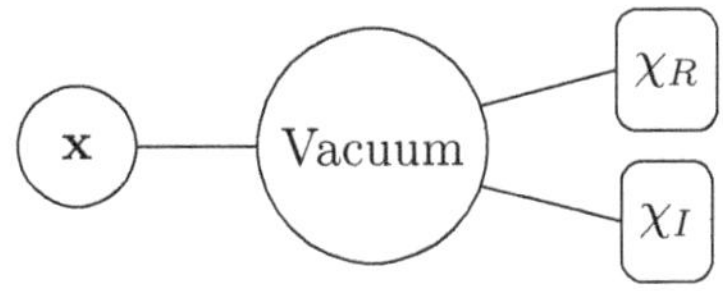

Fig. 8.4 The model formed by an `Input` layer, the `VacuumLayer`, and two output heads returning χ_R and χ_I

```
<tf.Tensor: shape=(5, 1), dtype=float32, numpy=
array([[0.],
       [0.],
       [0.],
       [0.],
       [0.]], dtype=float32)>]
```

We use two layers: the `VacuumLayer` and an `Input` layer. The latter is a standard layer of `tensorflow.keras.`. First, we create the layers, and then we connect them. Defining the model by the `tf.keras.Model` is needed for later use.

We represent the model as in Fig. 8.4; note that one has one input and two output heads. We will refer to these representations of models as "bug trains."

8.6 Pullback

To build arbitrary states or to emulate a quantum circuit, we organize the neural network representation of the density matrix as a layered sequence of gates. Each gate is a unitary operator that acts on the density matrix and transforms the latter into a new state.

To detail how it works, we start considering the transformation of the characteristic function under unitary operations.

We consider a linear transformation by the operator $\hat{U}$. The operators and the density matrix change as follows:

$$\hat{\tilde{\boldsymbol{a}}} = \hat{U}^\dagger \hat{\boldsymbol{a}} \hat{U} \ , \tag{8.6}$$

$$\tilde{\rho} = \hat{U} \rho \, \hat{U}^\dagger \ . \tag{8.7}$$

Correspondingly, for the transformed characteristic functions $\tilde{\chi}$, we have

$$\begin{aligned} \tilde{\chi} &= \mathrm{Tr}\left[\hat{\tilde{\rho}} e^{\mathrm{i}x\hat{\boldsymbol{R}}}\right] = \mathrm{Tr}\left[\hat{U}\rho\hat{U}^\dagger e^{\mathrm{i}x\hat{\boldsymbol{R}}}\right] = \mathrm{Tr}\left[\rho\hat{U}^\dagger e^{\mathrm{i}x\hat{\boldsymbol{R}}}\hat{U}\right] \\ &= \mathrm{Tr}\left[\rho e^{\mathrm{i}x\hat{U}^\dagger\hat{\boldsymbol{R}}\hat{U}}\right] = \mathrm{Tr}\left[\rho e^{\mathrm{i}x\hat{\tilde{\boldsymbol{R}}}}\right]. \end{aligned} \tag{8.8}$$

Recalling Eq. (7.140)

$$\hat{\tilde{\boldsymbol{R}}} = \mathbf{M}\boldsymbol{R} + \boldsymbol{d}', \tag{8.9}$$

we remark that the linear transformation is determined by the symplectic matrix $\mathbf{M}$ and the displacement $\boldsymbol{d}'$. We write $\tilde{\chi}$ in terms of $\mathbf{M}$ and $\boldsymbol{d}'$, by using Eq. (8.9):

$$\tilde{\chi} = \mathrm{Tr}\left[\rho e^{\mathrm{i}\boldsymbol{x}\mathbf{M}\hat{\boldsymbol{R}}}\right] e^{\mathrm{i}\boldsymbol{x}\boldsymbol{d}'}. \tag{8.10}$$

On the other hand, the expression for $\chi(\boldsymbol{x})$ is the following:

$$\chi = \mathrm{Tr}\left[\rho e^{\mathrm{i}\boldsymbol{x}\hat{\boldsymbol{R}}}\right]. \tag{8.11}$$

Hence, we have

$$\tilde{\chi}(\boldsymbol{x}) = \chi(\boldsymbol{x}\mathbf{M})e^{\mathrm{i}\boldsymbol{x}\boldsymbol{d}'} = \chi(\boldsymbol{x}\mathbf{M})e^{\mathrm{i}(\boldsymbol{x}\mathbf{M})(\mathbf{M}^{-1}\boldsymbol{d}')}. \tag{8.12}$$

From Eq. (8.12), we see that the modified characteristic function depends on the modified input vector $\boldsymbol{x}\mathbf{M}$ and has a modified displacement $\mathbf{M}^{-1}\boldsymbol{d}'$.

Introducing the new input vector $\boldsymbol{y} = \boldsymbol{x}\mathbf{M}$ and the bias $\boldsymbol{a} = \mathbf{M}^{-1}\boldsymbol{d}'$, we express the transformed characteristic function $\tilde{\chi}$ in terms of the original χ as follows:

$$\tilde{\chi}(\boldsymbol{x}) = \chi(\boldsymbol{y})e^{\mathrm{i}\boldsymbol{y}\boldsymbol{a}}. \tag{8.13}$$

8.7 The Pullback Layer

We give a graphical representation of the transformations in Eq. (8.13). We start considering a characteristic function with input vector $\boldsymbol{x}$ and multi-headed outputs χ_R and χ_I, as in Fig. 8.5.

We generalize the model by introducing the bias vector, $\boldsymbol{a}$, shown in Fig. 8.6. Figure 8.7 represents the transformation in Eq. (8.13).

We introduce a new layer, which we call "linear layer," having as parameters the symplectic matrix $\mathbf{M}$ and the displacement $\boldsymbol{d}'$. The linear layer is shown in Fig. 8.8, and it has two inputs $\boldsymbol{x}$ and $\boldsymbol{a}$, and two outputs: $\boldsymbol{y} = \boldsymbol{x}\mathbf{M}$, a row vector with the same size of $\boldsymbol{x}$, and a new displacement $\boldsymbol{b} = \mathbf{M}^{-1}(\boldsymbol{d}' + \boldsymbol{a})$ a column vector with the size of $\boldsymbol{d}$.

The linear layer enables representing different linear transformations. The bias vector is needed to cascade multiple gates, as detailed below.

By the linear layer, the transformation is a concatenation of two layers as in Fig. 8.9. We describe the model as the pullback of the linear layer from the

Fig. 8.5 Graphical representation of the model for the characteristic function

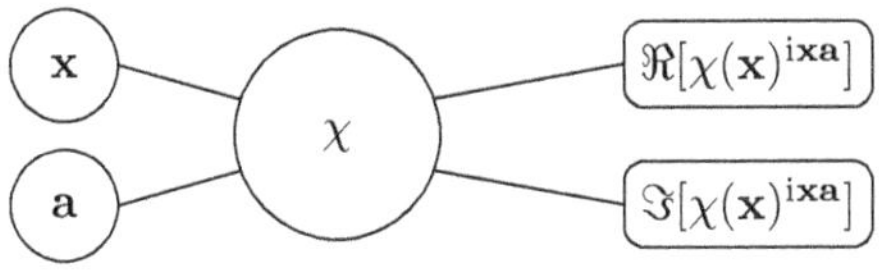

Fig. 8.6 Graphical representation of the generalized model for the characteristic function with bias $\boldsymbol{a}$, which is adopted in the linear transformations

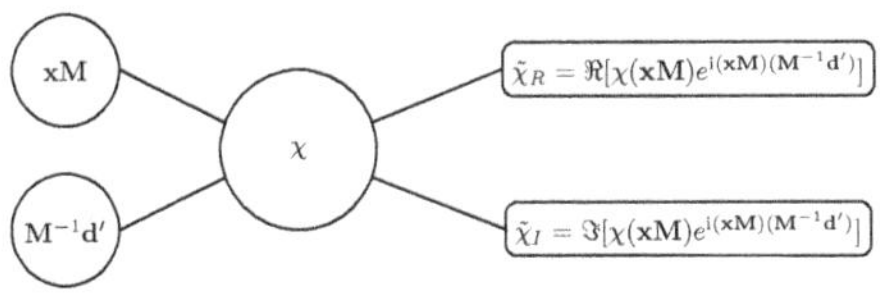

Fig. 8.7 Graphical representation of the transformed characteristic function in terms of the original characteristic function with modified input and bias

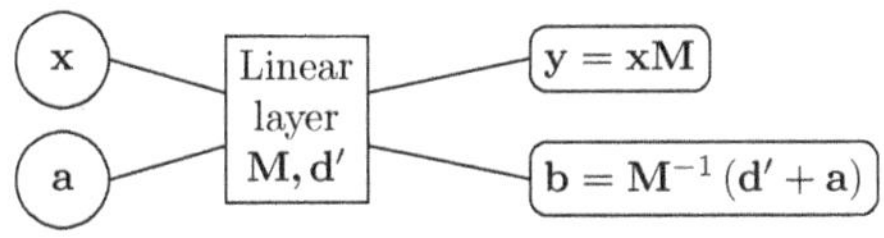

Fig. 8.8 Linear layer to transform the input variables to the χ layer

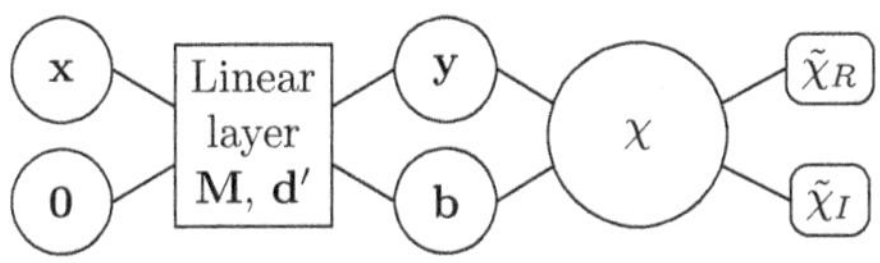

Fig. 8.9 Graphical representation of a linear transformation as a pullback of a linear layer from the original characteristic function. The resulting model corresponds to the transformed characteristic function in Fig. 8.7. Note that here $\boldsymbol{a} = \mathbf{0}$

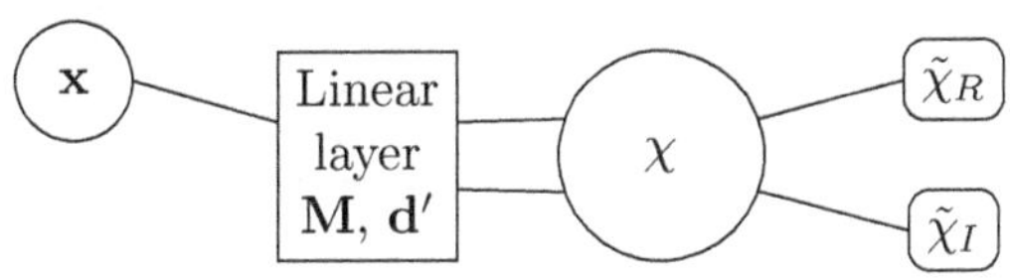

Fig. 8.10 Simplified model of Fig. 8.9, omitting the internal variables $\boldsymbol{y}$ and $\boldsymbol{b}$ and the zero input bias $\boldsymbol{a}$

characteristic function layer. The matrix $\mathbf{M}$ first acts on the input $\boldsymbol{x}$, and then the function χ is evaluated on the vector $\boldsymbol{y}$, plus all the other decorators due to the displacement. A simplified diagram of the mode is given in Fig. 8.10. This bug train is useful when considering multiple transformations on states.

8.8 Pullback of Gaussian States

For a Gaussian state [see Eq. (7.37)], the transformed characteristic function reads

$$\tilde{\chi}(\boldsymbol{x}) = e^{-\frac{1}{4}\boldsymbol{x}\mathbf{M}\boldsymbol{g}\mathbf{M}^\top \boldsymbol{x}^\top} e^{\mathrm{i}\boldsymbol{x}\mathbf{M}(\boldsymbol{d}+\mathbf{M}^{-1}\boldsymbol{d}')}. \tag{8.14}$$

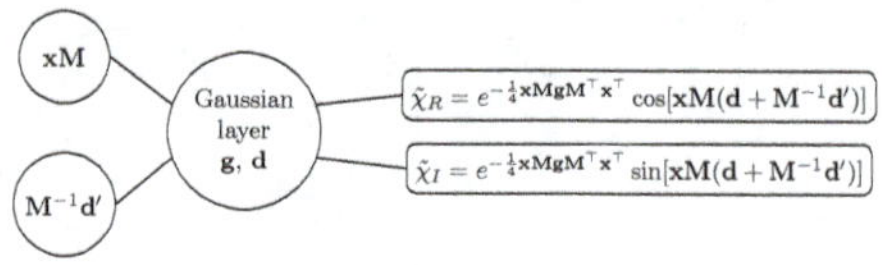

Fig. 8.11 Single-layer multi-head model for a linear transformation of a Gaussian state

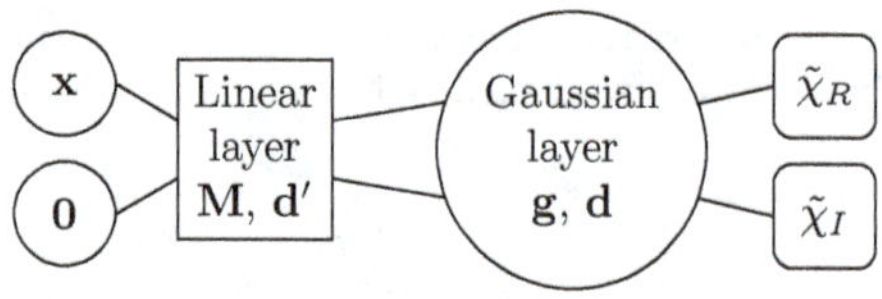

Fig. 8.12 Two-layer multi-head model equivalent to Fig. 8.11

Equation (8.14) is graphically represented in Fig. 8.11, by using the Gaussian layer in Fig. 8.2. This model is equivalent to Fig. 8.12 with the linear layer in Fig. 8.9.

As shown in Fig. 8.12, a linear transformation on the density matrix is a change of the variable $\boldsymbol{x}$. The linear transformation is a linear gate followed by the Gaussian gate, which is "pulling back" the linear operation from the Gaussian gate.

By using these two models, the linear transformation of the Gaussian state is a cascade of a linear pullback layer and a Gaussian state layer, as shown in Fig. 8.12.

8.9 Coding the Linear Layer

We implement the linear layer as a `Python` class. We define a `Keras` layer that takes as parameters the matrix $\mathbf{M}$, which must be symplectic (see Chap. 7.8) and the displacement $\boldsymbol{d}'$. We obtain the inverse of $\mathbf{M}$ by the properties of the symplectic matrices, i.e., $\mathbf{M}^{-1} = \mathbf{J}\,\mathbf{M}^\top\,\mathbf{J}^{-1} = \mathbf{J}\,\mathbf{M}^\top\,\mathbf{J}^\top$

We define a non-trainable linear layer, where $\mathbf{M}$ and $\boldsymbol{d}'$ are corresponding to `TensorFlow` constant tensors `M` and `d1`, with `IM` the inverse $\mathbf{M}^{-1}$ [5]

```
class LinearLayerConstant(layers.Layer):
""" Class for linear layer with constant parameters

    Examples
    --------
    L=(M,d1) # create linear layer with np matrix M and vector d1
    y, b = L(x) # call the layer with zero bias
    y, b = L(x,a) # call the layer with a bias vector
"""

    def __init__(self, M_np, d1_np, **kwargs):
        super(LinearLayerConstant, self).__init__(**kwargs)
        N,_ = M_np.shape
```

[5] The complete version of the function is in `phasespace.py`.

```
        self.N = N
        # Create the inverse symplectic matrix
        _, _, J = RXRP(n) # call to external function which
         gives J
        MI_np = np.matmul(J, np.matmul(M_np.transpose(),
         J.transpose()))

        self.M = tf.constant(M_np, dtype=self.dtype)
        self.MI = tf.constant(MI_np, dtype=self.dtype)
        self.d1 = tf.constant(d1_np, dtype=self.dtype)

    def call(self, x, a=None):
        y = tf.matmul(x,self.M)
        # check if a is in input
        if a is None:
            b = tf.matmul(self.MI, self.d1)
        else:
            b = tf.matmul(self.MI, self.d1+a)
        return [y, b]
```

The layer `LinearLayerConstant` is not trainable, as the parameters `self.M` (corresponding to $\mathbf{M}$) and `self.d1` (corresponding to $\boldsymbol{d}'$) are constant. To build complex states, we need to cascade multiple transformations, as done in quantum processors.

8.10 Pullback Cascading

The pullback is helpful with multiple transformations. A sequence of linear gates is equivalent to a sequence of pullbacks in reverse order as sketched in Fig. 8.13.

For example, we consider a system with the density matrix ρ and canonical observables $\hat{\boldsymbol{R}}$. First, the system is subject to a linear transformation with operator $\hat{U}_1$, such that the density matrix becomes

$$\hat{U}_1\rho\hat{U}_1^\dagger \tag{8.15}$$

and the new canonical vector is

$$\hat{R}_1 = \mathbf{M}_1\hat{R} + \boldsymbol{d}_1. \tag{8.16}$$

Then we consider a second transformation with unitary operator $\hat{U}_2$ and parameters $\mathbf{M}_2$ and $\boldsymbol{d}_2$, such that the final density matrix reads

$$\hat{U}_2\hat{U}_1\rho\hat{U}_1^\dagger\hat{U}_2^\dagger, \tag{8.17}$$

and the final vector of observables is

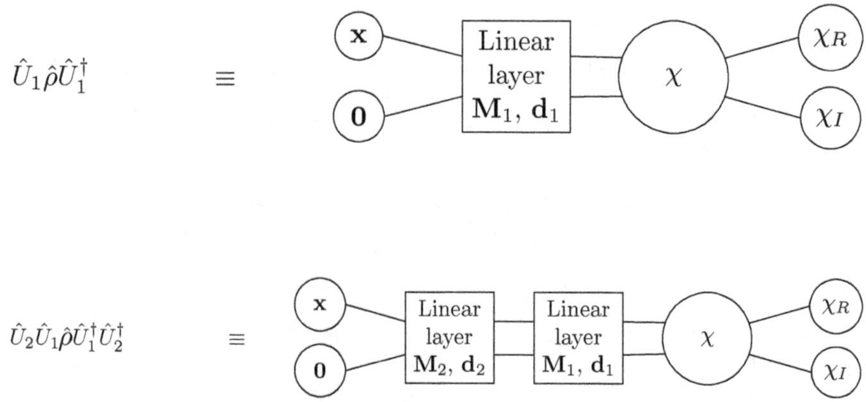

Fig. 8.13 Linear transformations and pullbacks. A single transformation with unitary operator $\hat{U}_1$ is a single pullback. A double transformation with unitary operators $\hat{U}_1$ and $\hat{U}_2$ is a double pullback. Note reversed order of the flow of data from $\boldsymbol{x}$ to the output through the network

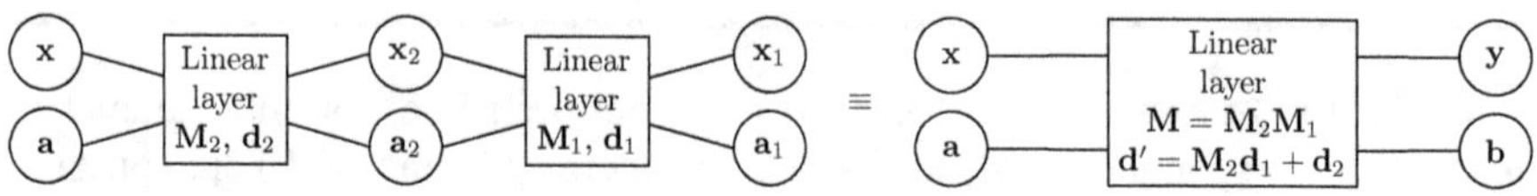

Fig. 8.14 Equivalence of two cascaded linear layers with a single linear layer

$$\hat{\boldsymbol{R}}_2 = \mathbf{M}_2\hat{\boldsymbol{R}}_1 + \boldsymbol{d}_2 = \mathbf{M}_2\mathbf{M}_1\hat{\boldsymbol{R}} + \boldsymbol{d}_2 + \mathbf{M}_2\boldsymbol{d}_1. \tag{8.18}$$

We remark that $\mathbf{M}_1\mathbf{M}_2 \neq \mathbf{M}_2\mathbf{M}_1$; hence, the order of the two transformations is relevant as the unitary operators $\hat{U}_1$ and $\hat{U}_2$ do not commute.

The sequence of the two linear transformations is equivalent to a single transformation with parameters

$$\begin{aligned} \mathbf{M} &= \mathbf{M}_2\mathbf{M}_1\ , \\ \boldsymbol{d}' &= \mathbf{M}_2\boldsymbol{d}_1 + \boldsymbol{d}_2\ . \end{aligned} \tag{8.19}$$

We have the following (Fig. 8.14).

Theorem 8.1 *The sequence of the two linear transformations $\hat{U}_1$ and $\hat{U}_2$ is equivalent to the pullback of the two linear layers with parameters* $(\mathbf{M}_1, \boldsymbol{d}_1)$ *and* $(\mathbf{M}_2, \boldsymbol{d}_2)$.

Proof As indicated in Fig. 8.14, the output of the linear layer corresponding to $\hat{U}_2$ is

$$\begin{aligned} \boldsymbol{x}_2 &= \boldsymbol{x}\mathbf{M}_2 \\ \boldsymbol{a}_2 &= \mathbf{M}_2^{-1}\left(\boldsymbol{d}_2 + \boldsymbol{a}\right) \end{aligned} \tag{8.20}$$

The output of the linear layer corresponding to $\hat{U}_1$ is

$$\begin{aligned} \boldsymbol{x}_1 &= \boldsymbol{x}_2\mathbf{M}_1 = \boldsymbol{x}\mathbf{M}_2\mathbf{M}_1 \\ \boldsymbol{a}_1 &= \mathbf{M}_1^{-1}\left(\boldsymbol{d}_1 + \boldsymbol{a}_2\right) = \mathbf{M}_1^{-1}\left(\boldsymbol{d}_1 + \mathbf{M}_2^{-1}\boldsymbol{d}_2 + \mathbf{M}_2^{-1}\boldsymbol{a}\right) \end{aligned} \tag{8.21}$$

For the layer with parameters $(\mathbf{M}, \boldsymbol{d}')$, we have (see (8.19))

$$\begin{aligned} \boldsymbol{y} &= \boldsymbol{x}\mathbf{M} = \boldsymbol{x}\mathbf{M}_2\mathbf{M}_1 \\ \boldsymbol{b} &= \mathbf{M}^{-1}\left(\boldsymbol{d}' + \boldsymbol{a}\right) \\ &= \mathbf{M}_1^{-1}\mathbf{M}_2^{-1}\left(\mathbf{M}_2\boldsymbol{d}_1 + \boldsymbol{d}_2 + \boldsymbol{a}\right) \\ &= \mathbf{M}_1^{-1}\left(\boldsymbol{d}_1 + \mathbf{M}_2^{-1}\boldsymbol{d}_2 + \mathbf{M}_2^{-1}\boldsymbol{a}\right) \end{aligned} \tag{8.22}$$

As $\boldsymbol{y} = \boldsymbol{x}_1$ and $\boldsymbol{b} = \boldsymbol{a}_1$, we have the proof. □

By extending the previous argument to three or more transformations, one realizes that the cascade of an arbitrary number of transformations is a cascade of pullbacks in reverse order.

For M transformations, reverse order means that one first makes a pullback of the operator 1, then the operator 2, etc. In the flow of data in the network, the data $\boldsymbol{x}$ first enter operator M, then $M-1$, and so forth to passing through the linear layer for the transformation 1. Figure 8.15 shows the case $M=3$ as example.

8.11 The Glauber Displacement Layer

Starting from the linear layer, we define specialized layers for significant unitary operators. The first we describe is the displacement operator, or Glauber operator, defined by

$$\hat{\mathcal{D}}(\alpha) = \exp\left(\alpha^*\hat{a} - \alpha\hat{a}^\dagger\right) . \tag{8.23}$$

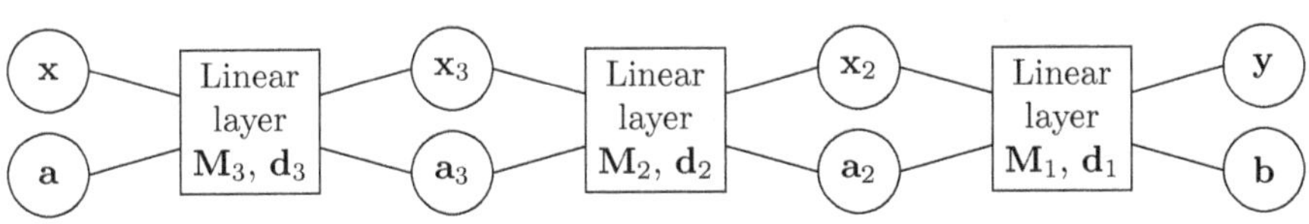

Fig. 8.15 Cascading three transformations. Note the reverse order of layer from right to left; this corresponds to a new state $|\psi\rangle$ obtained by an initial state $|\phi\rangle$ by three unitary operators, as in $|\psi\rangle = \hat{U}_3\hat{U}_2\hat{U}_1|\phi\rangle$

A coherent state is the displacement of vacuum:

$$|\alpha\rangle = \hat{\mathcal{D}}(\alpha)|0\rangle. \tag{8.24}$$

For a single mode, letting $\hat{U} = \hat{\mathcal{D}}(\alpha)$, we have [1]

$$\tilde{\hat{a}} = \hat{U}^\dagger \hat{a} \hat{U} = \hat{a} + \alpha \ , \tag{8.25}$$

which implies for the canonical vector $\boldsymbol{R}$

$$\tilde{\hat{\boldsymbol{R}}} = \hat{U}^\dagger \hat{\boldsymbol{R}} \hat{U} = \begin{pmatrix} \hat{x} \\ \hat{p} \end{pmatrix} + \begin{pmatrix} d_0 \\ d_1 \end{pmatrix} \tag{8.26}$$

with

$$\begin{aligned} d_0 &= \sqrt{2} \ \ (\alpha) \ , \\ d_1 &= \sqrt{2} \ \ (\alpha). \end{aligned} \tag{8.27}$$

For a many-body displacement $\hat{\mathcal{D}}(\boldsymbol{\alpha})$, with $\boldsymbol{\alpha} = (\alpha_0, \alpha_1, ..., \alpha_n)^\top$, we have

$$\tilde{\hat{\boldsymbol{a}}} = \hat{U}^\dagger \hat{\boldsymbol{a}} \hat{U} = \hat{\boldsymbol{a}} + \boldsymbol{\alpha}, \tag{8.28}$$

which is

$$\tilde{\hat{\boldsymbol{R}}} = \hat{U}^\dagger \hat{\boldsymbol{R}} \hat{U} = \hat{\boldsymbol{R}} + \boldsymbol{d} \tag{8.29}$$

with $(j = 0, 1, \ldots, n-1)$

$$\begin{aligned} d_{2j} &= \sqrt{2} \ \ (\alpha_j) \\ d_{2j+1} &= \sqrt{2} \ \ (\alpha_j) \end{aligned} \tag{8.30}$$

We remark that we have $\mathbf{M} = \mathbf{1}_N$ for the Glauber layer.

8.12 A Neural Network Representation of a Coherent State

To create a neural network model of a coherent state with a given displacement vector $\boldsymbol{d}$, one can start from the vacuum Gaussian state with $\boldsymbol{g} = \mathbf{1}_N$ and $\boldsymbol{d} = \mathbf{0}$ and pull back a linear gate with $\mathbf{M} = \mathbf{1}_N$ and displacement $\boldsymbol{d}$. No bias is needed ($\boldsymbol{a} = \mathbf{0}$).

We use the class `LinearLayerConstant` to define a function that returns a constant displacement layer for an input displacement vector $\boldsymbol{d}$ with dimensions $N \times 1$, corresponding to the `Python` array `dtarget` with shape `(N,1)`, as follows.

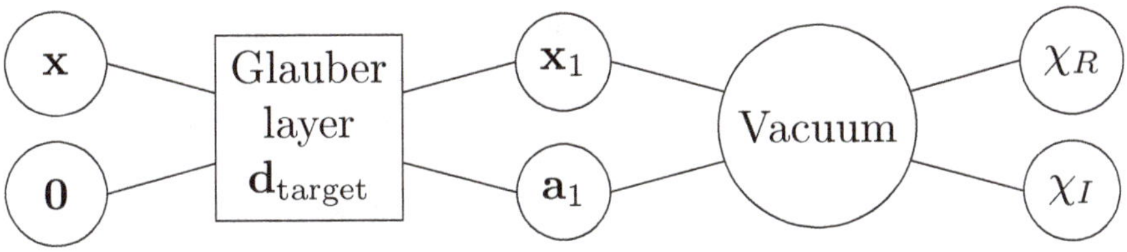

Fig. 8.16 Model for a coherent state obtained by pullback of a Glauber displacement operator (Glauber layer) from the vacuum

```
def DisplacementLayerConstant(dtarget_np):
    N = dtarget_np.shape[0] # dimension N of the input array
    D = LinearLayerConstant(np.eye(N),dtarget_np)
    return D
```

The following code gives a model for a coherent state obtained by a displacement operator applied on the vacuum layer. The model is made of three layers: input, pullback (displacement), and Gaussian layer (vacuum). Its simplified version is shown in the following[6] example with $\boldsymbol{d}_{\text{target}} = 3$ a constant vector with identical elements.

First, we create the layers, and then we connect them, following Fig. 8.16. `chir, chii` are the real and imaginary parts of the characteristic function, `x1, d1` give the internal status of the model, and `x` is the input. Defining the model by the `tf.keras.Model` is needed for later use.

```
# create a constant (non-trainable) coherent state
# Use a pullback of a displacement from the vacuum
N=20 # size
vacuum = VacuumLayer(N) # vacuum layer
dtarget = 3.0*np.ones((N,1)) # target displacement
D = DisplacementLayerConstant(dtarget) # displacement layer
x = tf.keras.layers.Input(N) # input layer
x1, a1 = D(xin) # connect the layers
chir, chii = vacuum(x1, d1) # connect the layers
# define the model
pullback = tf.keras.Model(inputs = xin, outputs=[chir, chii])
```

[6] The complete code is in `jupyternotebooks/phasespace/coherent.ipynb`.

8.13 A Linear Layer for a Random Interferometer

We want to model the scattering of one mode to another in a random interferometer, such as a random optical material, a disordered array of optical waveguides, or a multi-mode fiber. Modes are coupled linearly via random coefficients, and the system is represented by a unitary matrix **U**.

As detailed in Sect. 7.10, we use the matrices $\mathbf{W}_R$ and $\mathbf{W}_I$ to realize a trainable random medium with unitary operator **U** and symplectic matrix **M**.

The code is the following[7]. Note that we use the function `RQRP(N)`, which returns the matrices $\mathbf{R}_\mathrm{p}$ and $\mathbf{R}_\mathrm{q}$ detailed in Sect. 7.9. These matrices are stored as `tf.constant` in the constructor method of the layer.

```
class RandomLayer(layers.Layer):
    def __init__(self, N=10, trainable=True, **kwargs):
        super(RandomLayer, self).__init__(**kwargs)
        assert n % 2 == 0, " Dimension must be even "
        self.trainable = trainable
        self.N = N
        n = np.floor_divide(self.N, 2) # n=N/2
        # init the random weights
        wr_np = np.random.random((n, n))
        wi_np = np.random.random((n, n))
        self.WR = tf.Variable(wr_np,trainable=self.trainable)
        self.WI = tf.Variable(wi_np,trainable=self.trainable)
        # define the constant matrices for computing M
        Rx, Rp, J = RQRP(N)
        self.Rx = tf.constant(Rx)
        self.Rp = tf.constant(Rp)
        self.J = tf.constant(J)

    def call(self, x, di=None):
        # generate symmetric matrix
        HR = self.WR+tf.transpose(self.WR)
        # generate an antisymmetric matrix
        HI = self.WI-tf.transpose(self.WI)
        # exponentiate the Hermitian matrix time the imaginary
           unit
        U = tf.linalg.expm(tf.complex(-HI, HR))
        # return the real and immaginary part
        UR = tf.math.real(U)
        UI = tf.math.imag(U)
        # Build the symplectic matrix M
        M = \
           tf.matmul(self.Rx,
                       tf.matmul(UR, self.Rx, transpose_b=True)) +\
           tf.matmul(self.Rp,
                       tf.matmul(UR, self.Rp, transpose_b=True)) -\
```

[7] The complete code is in `phasespace.py`.

```
        tf.matmul(self.Rx,
                  tf.matmul(UI, self.Rp, transpose_b=True)) +\
        tf.matmul(self.Rp,
                  tf.matmul(UI, self.Rx, transpose_b=True))
    # Inverse of M
    MI = tf.matmul(tf.matmul(M, self.J),self.J, transpose_a=
      True)
    # no traspose here for M, as M is already transpose
    y = tf.matmul(x, M)
    if di is None:
        # zero bias output with zero bias input
        b = tf.constant(np.zeros((self.N, 1), dtype=np_real),
        dtype=self.dtype)
    else:
        b = tf.matmul(MI,di)
    return [y, b]
```

`RandomLayer` is by default a trainable layer with the `tf.variable` `M` trainable. However, during training, the layer must have a symplectic matrix **M**, corresponding to `M`. We cannot use directly `M` as weight. Indeed, during training **M** cannot vary arbitrarily, but must belong to the space of symplectic matrices. To achieve this goal, we use as weights the `tf.variables` `WR` and `WI`, which are properties of the object. `WR` and `WI` are randomly initiated in the constructor by the `NumPy` matrices `wr_np` and `wi_np`. Then `M` and its inverse `MI` are computed as in Sect. 7.9, when calling the layer by the method `call`. Note that `call` includes an optional bias input `di` to be cascaded with other layers.

In case we need a constant random layer, we specify `trainable=False` when creating the object or we define a function as in the following.

```
def RandomLayerConstant(N=10, **kwargs):
    """Return a class ResidualGaussianMultihead non-trainable
    with a random unitary (a constant RandomLayer)

    Parameters
    ----------
    N: size of the vector

    Returns
    -------
    a Constant RandomLayer
    """

    return RandomLayer(N, trainable=False, **kwargs)
```

We use the `RandomLayer` to build a model representing a multimode coherent state in a random interferometer. Figure 8.17 shows the pullback of a displacement

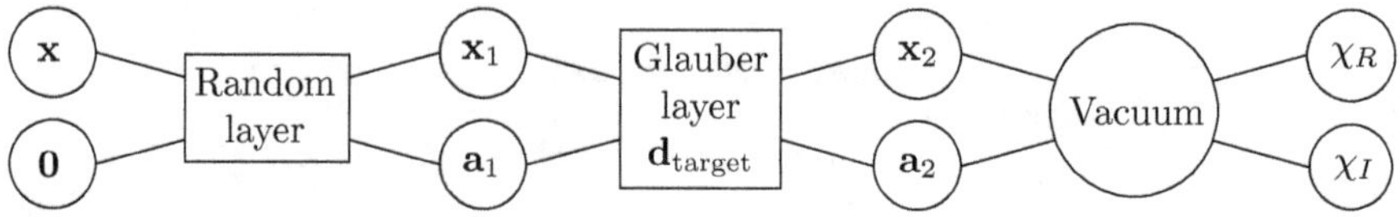

Fig. 8.17 A coherent state exiting a random interferometer obtained by pullback of a Glauber displacement operator (Glauber layer) from the vacuum, and a further pullback of a random layer

layer from the vacuum. A further pullback implements the random interferometer. The following code shows how to build the model.

```
# create a model representing the state at the output
#of a random interferometer seeded by a coherent state

#size
N=20

# create layers
vacuum = VacuumLayer(N) # vacuum layer
dtarget = 3.0*np.ones((N,1)) # target displacement
D = DisplacementLayerConstant(dtarget) # displacement layer
R = RandomLayer(N) # random interferometers
x = tf.keras.layers.Input(N) # input layer

# connect the layers in pullback order
x1, a1 = R(x)
x2, a2 = D(x1, a1) # connect the layers
chir, chii = vacuum(x2, d2) # connect the layers
# define the model
pullback = tf.keras.Model(inputs = xin, outputs=[chir, chii])
```

The resulting values for χ_R and χ_I for the tensors `chir` and `chii` give the state at the output of the random interferometer. In the next chapter, we will train the `RandomLayer` to obtain a target state.

Reference

1. S.M. Barnett, P.M. Radmore, *Methods in Theoretical Quantum Optics* (Oxford University Press, New York, 1997)

Chapter 9
Quantum Reservoir Computing

Harnessing disorder

Abstract We study models including random interferometers and trainable layers.We discuss different strategies for defining a cost function by using first and second derivatives of the characteristic function.We consider circuits with multiple interferometers and phase modulators.

9.1 Introduction

We consider examples of simple quantum reservoir computing, in which an interferometer, or a mode-mixer, is trained to perform a specific task, such as tailoring the output state.

We have to develop a feed-forward model for quantum circuits that synthesize a target state. We compute specific observables (e.g., the mean particle number) for which we need the derivatives of the characteristic function and hence of the model.

9.2 Observable Quantities as Derivatives of χ

To test if the neural network model returns the expected state, we exploit the properties of the characteristic function. We use Eq. (7.39), which is also written as

$$\langle \hat{\boldsymbol{R}} \rangle = -\mathrm{i}\nabla\chi|_{\boldsymbol{x}=0} = \nabla\chi_I|_{\boldsymbol{x}=0}\,, \tag{9.1}$$

$$\nabla\chi_R|_{\boldsymbol{x}=0} = 0\,, \tag{9.2}$$

by using the fact that $\langle \hat{\boldsymbol{R}} \rangle$ is real-valued.

C. Conti, *Quantum Machine Learning*, Quantum Science and Technology,
https://doi.org/10.1007/978-3-031-44226-1_9

The automatic differentiation `GradientTape` method enables to obtain the derivatives.[1] In the following code, we first define a model named `pullback` for a coherent state as in the previous chapter, and then we determine the derivatives of χ.

```
# define a model given by a
# pullback of a GlauberLayer from the vacuum
# the model corresponds to a coherent state
N = 10
xin = tf.keras.layers.input(N) # input layer
V = VacuumLayer(N) # vacuum layer
# Glauber layer dtarget=3.0
G = DisplacementLayerConstant(3.0*ones(N,1))
# connect the layers in the pullback order
x1, a1 = G(xin)
chir, chii = V(x1, a1)
# define a constant variable for evaluating the gradient at x=0
# note that x is a row vector as described above
x = tf.constant(np.zeros((1,N)))
# instruct the Gradient Tape for the derivative w.r.t. x
# we need persistent = True,
# as we will call two times the gradient
# when evaluating derivatives of chir and chii
with tf.GradientTape(persistent=True) as tape:
    tape.watch(x)
    chir, chii = pullback(x)
# print the derivative w.r.t. x of the real part of \chi
# return all zeros as the gradient of the real part
print(tape.gradient(chir,x))
# returns
tf.Tensor([[0. 0. 0. 0. 0. 0. 0. 0. 0. 0.]],
          shape=(1, 10), dtype=float32)
# print the derivative w.r.t. x of the imaginary part of \chi
# return the target displacement
print(tape.gradient(chii,x))
# returns
tf.Tensor([[3. 3. 3. 3. 3. 3. 3. 3. 3. 3.]],
          shape=(1, 10), dtype=float32)
# the output is the displacement vector as the mean value of R
```

We use `GradientTape` for the derivatives with respect to the input tensor `x`. We specify the derivation variable by the statement `tape.watch`, and then we report the target tensors as the output of the model that we named `pullback`, i.e., `chir, chii = pullback(x)`.

The `GradientTape` stores the derivatives in `tape.gradient`; we remark that they are computed at the value of the constant zero tensor `x`.

[1] The code is in `jupyternotebooks/phasespace/coherent.ipynb`.

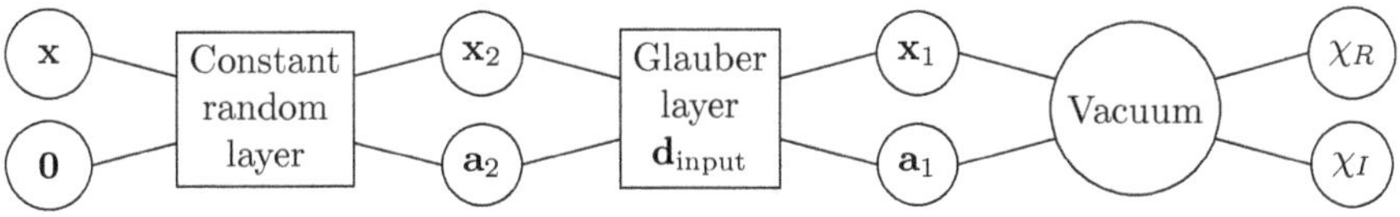

Fig. 9.1 Model for a coherent state with displacement vector $\boldsymbol{d}_{\text{input}}$ in a random interferometer. As in Fig. 8.17, the coherent state is a pullback of a displacement layer from the vacuum. The random interferometer is a pullback of a random linear layer from the Glauber layer. The input bias is not needed ($\boldsymbol{a} = \mathbf{0}$). The data in the model go from left to right. The pullback direction is from right to left and corresponds to the evolution of the state: first, a coherent state is created from the vacuum, and then the propagation through the random medium occurs. The output state displacement is $\boldsymbol{d}_{\text{output}}$

Note that we are deriving a coherent state characteristic matrix, so the output is the mean value $\langle \hat{\boldsymbol{R}} \rangle$, which has the values of displacement defined the Glauber layer. However, the shape of the displacement vector is `(N,1)`, while the shape of the mean observables is `(1, N)`.

9.3 A Coherent State in a Random Interferometer

We build a model for a coherent state with displacement $\boldsymbol{d}_{\text{input}}$ in a random interferometer with symplectic matrix $\mathbf{M}_{\text{random}}$, as shown in Fig. 9.1.[2]

The coherent state is a Gaussian layer with $\boldsymbol{g} = \mathbf{1}_N$ and displacement $\boldsymbol{d}_{\text{input}} = (3.0, 3.0, ..., 3.0)^\top$ representing a state with constant amplitude. The transmission through the random interferometer is a unitary matrix. The interferometer scrambles the modes and the result is a Gaussian state with a randomized displacement $\boldsymbol{d}_{\text{output}} = \mathbf{M}_{\text{random}} \boldsymbol{d}_{\text{input}}$.

To build the model, we cascade two linear layers as described in Sect. 8.6. We use the function `RandomLayer` to create the random interferometer.

```
N=20 # size
vacuum = VacuumLayer(N) # Vacuum Layer
dinput = 3.0*np.ones((N,1)) # input d vector
# define the linear layer for the pullback
L = DisplacementeLayerConstant(dinput) # Glauber operator
R = RandomLayer(N) # Constant Random Layer
# Build the model by connecting the layers
# define the input layer
xin = tf.keras.layers.Input(N)
# connect the layers in pullback order
```

[2] The code for this example is in `jupyternotebooks/phasespace/coherentcomplex.ipynb`.

```
x2, a2 = R(xin)
x1, a1 = L(x2,a2)
chir, chii = vacuum(x1, a1)
# build the model
pullback = tf.keras.Model(inputs = xin, outputs=[chir, chii])
```

We test the model by evaluating the derivatives of the characteristic function tensors `chir` and `chii` at $\boldsymbol{x} = 0$. The code below returns zeros for the derivatives of the real part of χ, and a random vector for the derivatives of the imaginary part χ_I, which is the displacement after the passage through the random medium.

```
# Test the model by the derivatives at x=0
x = tf.constant(np.zeros((1,N)))
# instruct the Gradient Tape for the derivative w.r.t. x
# here we need persistent = True,
# as we will call two times the gradient
# when evaluating derivatives of chir and chii
with tf.GradientTape(persistent=True) as tape:
    tape.watch(x)
    chir, chii = pullback(x)
# print the derivative w.r.t. x of the real part of \chi
# return all zeros
print(tape.gradient(chir,x))
# returns
[[0 0 0 ... 0 0 0]]
# print the derivative w.r.t. x of the imaginary part of \chi
# return the target displacement
print(tape.gradient(chii,x))
# returns (change at any run, as the interferometer is random)
[[0.407434523 -3.71409059 -1.13698542 ...
-3.94696689 1.48004842 -2.49121428]]
# check the number of bosons in output
doutput=tape.gradient(chii,x).numpy();
print(np.linalg.norm(doutput)**2/2)
# returns
44.99999656774207
# check the number of bosons in input
print(np.linalg.norm(dinput)**2/2)
# returns
45.0
# remark: identical within numerical precision
```

We check if the expected number of particles is conserved by comparing the modulus square of $\boldsymbol{d}_{\text{input}}$ and the modulus square of $\boldsymbol{d}_{\text{output}}$. They are (within a factor 1/2) the expectation value of $\hat{\boldsymbol{a}}^\dagger \cdot \hat{\boldsymbol{a}}$, as detailed in Sect. 7.4.2.

The corresponding model is shown in Fig. 9.1. We generate the coherent state by a pullback from the vacuum, and then we evolve (from right to left) to get the coherent state after the interferometer.

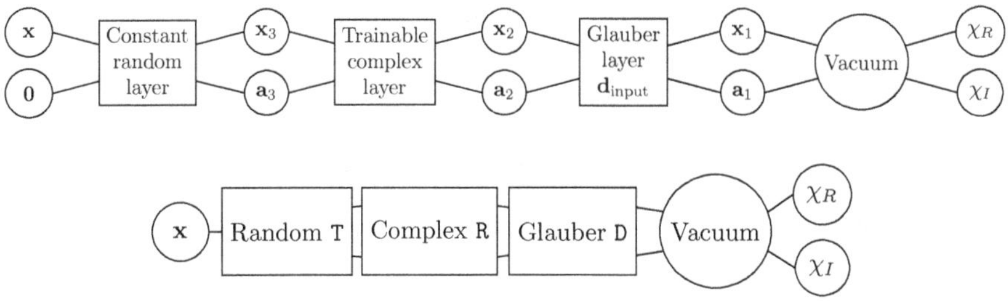

Fig. 9.2 Model for a coherent state with displacement vector $\boldsymbol{d}_{\text{input}}$ propagating in a constant random medium and a trainable interferometer. As in Fig. 9.1, the coherent state is obtained by a pullback of a displacement layer from the vacuum. The ancilla is not needed ($\boldsymbol{a} = \mathbf{0}$). The data in the model go from left to right. The pullback direction is from right to left and corresponds to the evolution of the state: first, a coherent state is created from vacuum, then the propagation through a trainable interferometer ("complex medium") and a constant random interferometer occurs, and the output state displacement is $\boldsymbol{d}_{\text{output}}$. In comparison with Fig. 9.1, we have a trainable interferometer to shape the output state after the propagation in the random medium. The bottom panel shows a simplified bug train

9.4 Training a Complex Medium for an Arbitrary Coherent State

We give an example of multiple cascading and training. We consider a trainable layer to generate a target coherent state. This example simulates the tailoring of light focusing through a random medium [1].

We consider a coherent state with displacement vector $\boldsymbol{d}_{\text{input}}$ that propagates in an interferometer. We use a trainable operator to generate a target coherent state with displacement $\boldsymbol{d}_{\text{target}}$.[3]

We remark that the displacement layer is un-trainable, i.e., the input $\boldsymbol{d}_{\text{input}}$ displacement is constant during the training.

Figure 9.2 shows the model. One starts from the vacuum state, and then a first pullback generates the input coherent state with displacement $\boldsymbol{d}_{\text{input}}$. Then the propagation through a trainable complex medium occurs, with symplectic operator $\mathbf{M}_{\text{hidden}}$, which returns a coherent state with displacement $\boldsymbol{d}_{\text{hidden}}$. And finally, a constant random layer is a linear layer with symplectic operator $\mathbf{M}_{\text{random}}$, which cannot be trained. The final output is a coherent state with displacement $\boldsymbol{d}_{\text{output}}$.

The following code generates the model in Fig. 9.2.[4]

---

[3] The code for this example is in `jupyternotebooks/phasespace/coherentcomplex.ipynb`.

[4] The code for this example is in `jupyternotebooks/phasespace/coherentcomplextraining.ipynb`.

```
N=20 # size
vacuum = VacuumLayer(N) # vacuum layer
dinput = 3.0*np.ones((N,1)) # constant vector for d
dinput = dinput/np.linalg.norm(dinput) #normalize dinput
D = DisplacementLayerConstant(dinput) # Glauber layer
R = RandomLayerConstant(N) # untrainable random medium
T = RandomLayer(N) # trainable random medium
# Build the model by connecting the layers
xin = tf.keras.layers.Input(N) # input layer
x3, a3 = R(xin)
x2, a2 = T(x3, a3)
x1, a1 = D(x2,a2)
chir, chii = vacuum(x1, a1)
# define the model
model  = tf.keras.Model(inputs = xin, outputs=[chir, chii])
```

First, we test the model by evaluating the derivatives of the characteristic function.

```
# Test the model by the derivatives
xzero = tf.constant(np.zeros((1,N)))
# instruct the Gradient Tape for the derivative w.r.t. x
# here we need persistent = True,
# as we will call two times the gradient
# when evaluating derivatives of chir and chii
with tf.GradientTape(persistent=True) as tape:
    tape.watch(xzero)
    chir, chii = model(xzero)
# print the derivative w.r.t. x of the real part of \chi
tf.print(tape.gradient(chir,xzero))
# outputs:
[[0 0 0 ... 0 0 0]]
# print the derivative w.r.t. x of the imaginary part of \chi
tf.print(tape.gradient(chii,xzero))
# outputs: (may change as random)
[[-0.069534920156002045 0.4104849100112915 ... ]]
```

We train to have the resulting expectation value of the canonical observables $\hat{\boldsymbol{R}}$ equal to a target displacement $\boldsymbol{d}_{\text{target}}$. This training optimizes the weight matrices $\mathbf{W}_R$ and $\mathbf{W}_I$ of the layer T.

It is important to stress that as the particle number is conserved, the training is successful only if the norms of the $\boldsymbol{d}_{\text{input}}$ and of $\boldsymbol{d}_{\text{target}}$ are equal.

We explore different approaches. First, we fit a target characteristic function, which is demanding but complete, as we know that—after the training—the final state will have all the statistical properties of a Gaussian state with a target displacement. As an alternative, we can train by using the expectation value of the observable or the covariance matrix. We will explore these possibilities in the following sections.

9.5 Training to Fit a Target Characteristic Function

Once we have the model, we need a cost function to train. Our goal is to tailor the trainable complex layer T to return a coherent state with a target $\boldsymbol{d}_{\text{target}}$ at the output of the random interferometer.

Our first strategy is fitting a known characteristic function $\chi_{\text{target}}(\boldsymbol{x})$ sampled in N_{batch} training vectors $\boldsymbol{x}_n$ with $n = 1, 2, ..., N_{\text{batch}}$, denoted as the training set.[5] The overall training set is stored in the Python array x with shape (N_{batch}, N).

The target state is a coherent state with $\boldsymbol{d} = \boldsymbol{d}_{\text{target}}$, with covariance matrix $\boldsymbol{g} = \mathbf{1}_N$, and

$$
\begin{aligned}
\chi_{R,\text{target}}(\boldsymbol{x}) &= \exp\left(-\frac{1}{4}\boldsymbol{x}\,\boldsymbol{x}^\top\right)\cos\left(\boldsymbol{x}\boldsymbol{d}_{\text{target}}\right) , \\
\chi_{I,\text{target}}(\boldsymbol{x}) &= \exp\left(-\frac{1}{4}\boldsymbol{x}\,\boldsymbol{x}^\top\right)\sin\left(\boldsymbol{x}\boldsymbol{d}_{\text{target}}\right) .
\end{aligned}
\tag{9.3}
$$

We recall that the target displacement $\boldsymbol{d}_{\text{target}}$ is a column vector with size $N \times 1$, with shape=(N,1). Correspondingly, the product $\boldsymbol{x}_n\,\boldsymbol{d}_{\text{target}}$ is scalar for a single sample in the training set.

We need to compute the target characteristic function at all the vectors $\boldsymbol{x}_n$ to generate the training set.

The target characteristic function is defined in the following code.

```
def chi_target(x, d):
    """ Characteristic function of a Gaussian coherent state
    Parameters
    ----------
    x : input
    d : displacement vector
    """
    yr = -0.25*np.matmul(x, x.transpose())
    yi = np.matmul(x, d)
    #return yr, yi
    return np.exp(yr)*np.cos(yi), np.exp(yr)*np.sin(yi)
```

[5] The code is in jupyternotebooks/phasespace/coherentcomplextraining .ipynb.

Training occurs by sampling $\chi_{R,\text{target}}$ and $\chi_{I,\text{target}}$ in the training set and adjusting the weights of the complex layer to minimize the cost function.

We define the normalized target displacement as follows.

```
dtarget = np.zeros((N,1));
dtarget[0]=3.14;
dtarget[1]=1.0;
# normalize the target displacement
dtarget = dtarget / np.linalg.norm(dtarget)
# print the target displacement
print(dtarget.transpose())
# outputs
[[0.9528 0.3035 0.     0.
   0.     0.     ... 0.   ]]
```

In this example, the target displacement has the first two elements as nonzeros and is normalized. The normalization to a unitary norm is the same as the input displacement to satisfy particle conservation. Otherwise, training is not possible.

We use random numbers as the training set and evaluate the target characteristic function as follows.

```
# size of training set
Nbatch = 100
# random input
xtrain = np.random.rand(Nbatch, N)-0.5
# corresponding training output as Nbatch values
chitestr = np.zeros(Nbatch)
chitesti = np.zeros(Nbatch)
for j in range(Nbatch):
    chitestr[j], chitesti[j] = chi_target(xtrain[j], dtarget)
```

For the cost function, one can use different error measures. For example, the average or the maximum between the two mean squared errors $||\chi_R(\boldsymbol{x}_n) - \chi_{R,\text{target}}(\boldsymbol{x}_n)||$ and $||\chi_I(\boldsymbol{x}_n) - \chi_{I,\text{target}}(\boldsymbol{x}_n)||$, where $\boldsymbol{x}_n$ is an input in the `Nbatch` values.

The following code compiles the model and performs the training for 100 epochs.

```
# compile the model with learning rate 0.01
model.compile(
              optimizer=tf.keras.optimizers.Adam(
                        learning_rate=0.01),
              loss='mean_squared_error')g
history = model.fit(x=xtrain,y=(chitestr, chitesti),epochs=100)
```

Figure 9.3 shows the training error versus the number of epochs.

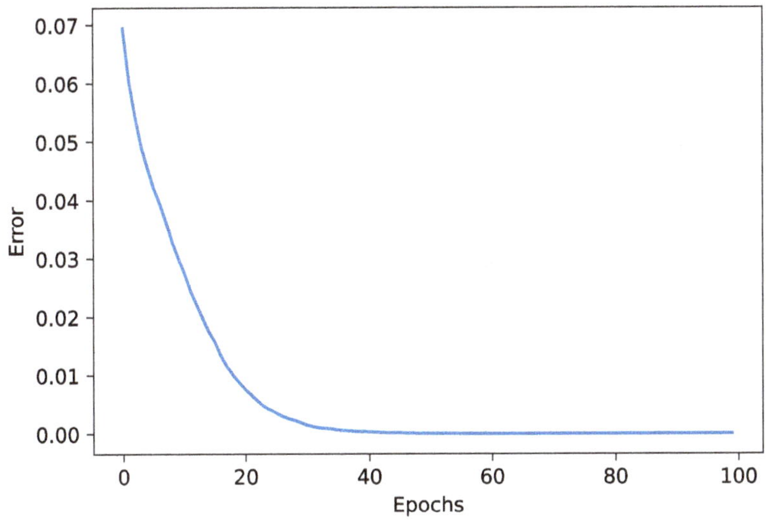

Fig. 9.3 Training error when fitting a target characteristic function

We recall that the weights in the complex layers are the matrices $\mathbf{W}_R$ and $\mathbf{W}_I$ in Eq. (7.181) used to generate the symplectic operator $\mathbf{M}$. After training, we test the derivatives to obtain the resulting expected values for the observables $\hat{\boldsymbol{R}}$, and check that they correspond to the target values (within numerical precision).[6]

```
with tf.GradientTape(persistent=True) as tape:
    tape.watch(xzero)
    chir, chii = model(xzero)
tf.print(tape.gradient(chii,xzero))
# outputs
[[0.95284205675125122 0.303466260433197
 -1.4115124940872192e-05 ...
5.0161033868789673e-05
4.32133674621582e-06 -1.4156103134155273e-05]]
```

[6] The code is in `jupyternotebooks/phasespace/coherentcomplextraining .ipynb`.

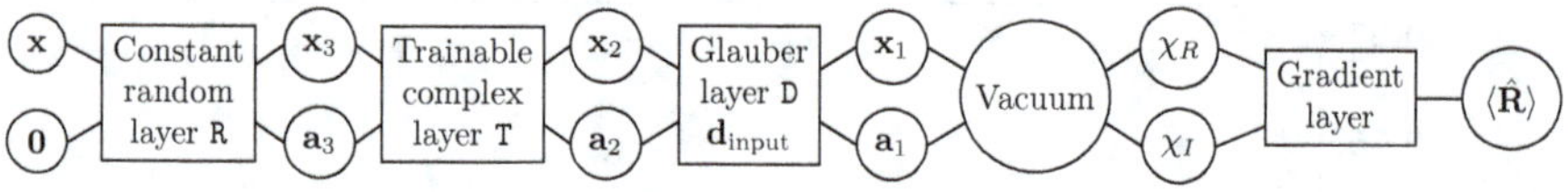

Fig. 9.4 Model for computing the derivatives of the characteristic function $\chi(x)$ for training a complex layer by the expectation value of the canonical variables

9.6 Training by First Derivatives

It is interesting to train the network by directly using observable quantities, as the expectation value of the displacement $\langle \hat{\boldsymbol{R}} \rangle$, which can be obtained by the derivatives of the model $\chi(\boldsymbol{x})$.

In this section, we train the network in a way such that

$$\langle \hat{\boldsymbol{R}} \rangle = \boldsymbol{d}_{\text{target}}. \tag{9.4}$$

We define a new model with additional "heads" that compute the derivatives, as in Fig. 9.4.

In order to return the derivatives of the model, we define a custom "gradient layer" that takes as input a characteristic function model and returns the derivative for its imaginary part evaluated at $\boldsymbol{x} = 0$. The code is the following.

```
class meanRLayer(layers.Layer):
    """
        Layer that returns the average of
        the expectation value of \hat{R}
        avg layer as derivative
        of the imaginary part
        of the model evaluated at x=0

        In the constructor as input:
        param: N: size of the of vector

        In the call as input
        param: c1: real part of chi(x) model
        param: c2: imag part of chi(x) model
        param: pullback: model to be derived wrt x
    """

    def __init__(self, N,
                 **kwargs):
        super(meanRlayer, self).__init__(**kwargs)
        self.N = N
        # constant x=0 for computing the derivative
        self.x0 = tf.constant(np.zeros((1,N)),
        ↪  dtype=self.type)

    def call(self, chir, chii, pullback):
        with tf.GradientTape() as tape:
```

```
            tape.watch(self.x0)
            _, ci = pullback(self.x0)
            chii_x = tape.gradient(ci, self.x0)
        return chii_x
```

The `meanRLayer` above enables to define a custom cost function that uses the expectation value of $\hat{\boldsymbol{R}}$, evaluated as the derivative of the model $\chi_I(\boldsymbol{x})$.

We build a new model by using the previous one for $\chi(\boldsymbol{x})$ by adding a layer that performs the derivatives (see Fig. 9.3). Note that in the `call` method, we specify the tensors `chir` and `chii` and also the name of the model to be derived, which is `pullback`. Even if `chir` and `chii` are not used in the call, they are needed by `TensorFlow` to create the graph for the computation.

The code for adding the layer to the model is the following.

```
der_layer = meanRLayer(N) # derivative layer
# connect the layer to the previous model
chii_x = der_layer(chir,chii, pullback)
# build the model
pullback_der = tf.keras.Model(inputs = xin, outputs=chii_x)
```

We are creating a new model `pullback_der` by using the previous one `pullback`.

Training occurs by imposing that for any $\boldsymbol{x}_n$ in the training set, the resulting $\langle\hat{\boldsymbol{R}}\rangle$ is equal to $\boldsymbol{d}_{\text{target}}$. We train by first compiling the model by the mean squared error as loss and then fitting. We adopt a target observable vector `Rtrain`, with `shape=(Nbatch,N)` for the input `xtrain`. [7]

```
pullback_der.compile(
optimizer=tf.keras.optimizers.Adam(learning_rate=0.01),
                loss='mean_squared_error')
```

Figure 9.5 shows the training error by using derivatives. Note that the output does not depend on the input $\boldsymbol{x}$. This is because different models, i.e., different states, can have the same expected values for observables. Hence, the optimization is only constrained to return a given statistical average. Different training processes may provide different states with the same mean values for the measurable quantities.

To completely determine Gaussian states, one must also impose the covariance matrix, i.e., the second derivatives of the model, as detailed in the following.

[7] The code is in `jupyternotebooks/phasespace/coherentcomplextrainingDER.ipynb`.

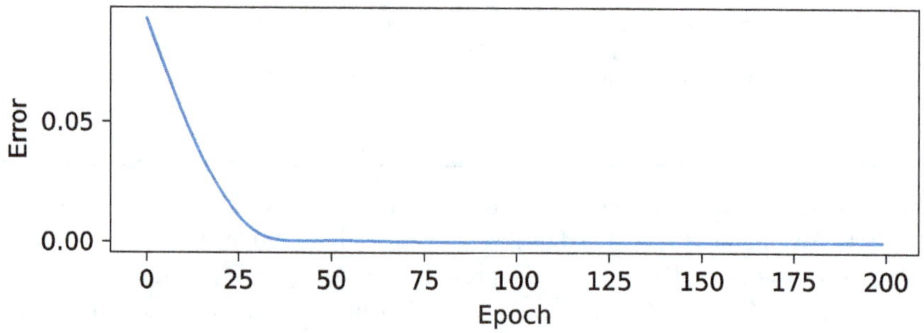

Fig. 9.5 Training error when fitting the transmitted state from a random interferometer with a target expectation value $\langle \hat{\boldsymbol{R}} \rangle$ obtained by the derivatives of the model

9.7 Training by Second Derivatives

For a Gaussian state, all the statistics are determined by the covariance symmetric real-valued matrix $\boldsymbol{g}$ and the real-valued displacement $\boldsymbol{d}$. Training by using the second derivatives is equivalent to fitting the entire characteristic function $\chi(\boldsymbol{x})$.

To compute the covariance matrix of $\chi(\boldsymbol{x})$, we use the following expression that holds for Gaussian states with a symmetric covariance matrix $\boldsymbol{g}$ as in Eq. (8.2):

$$g_{mn} = -2 \left. \frac{\partial^2 \chi_R}{\partial x_n \partial x_m} \right|_{x=0} - 2 \left. \frac{\partial \chi_I}{\partial x_n} \right|_{x=0} \left. \frac{\partial \chi_I}{\partial x_m} \right|_{x=0} \tag{9.5}$$

with $m, n = 0, 1, ..., N - 1$. The proof of Eq. (9.5) is given in Sect. 9.7.1.

We obtain g_{mn} by the `tf.gradient.jacobian` function with the following code, which defines a layer returning the covariance matrix and the expectation displacement.[8]

```
class CovarianceLayer(layers.Layer):
    """ Return the covariance of a characteristic function
        from the derivatives of the model
    """
    def __init__(self, N,
                 **kwargs):
            super(CovarianceLayer, self).__init__(**kwargs)
            # vector size
            self.N = N
            # constant x0 (x=0 for the derivatives)
            self.x0 = tf.constant(np.zeros((1,self.N)),
                         dtype = self.dtype)
    def call(self, c1, c2, chi):
```

[8] The full code for the `CovarianceLayer` is in `phasespace.py`.

```
    # c1 and c2 are dummy but needed for model building
    x = self.x0 # x=0 (must be constant as non-trainable)
    with tf.GradientTape() as t1:
        t1.watch(x) # watch constant
        # t2 is persistent as used two times
        with tf.GradientTape(persistent=True) as t2:
            t2.watch(x)
            cr, ci = chi(x)
        cr_x = t2.gradient(cr, x)
        ci_x = t2.gradient(ci, x)
    d2cr= t1.jacobian(cr_x,x) # hessian
    # reshape the Hessian to a matrix
    Hessian = tf.reshape(d2cr, [self.N,self.N])
    # covariance with the factor 2.0 for the definition of g
    cov = -2*(Hessian+tf.matmul(ci_x,ci_x,transpose_a=True))
    # free t2
    del t2
    return cov, ci_x, Hessian
```

In the function `call` of the `CovarianceLayer`, we have the vector `ci_x` with components $\frac{\partial \chi_I}{\partial x_n}$ at $\boldsymbol{x} = 0$ and `shape=(1,N)`.

We compute the $N \times N$ matrix $\frac{\partial \chi_I}{\partial x_n} \frac{\partial \chi_I}{\partial x_m}$ by multiplying with the transposed vector in `tf.matmul(ci_x,ci_x, transpose_a=True)`.

Once we have the layer for the covariance matrix, we proceed as above and, instead of the `meanRLayer`, we use the `CovarianceLayer`, which has two outputs including the $\langle \hat{\boldsymbol{R}} \rangle$.

The model is defined as follows (Fig. 9.5).[9]

```
cov_layer = CovarianceLayer(N)
g, Re = cov_layer(chir,chii, pullback)
pullback_cov = tf.keras.Model(inputs = xin, outputs=[g, Re])
```

Training occurs by specifying as targets the displacement vector $\boldsymbol{d}_{\text{target}}$ and the covariance matrix $\boldsymbol{g}$ (Fig. 9.6).

Note that the model has a list of two output tensors, `outputs=[g, Re]`. `Re` is the expected value for $\hat{\boldsymbol{R}}$, corresponding to $\boldsymbol{d}^\top$.

In the training set, we need to fix the output by the data `y=[g_train, Re_train]`. We define the training set as follows.

[9] The code is in `jupyternotebooks/phasespace/coherentcomplextrainingCOV.ipynb`.

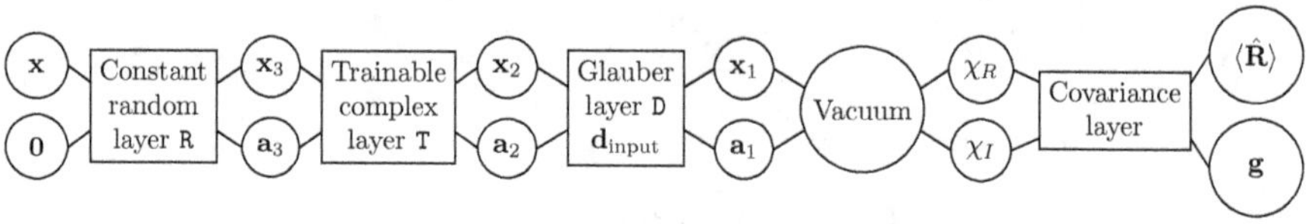

Fig. 9.6 Model for training a complex layer by the covariance matrix and the derivatives of the model

```
xtrain = np.random.rand(Nbatch, N)-0.5
dtrain = np.zeros((Nbatch,N))
gtrain = np.zeros((Nbatch,N,N))
for j in range(Nbatch):
    for i in range(N):
        dtrain[j,i]=dtarget[i]
        for k in range(N):
            gtrain[j,i,k]=gtarget[i,k]
```

`xtrain` are dummy `Nbatch` input random vectors with overall shape `(Nbatch, N)`. `Rtrain` is given by `Nbatch` vectors, which are all equals to the transpose of `dtarget` with shape `(N,1)`, thus forming an array with shape `(Nbatch, N)`. We also need `Nbatch` training matrices all equal to the target $\boldsymbol{g}$, as we needed the same output for any input vectors, such that `gtrain` has shape `(Nbatch, N, N)`.

In this example, we want to return a coherent state, such that the target matrix $\boldsymbol{g}$ is the identity $\boldsymbol{g} = \mathbf{1}_N$, corresponding to `g_train=np.eye(N)`. This is a simple example; higher-order derivatives will become helpful in a later section.

9.7.1 Proof of Eq. (9.5)*

Proof Equation (9.5) holds for a Gaussian characteristic model and can be derived by writing the Gaussian characteristic function in terms of the components of $\boldsymbol{x}$

$$\chi(\boldsymbol{x}) = \exp\left(-\frac{1}{4}\sum_{pq} g_{pq}x_p x_q + \mathrm{i}\sum_p x_p d_p\right), \tag{9.6}$$

with $p, q = 0, 1, ..., N-1$.

We have for the first derivative, by using the symmetry $g_{pq} = g_{qp}$,

$$\frac{\partial\chi}{\partial x_m} = \left(-\frac{1}{2}\sum_p g_{mp}x_p + \mathrm{i}d_m\right)\chi(\boldsymbol{x}) \tag{9.7}$$

with $m = 0, 1, ..., N-1$.

Evaluating Eq. (9.7) at $\boldsymbol{x} = 0$, we obtain as above, being $\chi(\mathbf{0}) = 1$,

$$\left.\frac{\partial \chi}{\partial x_m}\right|_{\boldsymbol{x}=0} = \mathrm{i} \left.\frac{\partial \chi_I}{\partial x_m}\right|_{\boldsymbol{x}=0} = \mathrm{i} d_m \tag{9.8}$$

For the second derivative of Eq. (9.6), we have

$$\begin{aligned}\frac{\partial^2 \chi}{\partial x_m \partial x_n} &= \\ -\frac{1}{2} g_{mn} \chi(\boldsymbol{x}) &+ \left(-\frac{1}{2}\sum_p g_{mp} x_p + \mathrm{i} d_m\right)\left(-\frac{1}{2}\sum_p g_{np} x_p + \mathrm{i} d_n\right)\chi(\boldsymbol{x})\end{aligned} \tag{9.9}$$

with $n = 0, 1, ..., N-1$. Evaluating Eq. (9.9) at $\boldsymbol{x} = 0$, we have

$$\begin{aligned}\left.\frac{\partial^2 \chi_R}{\partial x_m \partial x_n}\right|_{\boldsymbol{x}=0} &= -\frac{1}{2} g_{mn} - d_m d_n \\ \left.\frac{\partial^2 \chi_I}{\partial x_m \partial x_n}\right|_{\boldsymbol{x}=0} &= 0\end{aligned} \tag{9.10}$$

and using Eq. (9.8) in the first line of Eq. (9.10), we have the proof. □

9.8 Two Trainable Interferometers and a Reservoir

We consider a multilayer device, with two trainable complex media.

Figure 9.7 shows the model, including the two trainable complex layers and the untrainable random interferometer (the "reservoir").

Proceeding in the pullback direction, from right to left, the vacuum state is turned into a coherent state by the Glauber layers, then enters the input complex layer, and passes through the random interferometer, and the output complex layer.

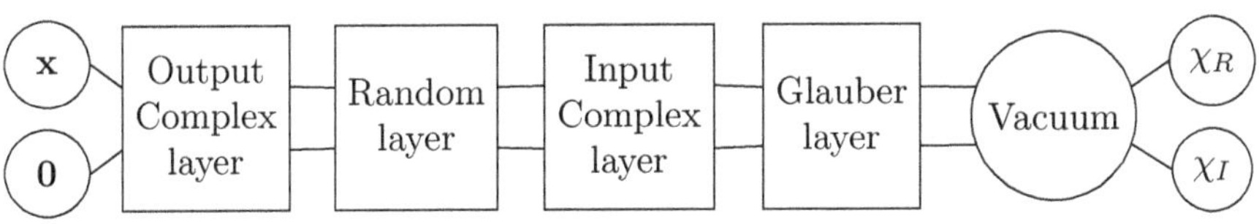

Fig. 9.7 Bug train of the model with two trainable complex media, at the input and output of a random interferometer (the reservoir). The order of the layers is inverted with respect to the initial and final state in the transformation. The input layer has symplectic matrix $\mathbf{M}_i$, and the output is $\mathbf{M}_o$. The bottom panel shows the training error vs the number of iterations

We train the model with a target coherent state. A derivative layer gives the cost function. There is no need to use the covariance matrix as the input and output are coherent states with covariance matrix $\mathbf{1}_N$. As far as there is no squeezer or generator of non-Gaussian states (as detailed in later chapters), the covariance matrix is not varied upon evolution and all the states are coherent with $\boldsymbol{g} = \mathbf{1}_N$.

The code for the model is as follows.

```
# define the layers give dinput and N
D = DisplacementLayerConstant(dinput)
R = RandomLayerConstant(N)
Ci= RandomLayer(N)
Co= RandomLayer(N)
# connect the layers
xin = tf.keras.layers.Input(N);print(xin) # input layer
x4, a4 = Co(xin) # output trainable layer
x3, a3 = R(x4, a4) # reservoir layer (untrainable)
x2, a2 = Ci(x3,a3) # input trainable layer
x1, a1 = D(x2,a2) # linear layer input coherent state
chir, chii = vacuum(x1, a1)  # vacuum
pullback = tf.keras.Model(inputs = xin, outputs=[chir, chii])
# Add the derivative layer for training
der_layer = ps.meanRLayer(N)
Re= der_layer(chir,chii, pullback)
pullback_der = tf.keras.Model(inputs = xin, outputs=Re)
```

Training occurs as described in the previous examples.[10]

9.9 Phase Modulator

We consider a specific complex trainable gate corresponding to phase modulation (PM). This kind of gate intervenes in optics when considering trainable interferometer or spatial light modulators. Phase modulation is effective if followed by propagation, which turns the phase into amplitude modulation. We consider the propagation of a phase modulated state through a complex medium.

In a phase modulator, the components of the state are modulated by a phase, and the classical expected value is hence multiplied by a phase factor. This gate is a unitary operator such that

$$\tilde{\hat{a}}_k = e^{i\theta_k}\hat{a}_k \tag{9.11}$$

[10] The code is in `jupyternotebooks/phasespace/twolayersreservoir.ipynb`.

where θ_k is the modulation angle of the field component k, with $k = 0, 1, ..., n-1$, and it is a particular case of Eq. (7.138).

In matrix form,

$$\hat{\tilde{\boldsymbol{a}}} = \mathbf{U}_{\mathrm{PM}} \hat{\boldsymbol{a}} \tag{9.12}$$

where

$$\mathbf{U}_{\mathrm{PM}} = \begin{pmatrix} e^{\mathrm{i}\theta_0} & & \\ & \ddots & \\ & & e^{\mathrm{i}\theta_{n-1}} \end{pmatrix}. \tag{9.13}$$

We let $\mathbf{U}_{\mathrm{PM}} = \mathrm{diag}\left(e^{\mathrm{i}\theta_0}, e^{\mathrm{i}\theta_1}, ..., e^{\mathrm{i}\theta_{N/2-1}}\right) = \mathbf{U}_{\mathrm{PM},R} + \mathrm{i}\mathbf{U}_{\mathrm{PM},I}$ and

$$\mathbf{U}_{\mathrm{PM,R}} = \mathrm{diag}\left[\cos(\theta_0), \cos(\theta_1), ..., \cos(\theta_{N/2-1})\right], \tag{9.14}$$

$$\mathbf{U}_{\mathrm{PM,I}} = \mathrm{diag}\left[\sin(\theta_0), \sin(\theta_1), ..., \sin(\theta_{N/2-1})\right]. \tag{9.15}$$

The variable q_k transforms as

$$\begin{aligned} \hat{\tilde{q}}_k = \frac{\hat{\tilde{a}}_k + \hat{\tilde{a}}_k^\dagger}{\sqrt{2}} = \frac{\hat{a}_k\, e^{\mathrm{i}\theta_k} + \hat{a}_k^\dagger\, e^{-\mathrm{i}\theta_k}}{\sqrt{2}} = \\ \frac{1}{2}\left[\left(\hat{q}_k + \mathrm{i}\hat{p}_k\right) e^{\mathrm{i}\theta_k} + \left(\hat{q}_k - \mathrm{i}\hat{p}_k\right) e^{-\mathrm{i}\theta_k}\right] = \\ \hat{q}_k \cos(\theta_k) - \hat{p}_k \sin(\theta_k). \end{aligned} \tag{9.16}$$

Seemingly,

$$\hat{\tilde{p}}_k = \hat{q}_k \sin(\theta_k) + \hat{p}_k \cos(\theta_k). \tag{9.17}$$

Hence, the symplectic operator $\mathbf{M}_{\mathrm{PM}}$ is a block diagonal matrix such that

$$\mathbf{M}_{\mathrm{PM}} = \begin{pmatrix} \cos(\theta_0) & -\sin(\theta_0) & 0 & 0 & \cdots & 0 & 0 \\ \sin(\theta_0) & \cos(\theta_0) & 0 & 0 & \cdots & 0 & 0 \\ 0 & 0 & \cos(\theta_1) & -\sin(\theta_1) & \cdots & 0 & 0 \\ 0 & 0 & \sin(\theta_1) & \cos(\theta_1) & \cdots & 0 & 0 \\ \vdots & \vdots & \vdots & \vdots & \ddots & \vdots & \vdots \\ 0 & 0 & 0 & 0 & \cdots & \cos(\theta_{n-1}) & -\sin(\theta_{n-1}) \\ 0 & 0 & 0 & 0 & \cdots & \sin(\theta_{n-1}) & \cos(\theta_{n-1}) \end{pmatrix}. \tag{9.18}$$

For the linear transformation corresponding to $\mathbf{U}_{\mathrm{PM}}$, we have also $\boldsymbol{d}'_{\mathrm{PM}} = \mathbf{0}$.

In our Python implementation, the matrix $\mathbf{M}_{\mathrm{PM}}$ is realized by using Eq. (7.182), by the diagonal matrices $\mathbf{U}_{\mathrm{PM},R}$ and $\mathbf{U}_{\mathrm{PM},I}$ above, which we build in TensorFlow by the `tf.linalg.diag` method.[11]

The code for the PM layer is similar to the code for the random interferometer in Sec. 8.13, but the symplectic matrix is diagonal and the layer has only $n = N/2$ weights corresponding to the values of the phases $\theta_0, \theta_1, ..., \theta_{n-1}$.

The following is the resulting code, with[12] the variable `theta_np` randomly initialized.

```
class PhaseModulatorLayer(layers.Layer):
    def __init__(self, N=10,
                 **kwargs):
        super(PhaseModulatorLayer, self).__init__(**kwargs)
        self.N = N
        n = np.floor_divide(self.n, 2)
        theta_np = np.random.random((n, 1)) #init random phases
        self.theta = tf.Variable(theta_np,
                     trainable=self.trainable)
        Rq, Rp, J = RQRP(N)
        self.Rq = tf.constant(Rq)
        self.Rp = tf.constant(Rp)
        self.J = tf.constant(J)

    def call(self, x, di=None):
        UR=tf.cos(self.theta)
        UI=tf.sin(self.theta)
        # Build the symplectic matrix M
        M = \
            tf.matmul(self.Rq,
                      tf.matmul(UR, self.Rq, transpose_b=True))
                      ↪  +\
            tf.matmul(self.Rp,
                      tf.matmul(UR, self.Rp, transpose_b=True))
                      ↪  -\
            tf.matmul(self.Rq,
                      tf.matmul(UI, self.Rp, transpose_b=True))
                      ↪  +\
            tf.matmul(self.Rp,
                      tf.matmul(UI, self.Rq, transpose_b=True))
        # Inverse of M
        MI = tf.matmul(tf.matmul(M, self.J),self.J,
              transpose_a=True)
        # no transpose here for M, as M is already transpose
        if di is None:
```

[11] A more efficient implementation can be done by using a vector instead of a diagonal matrix, but for the sake of simplicity and uniformity with the other layers, we retain the diagonal matrix implementation.

[12] The full code is in `phasespace.py`.

```
            d2 = tf.constant(np.zeros((self.n, 1))
        else:
            d2 = di
        return [tf.matmul(x, M), tf.matmul(MI, d2)]
```

9.10 Training Phase Modulators

We use the phase modulator and the random interferometer for building an arbitrary coherent state. We follow the example in Sect. 9.4 by replacing the trainable complex random interferometer with the phase modulator, as shown in Fig. 9.8. The code for the model is the following.[13]

```
PM_train = PhaseModulator(N) # define the PM layers
# input layer and connect the layers
xin = tf.keras.layers.Input(N);
x3, a3 = R(xin)
x2, a2 = PM_train(x3,d3)
x1, a1 = D(x2,d2)
chir, chii = vacuum(x1, d1)
pullback = tf.keras.Model(inputs = xin, outputs=[chir, chii])
# Add the derivative layer
der_layer = ps.avgR(N)
Re = der_layer(chir,chii, pullback)
pullback_der = tf.keras.Model(inputs = xin, outputs=Re)
```

We train as before by the first derivative, which requires adding the derivative layer.

In the specific example, we used $N = 10$ and $N_{\text{batch}} = 100$. After 100 epochs, the cost function reaches the minimum value of the order of 10^{-2}. With the adopted complex trainable medium in the previous examples, the final cost function is as small as the numerical precision (e.g., 10^{-12} after a sufficiently large number of epochs).

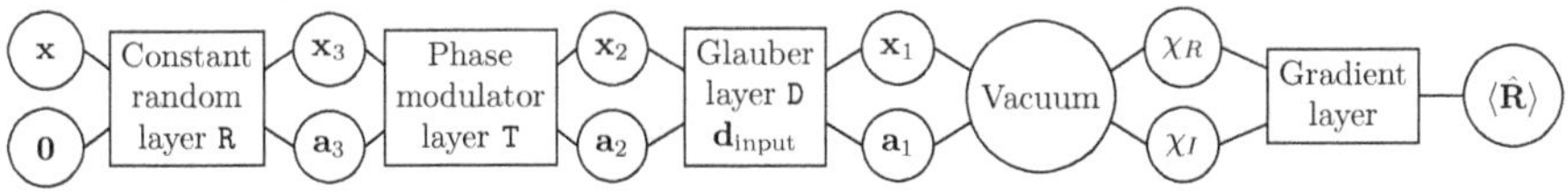

Fig. 9.8 Bug train for a model with a phase modulator and a random interferometer, trained by using the expectation value of the canonical vector

[13] The full example is in `jupyternotebooks/phasespace/phasemodulator.ipynb`.

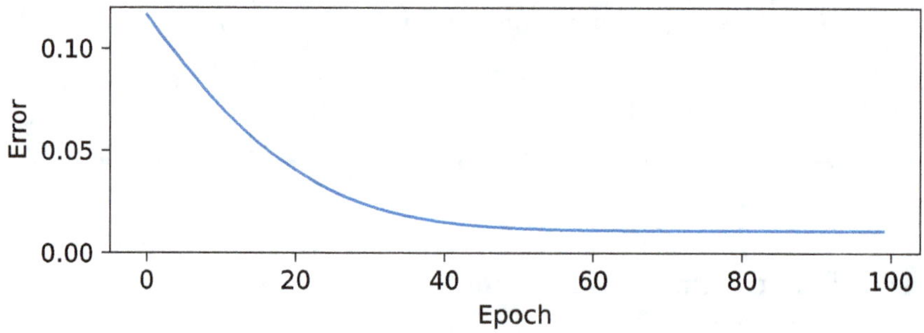

Fig. 9.9 Cost function w.r.t. number of epochs. Note that the residual error cannot be arbitrarily reduced

As one can see by printing the model summary, the number of trainable parameters for the phase modulator is much smaller than the case of the complex trainable medium. Indeed, the trainable parameters are only the n phases θ_k. As a result, when training the model, one finds that the cost function is bounded from below (see Fig. 9.9). Improvements are obtained by increasing n or adding further phase modulators at output to increase the number of trainable parameters.

9.11 Further Reading

- Optical quantum gates by random media: In recent years, innovative experiments have shown how to realize multilevel quantum gates with optical systems and random reservoir. Among the related publications, the interested reader may consider [2–4].
- Using reservoir computing and `tensorwflow` to program multilevel gates: The manuscript [5] reports on training random interferometers for arbitrary gates with `TensorFlow`.

References

1. I. Vellekoop, A. Mosk, Opt. Commun. **281**(11), 3071 (2008). https://doi.org/10.1016/j.optcom.2008.02.022. http://linkinghub.elsevier.com/retrieve/pii/S0030401808001430
2. S.R. Huisman, T.J. Huisman, T.A.W. Wolterink, A.P. Mosk, P.W.H. Pinkse, Opt. Express **23**(3), 3102 (2015). https://doi.org/10.1364/OE.23.003102. http://www.opticsexpress.org/abstract.cfm?URI=oe-23-3-3102
3. S. Leedumrongwatthanakun, L. Innocenti, H. Defienne, T. Juffmann, A. Ferraro, M. Paternostro, S. Gigan, Nat. Photon. **14**, 139 (2020)
4. S. Goel, S. Leedumrongwatthanakun, N.H. Valencia, W. McCutcheon, C. Conti, P.W.H. Pinkse, M. Malik, arXiv:2204.00578 (2022). http://arxiv.org/abs/2204.00578v1
5. G. Marcucci, D. Pierangeli, P.W.H. Pinkse, M. Malik, C. Conti, Opt. Express **28**(9), 14018 (2020). https://doi.org/10.1364/OE.389432. http://www.opticsexpress.org/abstract.cfm?URI=oe-28-9-14018

Chapter 10
Squeezing, Beam Splitters, and Detection

Make quantum great again

Abstract We introduce the squeezing layer by generalizing the symplectic operator. We study squeezed states including the vacuum and the coherent states. We detail multi-mode squeezing operators and beam splitters. We introduce the photon counting layer and homodyne detection.

10.1 Introduction

The previous chapters introduced the basic notions for the representation of the characteristic function and its transformations by neural networks. But we have been mainly dealing with coherent states characterized by an identity covariance matrix. The phase-space approach allows to deal with much more general states. While still limiting the analysis to Gaussian states, we will consider here the representation of nonclassical states as squeezed states and applications as optimization of the degree of entanglement.

To generate nonclassical states, we consider a wider class of linear transformations.

10.2 The Generalized Symplectic Operator

Different nonclassical states are generated from vacuum by unitary operators resulting into the following linear relation:

$$\hat{\tilde{a}} = \mathbf{U}\hat{a} + \mathbf{W}\hat{a}^{\dagger}, \tag{10.1}$$

which generalizes Eq. (7.138). Using the matrices $\mathbf{U}$ and $\mathbf{W}$, one obtains the corresponding symplectic matrix $\mathbf{M}$.

C. Conti, *Quantum Machine Learning*, Quantum Science and Technology,
https://doi.org/10.1007/978-3-031-44226-1_10

Let

$$\mathbf{U} = \mathbf{U}_R + \mathrm{i}\mathbf{U}_I \ , \tag{10.2}$$

and

$$\mathbf{W} = \mathbf{W}_R + \mathrm{i}\mathbf{W}_I \ , \tag{10.3}$$

we write the $\widehat{\tilde{a}}$ as follows:

$$\widehat{\tilde{a}} = \frac{\widehat{\tilde{q}} + \mathrm{i}\widehat{\tilde{p}}}{\sqrt{2}} = \mathbf{U}\hat{a} + \mathbf{W}\hat{a}^\dagger \ , \tag{10.4}$$

$$\widehat{\tilde{a}}^\dagger = \frac{\widehat{\tilde{q}} - \mathrm{i}\widehat{\tilde{p}}}{\sqrt{2}} = \mathbf{U}^*\hat{a}^\dagger + \mathbf{W}^*\hat{a} \ . \tag{10.5}$$

We have

$$\begin{aligned} \widehat{\tilde{q}} &= (\mathbf{U}_R + \mathbf{W}_R)\hat{q} + (-\mathbf{U}_I + \mathbf{W}_I)\hat{p} \ , \\ \widehat{\tilde{p}} &= (\mathbf{U}_I + \mathbf{W}_I)\hat{q} + (\mathbf{U}_R - \mathbf{W}_R)\hat{p} \ , \end{aligned} \tag{10.6}$$

which generalize Eq. (7.166). Following the same arguments for Eqs. (7.168) and (7.182), we have for the symplectic matrix

$$\begin{aligned} \mathbf{M} &= \mathbf{M}_1 + \mathbf{M}_2, \\ \mathbf{M}_1 &= \mathbf{R}_\mathrm{q}(\mathbf{U}_R + \mathbf{W}_R)\mathbf{R}_\mathrm{q}^\top + \mathbf{R}_\mathrm{p}(\mathbf{U}_R - \mathbf{W}_R)\mathbf{R}_\mathrm{p}^\top \ , \\ \mathbf{M}_2 &= \mathbf{R}_\mathrm{p}(\mathbf{U}_I + \mathbf{W}_I)\mathbf{R}_\mathrm{q}^\top + \mathbf{R}_\mathrm{q}(-\mathbf{U}_I + \mathbf{W}_I)\mathbf{R}_\mathrm{p}^\top \ . \end{aligned} \tag{10.7}$$

Equation (10.7) is used for generating layers for a linear transformation.

10.3 Single-Mode Squeezed State

We consider a single-mode squeezed state, so that $N = 2$, and

$$\hat{\boldsymbol{R}} = \begin{pmatrix} \hat{x} \\ \hat{p} \end{pmatrix} \ . \tag{10.8}$$

By using the squeezing operator with parameters r and θ [1]

$$\widehat{\tilde{a}} = \hat{S}^\dagger \hat{a} \hat{S} = \cosh(r)\hat{a} - e^{\mathrm{i}\theta}\sinh(r)\hat{a}^\dagger \ , \tag{10.9}$$

we have that the matrix $\mathbf{U}_{R,I}$ and $\mathbf{W}_{R,I}$ are complex scalars as follows:

$$\begin{aligned} U_R &= \cosh(r) \\ U_I &= 0 \\ W_R &= -\cos(\theta)\sinh(r) \\ W_I &= -\sin(\theta)\sinh(r) \end{aligned} . \tag{10.10}$$

By (10.7) we find the symplectic operator for the squeezing:[1]

$$\mathbf{M}_s(r,\theta) = \begin{pmatrix} \cosh(r) - \cos(\theta)\sinh(r) & -\sin(\theta)\sinh(r) \\ -\sin(\theta)\sinh(r) & \cosh(r) + \cos(\theta)\sinh(r) \end{pmatrix} \tag{10.11}$$

One can verify by direct calculation that $\mathbf{M}$ is symplectic, i.e., $\mathbf{M}^\top \mathbf{J}\mathbf{M} = \mathbf{J}$.

In the special case, $\theta = 0$, i.e., for a real squeezing parameter, we have

$$\mathbf{M}_s(r,0) = \begin{pmatrix} \exp(-r) & 0 \\ 0 & \exp(r) \end{pmatrix} . \tag{10.12}$$

10.4 Multimode Squeezed Vacuum Model

We are interested to multimode systems, so we consider a quantum n-body state and apply the squeezing operator to one mode.[2]

$\hat{\boldsymbol{R}}$ has dimension $N = 2n$. The single-mode squeezing operator is obtained by a linear gate with $\boldsymbol{d}' = 0$ and $\mathbf{M}$ given by Eq. (10.11) for the mode to be squeezed corresponding to an index n_{squeezed} in the range 0 to $n-1$.

For example, for $n = 4$, and $n_{\text{squeezed}} = 0$, we have

$$\mathbf{M} = \begin{pmatrix} M_{s,11} & M_{s,12} & 0&0&0&0&0&0 \\ M_{s,21} & M_{s,22} & 0&0&0&0&0&0 \\ 0 & 0 & 1&0&0&0&0&0 \\ 0 & 0 & 0&1&0&0&0&0 \\ 0 & 0 & 0&0&1&0&0&0 \\ 0 & 0 & 0&0&0&1&0&0 \\ 0 & 0 & 0&0&0&0&1&0 \\ 0 & 0 & 0&0&0&0&0&1 \end{pmatrix} . \tag{10.13}$$

Seemingly, for $n = 4$, $n_{\text{squeezed}} = 2$,

[1] See the MATLAB symbolic file `matlabsymbolic/squeezedoperator.m`.

[2] See the notebook `jupyternotebooks/phasespace/singlemodesqueezer.ipynb`.

$$\mathbf{M} = \begin{pmatrix} 1\,0\,0\,0 & 0 & 0 & 0\,0 \\ 0\,1\,0\,0 & 0 & 0 & 0\,0 \\ 0\,0\,1\,0 & 0 & 0 & 0\,0 \\ 0\,0\,0\,1 & 0 & 0 & 0\,0 \\ 0\,0\,0\,0 & M_{s,11} & M_{s,12} & 0\,0 \\ 0\,0\,0\,0 & M_{s,21} & M_{s,22} & 0\,0 \\ 0\,0\,0\,0 & 0 & 0 & 1\,0 \\ 0\,0\,0\,0 & 0 & 0 & 0\,1 \end{pmatrix} \tag{10.14}$$

Coding these matrices in an efficient way is critical for the compatibility of the various training functions and methods. The matrices are realized in `TensorFlow` by the following algorithm:

- Define four variables corresponding to the elements of the 2×2 matrix $M_s(r, \theta)$.
- Subtract 1 from the diagonal elements of $M_s(r, \theta)$.
- Build a 2×2 matrix.
- Increase the dimensions to $N \times N$ by tabbing with zero values by the method `tf.pad`.
- Add the identity matrix of dimension N by `tf.eye(N)`.

Subtracting 1 at the beginning is needed to use a simple matrix addition to fill the diagonal elements instead of using cycles or similar. Other smart solutions by using the various tensor manipulation routines of `TensorFlow` can be figured out.

The code for the resulting function is the following.

```
def get_M():
""" Return the M matrix for single mode squeezing operator """
    # Build the elements of the single mode squeezing matrix
    # and subtract ones to the diagonals
        M11 = tf.math.cosh(self.r)-\\
            tf.math.cos(self.theta)*tf.math.sinh(self.r)-1
        M12 = tf.math.sin(-self.theta)*tf.math.sinh(self.r)
        M21 = tf.math.sin(-self.theta)*tf.math.sinh(self.r)
        M22 = tf.math.cosh(self.r)+\\
            tf.math.cos(self.theta)*tf.math.sinh(self.r)-1
        # Form a tensor with the scalar elements
        L1 = tf.stack([M11, M12, M21, M22],0)
        # Reshape the tensor as a 2x2 matrix
        L2 = tf.reshape(L1, (2,2))
        # Pad the tensor to reach NxN size with zeros
        L3 = tf.pad(L2, self.paddings, constant_values=0)
        # Sum the indentity matrix
        M = L3+tf.eye(self.N)
        return M
```

This function is used below to define a multimode squeezing layer. Note that the `tf.pad` function adopts the 2×2 tensor `paddings`, which contains *four* numbers representing the number of zero rows to be added above and below and the columns to be added on the left and on the right to the 2×2 matrix. For example, for $n = 4$ and $n_{\text{squeezed}} = 2$ as in (10.14), we have paddings=[[4,2] [4,2]].

The code for defining the single-mode squeezing layer is the following.[3]

```
class SingleModeSqueezerLayer(layers.Layer):
    def __init__(self, N=10, r_np=1.0,
                theta_np=0.0,
                n_squeezed=0,
                trainable=True,
               **kwargs):
        super(SingleModeSqueezerLayer, self).__init__(**kwargs)
        self.trainable = trainable
        self.N = N
        self.n_squeezed = n_squeezed
        # squeezing parameters
        self.theta = tf.Variable(theta_np,
                    trainable=self.trainable)
        self.r = tf.Variable(r_np,
                trainable=self.trainable)
        Rq, Rp, J = RQRP(N)
        self.Rq = tf.constant(Rq)
        self.Rp = tf.constant(Rp)
        self.J = tf.constant(J)
        # padding vector for the matrix
        self.paddings = tf.constant([[2*n_squeezed,
                    self.N-2*(n_squeezed+1)],
                    [2*n_squeezed, self.N-2*(n_squeezed+1)]])

    def get_M(self):
        # return the M matrix and its inverse MI
        # Build the symplectic matrix M
        # subtract 1 in the diag, as then sum identity
        M11 = tf.math.cosh(self.r)- \
              tf.math.cos(self.theta)*tf.math.sinh(self.r)-1
        M12 = tf.math.sin(-self.theta)*tf.math.sinh(self.r)
        M21 = tf.math.sin(-self.theta)*tf.math.sinh(self.r)
        M22 = tf.math.cosh(self.r)+ \
              tf.math.cos(self.theta)*tf.math.sinh(self.r)-1
        L1 = tf.stack([M11, M12, M21, M22],0)
        L2 = tf.reshape(L1, (2,2))
        L3 = tf.pad(L2, self.paddings, constant_values=0)
        M = L3+tf.eye(self.N)
        # Inverse of M
        MI = tf.matmul(tf.matmul(M, self.J), self.J,
              transpose_a=True)
        return M, MI
```

[3] This code with some extension is in the function phasespace.py.

```
    def call(self, x, di=None):
        # build M and its inverse
        M, MI=self.get_M()
        # build
        if di is None:
            d2 = tf.constant(np.zeros((self.N, 1),
                 dtype=np_real), dtype=tf_real)
        else:
            d2 = di
        return [tf.matmul(x, M), tf.matmul(MI, d2)]
```

Note that we define a method `get_M` for the class `SingleModeSqueezer`, and use the method in the `call` function.

A model for squeezed vacuum is built by pulling back from the vacuum a linear layer as in Fig. 10.1, as in the following.[4]

```
# define the layers
vacuum = VacuumLayer(N)
squeezer=SingleModeSqueezer(N,
      r_np=1.0, theta_np=1.52, n_squeezed=0)
# connects the layers
xin = tf.keras.layers.Input(N)
x1, a1 = squeezer(xin)
chir, chii = vacuum(x1, a1)
single_mode_squeezed =
        tf.keras.Model(inputs = xin, outputs=[chir, chii])
```

10.5 Covariance Matrix and Squeezing

One verifies that the model corresponds to a squeezed vacuum state by the covariance matrix. For a single mode in a classical coherent state, the covariance matrix has two unitary eigenvalues. If the mode is squeezed, the two eigenvalues are $\exp(2r)$ and $\exp(-2r)$.

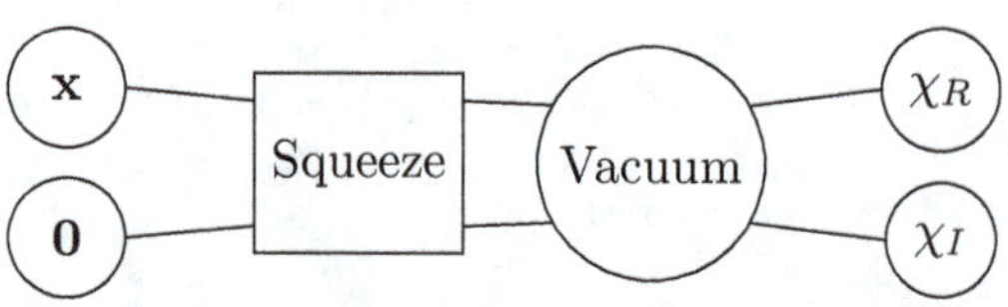

Fig. 10.1 A squeezed coherent state obtained by pullback with a single-mode squeezing layer from the vacuum

[4] See the notebook `jupyternotebooks/phasespace/singlemodeSqueezer.ipynb`.

We use the covariance layer to obtain the covariance matrix and the mean displacement $\langle \hat{\boldsymbol{R}} \rangle$, as follows.

```
cov_layer = CovarianceLayer(N)
covariance_matrix, Re = cov_layer(chir,chii,
                         single_mode_squeezed)
squeezed_cov = tf.keras.Model(inputs = xin,
                              outputs=[covariance_matrix, Re])
```

Note that, as we are dealing with a vacuum state, the displacement is zero. This is found by calling the covariance layer and inspecting the `Re` output tensor.

```
xtrain = np.random.rand(1, N)-0.5 # training point
cov0,d0=squeezed_cov(xtrain);
print(d0) # d0 is the value stored  in Re
# output:
# tf.Tensor([[0. 0. 0. 0.]], shape=(1, 4), dtype=float64)
```

The squeezing is found by the eigenvalues and eigenvectors of the `cov0` matrix using `np.linalg.eig(cov0.numpy())`, and one finds that the eigenvalues are $\exp(-2r) = 0.14$ and $\exp(2r) = 7.4$ for $r = 1.0$. The squeezed mode is a superposition of the `cov0` eigenvectors, while the other modes have unitary eigenvalues and are spanned by the orthogonal eigenvectors.

Note that realizing a gate that squeezes different modes may be done by simply cascading single-mode squeezing operators. This will be reported below.

10.6 Squeezed Coherent States

10.6.1 Displacing the Squeezed Vacuum

The squeeze operator is represented by a linear gate, which we cascade with other layers, as the Glauber displacement layer. Figure 10.2a shows a model to generate a squeezed coherent state from the squeezed vacuum by a displacement layer.

The squeezed coherent states $|\alpha, \zeta\rangle$ are built by applying the displacement operator $\hat{\mathcal{D}}(\alpha)$ to a squeezed vacuum $\hat{\mathcal{S}}(\zeta)|0\rangle$, where $\zeta = r\,e^{i\theta}$ is the complex squeezing parameter, that is,

$$|\alpha, \zeta\rangle = \hat{\mathcal{D}}(\alpha)\hat{\mathcal{S}}(\zeta)|0\rangle \,. \tag{10.15}$$

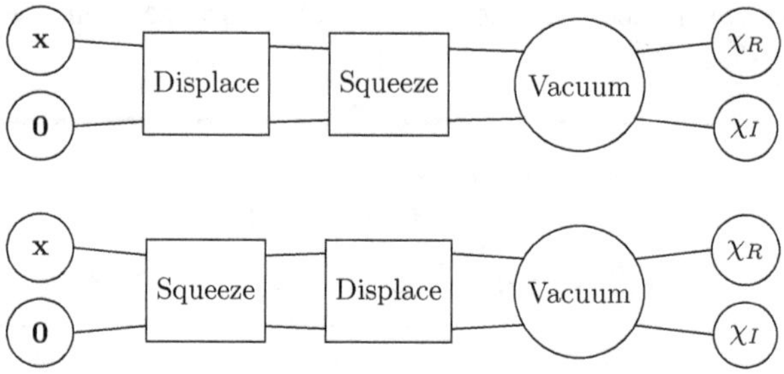

Fig. 10.2 Pullback model for squeezed coherent state by displacing a squeezed vacuum (top panel). Pullback model for squeezed coherent state by squeezing a displaced vacuum (bottom panel). Here all the layers act on the same mode

The resulting state has $\langle \hat{a} \rangle = \alpha$, and it is squeezed. In the following coding example, we first define a displacement layer with a constant displacement $d_j = 3.0$ for $j = 0, 1, \ldots, N-1$, and then we concatenate the layers.

```
N=2
dtarget = 3.0*np.ones((N,1))
D = DisplacementLayerConstant(dtarget)
xin = tf.keras.layers.Input(N)
x2, a2 = D(xin)
x1, a1 = squeezer(x2,a2 )
chir, chii = vacuum(x1, a1)
single_mode_squeezed_1 = tf.keras.Model(inputs = xin,
                                        outputs=[chir, chii])
cov_layer = CovarianceLayer(N)
cova, Re = cov_layer(chir,chii, single_mode_squeezed_1)
squeezed_cov_1 = tf.keras.Model(inputs = xin,
                                outputs=[cova, Re])
# we run the model to obtain the mean displacement
# and eigenvalues of the covariance matrix
cov1, d1=squeezed_cov_1(xtrain);
```

The model includes a `CovarianceLayer`. When running the model, we obtain a displacement `d1=dtarget`, and the eigenvalues 1, 1, 0.14, 7.39, corresponding to a squeezed mode. Inspecting the eigenvectors, one finds that the squeezed mode has index $n_{\text{squeezed}} = 0$, as expected.

10.6.2 *Squeezing the Displaced Vacuum*

A different squeezed coherent state is obtained by squeezing a coherent state. We first pullback a displacement layer from the vacuum to create the coherent state, and then pullback a squeezing layer, as shown in Fig. 10.2b. The bug train corresponds to the following Eq. (10.16):

$$|\alpha \cosh(r) - \alpha^* e^{i\theta} \sinh(r), \zeta\rangle = \hat{\mathcal{S}}(\zeta)\hat{\mathcal{D}}(\alpha)|0\rangle \ . \tag{10.16}$$

The result is a squeezed coherent state with the same eigenvalues for the covariance matrix as above, but the mean value of the displacement is changed [1], i.e.,

$$\langle \hat{a} \rangle = \alpha \cosh(r) - \alpha^* e^{i\theta} \sinh(r). \tag{10.17}$$

As an example, let us start with a non-squeezed coherent state obtained by displacing with `dtarget` above, i.e., a constant vector elements all equal to 3. Letting $n_{\text{squeezed}} = 0$, the expected value of the displacement is $d_0 = \langle \hat{x}_0 \rangle = 3$ and $d_1 = \langle \hat{p}_0 \rangle = 3$. Correspondingly, before squeezing,

$$\langle \hat{a}_0 \rangle = \frac{3.0 + i3.0}{\sqrt{2}}. \tag{10.18}$$

After squeezing, we have ($r = 1$ and $\theta = \pi/2$)

$$\langle \hat{\tilde{a}}_0 \rangle = \alpha \cosh(r) - \alpha^* e^{i\theta} \sinh(r), \tag{10.19}$$

which gives $d_0 = \langle \hat{R}_0 \rangle \simeq 1.1$ and $d_1 = \langle \hat{R}_1 \rangle \simeq 1.1$. The values of the other displacements and the eigenvalues of the covariance matrix are unchanged. [5]

This may be verified by the following code, which follows the pullback in Fig. 10.2b.

```
xin = tf.keras.layers.Input(N)
x2, a2 = squeezer(xin )
x1, a1 = D(x1,a1 )
chir, chii = vacuum(x1, a1)
single_mode_squeezed_2 =
tf.keras.Model(inputs = xin, outputs=[chir, chii])
cov_layer = ps.CovarianceLayer(N)
cova, Re = cov_layer(chir,chii, single_mode_squeezed_2)
squeezed_cov_2 = tf.keras.Model(inputs = xin,
                 outputs=[cova, Re])
```

When running the model by

```
cov2, d2=squeezed_cov_2(xtrain)
print(d2)
```

[5] See the notebook `jupyternotebooks/phasespace/singlemodesqueezer.ipynb`.

```
# output is
# tf.Tensor([1.1 1.1 3. 3.]], shape(1,4), dtype=float64)
```

we obtain the expected theoretical values of the displacement.

10.7 Two-Mode Squeezing Layer

We consider the two-mode squeezing operator [1]. Given two modes with indices A and B, in the range $0, 1, 2 \ldots, n-1$, the two-mode squeezing operator is

$$\hat{S}_{AB}(\zeta) = \exp(-\zeta \hat{a}_A^\dagger \hat{a}_B^\dagger + \zeta^* \hat{a}_B \hat{a}_A). \tag{10.20}$$

Letting $\zeta = r\, e^{i\theta}$, we have for the transformed operators [1]

$$\begin{aligned} \tilde{\hat{a}}_A &= \hat{S}_{AB}^\dagger \hat{a}_A \hat{S}_{AB} = \cosh(r)\hat{a}_A - e^{i\theta}\sinh(r)\hat{a}_B^\dagger \\ \tilde{\hat{a}}_B &= \hat{S}_{AB}^\dagger \hat{a}_B \hat{S}_{AB} = \cosh(r)\hat{a}_B - e^{i\theta}\sinh(r)\hat{a}_A^\dagger \end{aligned} . \tag{10.21}$$

The corresponding 2×2 complex matrices $\mathbf{U}$ and $\mathbf{W}$ in Eq. (10.1) are

$$\mathbf{U} = \begin{pmatrix} \cosh(r) & 0 \\ 0 & \cosh(r) \end{pmatrix} \tag{10.22}$$

and

$$\mathbf{W} = \begin{pmatrix} 0 & -e^{i\theta}\sinh(r) \\ -e^{i\theta}\sinh(r) & 0 \end{pmatrix}. \tag{10.23}$$

We have for the real and imaginary parts of $\mathbf{U}$ and $\mathbf{W}$

$$\begin{aligned} \mathbf{U}_R &= \begin{pmatrix} \cosh(r) & 0 \\ 0 & \cosh(r) \end{pmatrix} \\ \mathbf{U}_I &= \begin{pmatrix} 0 & 0 \\ 0 & 0 \end{pmatrix} \\ \mathbf{W}_R &= \begin{pmatrix} 0 & -\cos(\theta)\sinh(r) \\ -\cos(\theta)\sinh(r) & 0 \end{pmatrix} \\ \mathbf{W}_I &= \begin{pmatrix} 0 & -\sin(\theta)\sinh(r) \\ -\sin(\theta)\sinh(r) & 0 \end{pmatrix}. \end{aligned} \tag{10.24}$$

For $n = 2$ states, (10.7) gives[6]

$$\mathbf{M}_{2s} = \begin{pmatrix} \cosh(r) & 0 & -\cos(\theta)\sinh(r) & -\sin(\theta)\sinh(r) \\ 0 & \cosh(r) & -\sin(\theta)\sinh(r) & \cos(\theta)\sinh(r) \\ -\cos(\theta)\sinh(r) & -\sin(\theta)\sinh(r) & \cosh(r) & 0 \\ -\sin(\theta)\sinh(r) & \cos(\theta)\sinh(r) & 0 & \cosh(r) \end{pmatrix} \tag{10.25}$$

For n-body states, the elements of the matrix $\mathbf{M}_{2s}$ can be embedded in the $N \times N$ matrix $\mathbf{M}$. For example, when $n = 4$ ($N = 8$), and assuming that the two squeezed modes have index $A = 1$ and $B = 3$, we have

$$\mathbf{M} = \begin{pmatrix} 1 & 0 & 0 & 0 & 0 & 0 & 0 & 0 \\ 0 & 1 & 0 & 0 & 0 & 0 & 0 & 0 \\ 0 & 0 & M_{2s,00} & M_{2s,01} & 0 & 0 & M_{2s,02} & M_{2s,03} \\ 0 & 0 & M_{2s,10} & M_{2s,11} & 0 & 0 & M_{2s,12} & M_{2s,13} \\ 0 & 0 & 0 & 0 & 1 & 0 & 0 & 0 \\ 0 & 0 & 0 & 0 & 0 & 1 & 0 & 0 \\ 0 & 0 & M_{2s,20} & M_{2s,21} & 0 & 0 & M_{2s,22} & M_{2s,23} \\ 0 & 0 & M_{2s,30} & M_{2s,31} & 0 & 0 & M_{2s,32} & M_{2s,33} \end{pmatrix} \tag{10.26}$$

where $M_{2s,pq}$ are the elements of the matrix in Eq. (10.25) with $p, q = 0, 1, 2, 3$.

This embedding is made in tensorflow as above by padding with zero matrices. We define a layer named `TwoModeSqueezerLayer`. The code is as follows;[7] note the building of the matrix $\mathbf{M}$, which requires a bit of matrix manipulation by using the methods of `TensorFlow`.

```
class TwoModeSqueezerLayer(layers.Layer):
    """
        Define a multi head linear random layer corresponding to
        two-mode squeezing opeator

        :param N: dimension (n=N/2)
        :param r_np : squeezing parameter
        :param theta_np : angle squeezing parameter
        :param n_s1 : index of the squeezed mode 1 (from 0 to
↪   N/2-1)
        :param n_s2 : index of the squeezed mode 2 (from 0 to
↪   N/2-1)
        :param trainable : r and theta are trainable (default is
↪   True)
```

[6] The MATLAB file `matlabsymbolic/twomodesqueezedoperator.m` reports this symbolic derivation.

[7] The code is in the library `phasespace.py`.

```
    :output x M
    :output b = M^(-1) d

"""

def __init__(self, N=10, r_np=1.0,
             theta_np=0.0,
             n_s1=0,
             n_s2=1,
             trainable=True,
          **kwargs):
    super(TwoModeSqueezerLayer, self).__init__(**kwargs)
    self.trainable = trainable
    self.N = N
    self.n_s1 = n_s1
    self.n_s2 = n_s2
    # squeezing parameters
    self.theta = tf.Variable(theta_np, dtype=tf_real,
                              trainable=self.trainable)
    self.r = tf.Variable(r_np, dtype=tf_real,
                              trainable=self.trainable)
    Rq, Rp, J = RQRP(N)
    self.Rq = tf.constant(Rq)
    self.Rp = tf.constant(Rp)
    self.J = tf.constant(J)
    # padding vectors for the matrix
    self.paddings11 = tf.constant([[2*n_s1, \
            self.N-2*(n_s1+1)],[2*n_s1,
            ↪  self.N-2*(n_s1+1)]])
    self.paddings12 = tf.constant([[2*n_s1, \
            self.N-2*(n_s1+1)],[2*n_s2,
            ↪  self.N-2*(n_s2+1)]])
    self.paddings21 = tf.constant([[2*n_s2, \
            self.N-2*(n_s2+1)],[2*n_s1,
            ↪  self.N-2*(n_s1+1)]])
    self.paddings22 = tf.constant([[2*n_s2, \
            self.N-2*(n_s2+1)],[2*n_s2,
            ↪  self.N-2*(n_s2+1)]])

@tf.function
def get_M(self):
    # return the M matrix and its inverse MI
    # Build the symplectic matrix M
    # fix the following elements by correct expressions
    M00 = tf.math.cosh(self.r)-1 # subtract 1 in the diag
    M01 = tf.constant(0.0)
    M02 = tf.math.sinh(-self.r)*tf.math.cos(self.theta) #
    ↪  note minus
    M03 = tf.math.sin(self.theta)*tf.math.sinh(-self.r)

    M10 = tf.constant(0.0)
    M11 = tf.math.cosh(self.r)-1 # subtract 1 in the diag
    M12 = tf.math.sin(self.theta)*tf.math.sinh(-self.r)
```

```
        M13 = tf.math.cos(self.theta)*tf.math.sinh(self.r)

        M20 = tf.math.cos(self.theta)*tf.math.sinh(-self.r)
        M21 = tf.math.sin(self.theta)*tf.math.sinh(-self.r)
        M22 = tf.math.cosh(self.r)-1 # subtract 1 in the diag
        M23 = tf.constant(0.0)

        M30 = tf.math.sin(self.theta)*tf.math.sinh(-self.r)
        M31 = tf.math.cos(self.theta)*tf.math.sinh(self.r)
        M32 = tf.constant(0.0)
        M33 = tf.math.cosh(self.r)-1  # subtract 1 in the diag

        La = tf.stack([M00, M01, M10, M11],0)
        L11 = tf.reshape(La, (2,2))

        Lb = tf.stack([M02, M03, M12, M13],0)
        L12 = tf.reshape(Lb, (2,2))

        Lc = tf.stack([M20, M21, M30, M31],0)
        L21 = tf.reshape(Lc, (2,2))

        Ld = tf.stack([M22, M23, M32, M33],0)
        L22 = tf.reshape(Ld, (2,2))

        Ma = tf.pad(L11, self.paddings11, constant_values=0)
        Mb = tf.pad(L12, self.paddings12, constant_values=0)
        Mc = tf.pad(L21, self.paddings21, constant_values=0)
        Md = tf.pad(L22, self.paddings22, constant_values=0)


        M = Ma+Mb+Mc+Md+tf.eye(self.N) # add identity to restore
        ↪ diag
        # Inverse of M
        MI = tf.matmul(tf.matmul(M, self.J), self.J,
        ↪ transpose_a=True)
        return M, MI



    def call(self, x, di=None):
        # build M and its inverse
        M, MI=self.get_M()
        # build
        if di is None:
            d2 = tf.constant(np.zeros((self.N, 1),
            ↪ dtype=np_real), dtype=tf_real)
        else:
            d2 = di
        return [tf.matmul(x, M), tf.matmul(MI, d2)]
```

A simple two-mode squeezed vacuum example is in.[8]

[8] The code is in `jupyternotebooks/phasespace/twomodesqueezer.ipynb`.

10.8 Beam Splitter

The beam splitter is a special interferometer that mixes modes. For many optical experiments, it is used to create specific states or to synthesize arbitrary unitary operators.

Given two modes, with amplitudes $\hat{a}_1$ and $\hat{a}_2$, the equations for a beam splitter are written as [2]

$$\hat{\tilde{a}}_1 = \cos(\theta)e^{-i\phi_0}\hat{a}_1 - \sin(\theta)e^{-i\phi_1}\hat{a}_2 \tag{10.27}$$

$$\hat{\tilde{a}}_2 = \sin(\theta)e^{+i\phi_1}\hat{a}_1 + \cos(\theta)e^{+i\phi_0}\hat{a}_2 \,. \tag{10.28}$$

which correspond to the 2×2 matrices $\mathbf{U}$ and $\mathbf{W}$

$$\mathbf{U} = \begin{pmatrix} \cos(\theta)e^{-i\phi_0} & -\sin(\theta)e^{-i\phi_1} \\ \sin(\theta)e^{i\phi_1} & \cos(\theta)e^{i\phi_0} \end{pmatrix} \tag{10.29}$$

and

$$\mathbf{W} = \mathbf{0} \,. \tag{10.30}$$

We have for the real and imaginary parts

$$\mathbf{U}_R = \begin{pmatrix} \cos(\theta)\cos(\phi_0) & -\sin(\theta)\cos(\phi_1) \\ \sin(\theta)\cos(\phi_1) & \cos(\theta)\cos(\phi_0) \end{pmatrix} \tag{10.31}$$

$$\mathbf{U}_I = \begin{pmatrix} -\cos(\theta)\sin(\phi_0) & \sin(\theta)\sin(\phi_1) \\ \sin(\theta)\sin(\phi_1) & \cos(\theta)\sin(\phi_0) \end{pmatrix} , \tag{10.32}$$

and

$$\mathbf{W}_R = \mathbf{W}_I = \mathbf{0} \,. \tag{10.33}$$

For $n = 2$ states, Eq. (10.7) gives[9]

$$\mathbf{M}_{bs} = \begin{pmatrix} \cos(\phi_0)\cos(\theta) & \cos(\theta)\sin(\phi_0) & -\cos(\phi_1)\sin(\theta) & -\sin(\phi_1)\sin(\theta) \\ -\sin(\phi_0)\cos(\theta) & \cos(\phi_0)\cos(\theta) & \sin(\phi_1)\sin(\theta) & -\cos(\phi_1)\sin(\theta) \\ \cos(\phi_1)\sin(\theta) & -\sin(\phi_1)\sin(\theta) & \cos(\phi_0)\cos(\theta) & -\sin(\phi_0)\cos(\theta) \\ \sin(\phi_1)\sin(\theta) & \cos(\phi_1)\sin(\theta) & \sin(\phi_0)\cos(\theta) & \cos(\phi_0)\cos(\theta) \end{pmatrix} \tag{10.34}$$

[9] See the `MATLAB` symbolic file `matlabsymbolic/beamsplitter.m`.

For a 50 : 50 beam splitter, we have $\theta = \pi/4$ and $\phi_0 = \phi_1 = 0$, such that

$$\mathbf{M}_{bs} = \frac{1}{\sqrt{2}} \begin{pmatrix} 1 & 0 & -1 & 0 \\ 0 & 1 & 0 & -1 \\ 1 & 0 & 1 & 0 \\ 0 & 1 & 0 & 1 \end{pmatrix} . \tag{10.35}$$

10.9 The Beam Splitter Layer

We can split one input mode into two output modes by a beam splitter and leave the other modes unaffected. As for the two-mode squeezing layer, $\mathbf{M}_{bs}$ can be embedded in a many-mode symplectic matrix with $n > 2$ [see (10.26)]. We define the $n-$mode beam splitter layer as in the following code.[10]

```
class BeamSplitterLayer(layers.Layer):
    """
    Define a many-mode beam splitter

    Parameters
    -----------
    N: dimension
    n_0 : index of the mode 1 (from 0 to N/2)
    n_1 : index of the mode 2 (from 0 to N/2)
    theta_np : angle parameter (default pi/4 to 50:50)
    phi0_np : delay parameter (default 0)
    phi1_np : delay parameter (default 0)
    trainable_theta : is theta is trainable (default is True)
    trainable_phi0 : is phi0 is trainable (default is True)
    trainable_phi1 : is phi1 is trainable (default is True)

    Returns:
    --------
    y =  x M
    b = M^(-1) d

    """

    def __init__(self, N=10,
                        n_0=0,
                        n_1=1,
                        theta_np=np.pi/4,
                        phi0_np = 0.0,
                        phi1_np = 0.0,
                        trainable_theta=True,
```

[10] The complete code is in the library `phasespace.py`.

```
                 trainable_phi0=True,
                 trainable_phi1=True,
              **kwargs):
        super(BeamSplitter, self).__init__(**kwargs)
        self.N = N
        self.n_0 = n_0
        self.n_1 = n_1
        # squeezing parameters
        self.theta = tf.Variable(theta_np, dtype=tf_real,
                                 trainable=trainable_theta)
        self.phi0 = tf.Variable(phi0_np, dtype=tf_real,
                                 trainable=trainable_phi0)
        self.phi1 = tf.Variable(phi1_np, dtype=tf_real,
                                 trainable=trainable_phi1)
        Rq, Rp, J = RQRP(N)
        self.Rq = tf.constant(Rq)
        self.Rp = tf.constant(Rp)
        self.J = tf.constant(J)
        # padding vector for the matrix
        self.paddings11 = tf.constant([[2*n_0, \
                    self.N-2*(n_0+1)],
                    [2*n_0, self.N-2*(n_0+1)]])
        self.paddings12 = tf.constant([[2*n_0, \
                    self.N-2*(n_0+1)],
                    [2*n_1, self.N-2*(n_1+1)]])
        self.paddings21 = tf.constant([[2*n_1, \
                    self.N-2*(n_1+1)],
                    [2*n_0, self.N-2*(n_0+1)]])
        self.paddings22 = tf.constant([[2*n_1, \
                    self.N-2*(n_1+1)],
                    [2*n_1, self.N-2*(n_1+1)]])

    def call(self, x, di=None):
        # build M and its inverse
        # return the M matrix and its inverse MI
        # Build the symplectic matrix M
        ## subtract 1 diag
        M11 = tf.math.cos(self.phi0)*tf.math.cos(self.theta)-1
        M12 = tf.math.sin(self.phi0)*tf.math.cos(self.theta)
        ## note -1 in sin
        M13 = tf.math.cos(self.phi1)*tf.math.sin(-self.theta)
        ## note -1 in sin
        M14 = tf.math.sin(self.phi1)*tf.math.sin(-self.theta)

        ## note -1 in sin
        M21 = tf.math.sin(-self.phi0)*tf.math.cos(self.theta)
        ## subtract 1 diag
        M22 = tf.math.cos(self.phi0)*tf.math.cos(self.theta)-1
        M23 = tf.math.sin(self.phi1)*tf.math.sin(self.theta)
        ## note -1 in sin
        M24 = tf.math.cos(self.phi1)*tf.math.sin(-self.theta)

        M31 = tf.math.cos(self.phi1)*tf.math.sin(self.theta)
        ## note -1 in sin
```

```
M32 = tf.math.sin(self.phi1)*tf.math.sin(-self.theta)
## subtract 1 diag
M33 = tf.math.cos(self.phi0)*tf.math.cos(self.theta)-1
## note -1 in sin
M34 = tf.math.sin(-self.phi0)*tf.math.cos(self.theta)

M41 = tf.math.sin(self.phi1)*tf.math.sin(self.theta)
M42 = tf.math.cos(self.phi1)*tf.math.sin(self.theta)
## note -1 in sin
M43 = tf.math.sin(self.phi0)*tf.math.cos(self.theta)
## subtract 1 diag
M44 = tf.math.cos(self.phi0)*tf.math.cos(self.theta)-1

La = tf.stack([M11, M12, M21, M22],0)
L11 = tf.reshape(La, (2,2))

Lb = tf.stack([M13, M14, M23, M24],0)
L12 = tf.reshape(Lb, (2,2))

Lc = tf.stack([M31, M32, M41, M42],0)
L21 = tf.reshape(Lc, (2,2))

Ld = tf.stack([M33, M34, M43, M44],0)
L22 = tf.reshape(Ld, (2,2))

Ma = tf.pad(L11, self.paddings11, constant_values=0)
Mb = tf.pad(L12, self.paddings12, constant_values=0)
Mc = tf.pad(L21, self.paddings21, constant_values=0)
Md = tf.pad(L22, self.paddings22, constant_values=0)


M = Ma+Mb+Mc+Md+tf.eye(self.N)
# Inverse of M
MI = tf.matmul(tf.matmul(M, self.J), self.J,
     transpose_a=True)
# build
if di is None:
    d2 = tf.constant(np.zeros((self.N, 1)),
         dtype=self.dtype)
else:
    d2 = di
return [tf.matmul(x, M), tf.matmul(MI, d2)]
```

Multiple beam splitters acting on different modes can be cascaded for quantum optics experiments with multimode interferometers.[11]

[11] The code for this example is in `jupyternotebooks/phasespace/beamsplitter.ipynb`.

10.10 Photon Counting Layer

In many applications, we need to measure the expectation value of the number of photons (or bosonic particles) in each channel, which corresponds to the current measured in a photodetector. We introduce a new layer, which evaluates the expected number of photons:

$$\langle \hat{n}_j \rangle = \langle \hat{a}_j^\dagger a_j \rangle \tag{10.36}$$

in each mode by using the derivatives of the $\chi(\boldsymbol{x})$ modes.

We use Eqs. (7.14) and (7.33) for $\mathcal{P} = 0$, as the $\chi(\boldsymbol{x})$ corresponds to symmetric ordering and we have

$$\langle \hat{n}_j \rangle = \langle \hat{a}_j^\dagger \hat{a}_j \rangle_{\mathcal{P}} - \frac{1}{2} = -\frac{1}{2}\left(\frac{\partial^2}{\partial x_{2j}^2} + \frac{\partial^2}{\partial x_{2j+1}^2}\right)\chi_R\bigg|_{\boldsymbol{x}=0} - \frac{1}{2} \tag{10.37}$$

This equation is used in the following layer, which returns the average number of photons (or bosonic particles) in each mode in a vector with dimension $(n, 1)$, denoted `nphoton`. First derivatives are computed by the `tf.gradient` methods. Second derivatives are given by the `tf.jacobian`.[12]

```
class PhotonCountingLayer(layers.Layer):
    """ Photon counting layer

        Return the expected photon number in each mode
        by using the second derivatives of the model

        Given
        x as 1xN vector
        d as Nx1 vector
        g as NxX symmetric matrix

        Build the (1xN) vector of second derivatives of chi
        chiR_xx chiR_yy chiR_zz

        Multiply by (Rq+Rp)/2 and add 1/2

        In the constructor:
        Parameters
        ----------
        N: size of the of vector

        In the call
```

[12] The class definition is in the file `phasespace.py`.

```
        Parameters
        ----------
        c1: real part of chi
        c2: imag part of chi
        pullback: model to be derived wrt x

        Returns
        -------
        nphotons: vector of expected values of the photon number
    """

    def __init__(self, N, **kwargs):
            super(PhotonCountingLayer, self).__init__(**kwargs)
            # vector size
            self.N = N
            # constant matrix to extract the modes
            Rq, Rp, _ = RQRP(N)
            self.Rq = tf.constant(Rq)
            self.Rp = tf.constant(Rp)

    def call(self, c1, c2, chi):
        # c1 and c2 are dummy here, needed to make a model
        # chi is the model
        x = tf.Variable(np.zeros((1, self.N)), dtype=self.dtype)
        with tf.GradientTape() as t1:
            with tf.GradientTape() as t2:
                cr, _ = chi(x)
            cr_x = t2.gradient(cr, x)
        cr_xx=t1.jacobian(cr_x,x)
        # extract the diagonal part
        tmp1  =tf.reshape(cr_xx, [self.N,self.N])
        tmp2 = tf.linalg.diag_part(tmp1)
        lapls = tf.reshape(tmp2,[1,self.N])
        # sum the derivatives to have the photon number
        nqq = tf.matmul(lapls,self.Rq)
        npp = tf.matmul(lapls,self.Rp)
        nboson = -0.5*(nqq+npp)-0.5
        return nboson
```

It is important to understand how the methods `tf.gradients` and `tf.jacobian` work. `tf.gradient` returns the derivatives of a scalar with respect to the components of the tensor x.[13]

Derivatives are evaluated by using the methods `tf.GradientTape`, which records all the coefficients in the graph, used for the derivatives in the neural network. As we are computing second derivatives, we need *two* records, i.e., two "tapes" denoted as `t1` and `t2`.

[13] A detailed example for the `PhotonCountingLayer` is in `jupyternotebooks/phasespace/photoncountinglayer.ipynb`.

As `x` has N components corresponding to the components of $\boldsymbol{x}$, and `tf.gradient` applies to the scalar variable `cr`, corresponding to χ_R, the output tensor `cr_x` is an array with shape $(1, N)$, i.e., it has N components corresponding to the derivatives

$$\left(\frac{\partial \chi_R}{\partial x_0}, \frac{\partial \chi_R}{\partial x_1}, \ldots, \frac{\partial \chi_R}{\partial x_{N-1}},\right) \tag{10.38}$$

evaluated at $\boldsymbol{x} = 0$, which is the value for the variable `x`.

Let us consider the case $N = 4$. As we apply the `tf.jacobian`, the components of `cr_x` are differentiated with respect to the components of the variable `x`. The resulting tensor `cr_xx` has shape $[1, 4, 1, 4]$. The first two indices refer to the elements of `cr_x`. For example, $\partial \chi_R / \partial x_1$ corresponds to the first indices $[0, 1]$: 0 because we have only one element in the first component, and 1 is for x_1 (0 for x_0, 2 for x_2, and so forth). The second two indices are the differentiation variable; for example, the second derivative $\frac{\partial^2 \chi}{\partial x_1 \partial x_2}$ has indices $[0, 1, 0, 2]$. Seemingly, $\frac{\partial^2 \chi}{\partial x_3^2}$ is $[0, 3, 0, 3]$. The first and third indices are redundant here, but this is because this approach is valid for any tensor shape to be derived, and these redundant indices are helpful for higher-dimensional tensors.

The second derivatives of interest to us are the diagonal ones, with components $\partial^2 \chi / \partial x_j^2$ evaluated at $\boldsymbol{x} = 0$ and $j = 0, 1, 2, \ldots, N - 1$. We extract them by first eliminating the redundant indices by using `tf.reshape`, which returns the $N \times N$ matrix of the Hessian of χ_R, i.e., the matrix of the second derivatives. Then we extract the diagonal part by `tf.linalg.diag_part`, and the resulting variable is `lapls`. Finally, we reshape to the shape $[1, N]$, which is needed for the next steps. Other extractions are possible as, e.g., using the method `tf.gather_nd`.

We remark that the described approach is not the most computationally efficient, as we determine all the derivatives and only use some of them. We keep this methodology as simple as is and becomes helpful in later chapters.

To obtain the quantities $\partial_q^2 \chi_R + \partial_p^2 \chi_R$, we right multiply `lapls` with the $N \times n$ matrix $\boldsymbol{R}_q$ to have the vector `nqq` with `shape=(1, n)`:

$$\left(\frac{\partial^2 \chi}{\partial x_0^2}, \frac{\partial^2 \chi}{\partial x_2^2}, \ldots, \frac{\partial^2 \chi}{\partial x_{N-2}^2}\right). \tag{10.39}$$

Seemingly, by right multiplying by the matrix $\boldsymbol{R}_p$ to have the vector `npp` with `shape=(1,n)`

$$\left(\frac{\partial^2 \chi}{\partial x_1^2}, \frac{\partial^2 \chi}{\partial x_3^2}, \ldots, \frac{\partial^2 \chi}{\partial x_{N-1}^2}\right). \tag{10.40}$$

evaluated at $\boldsymbol{x} = 0$. Following Eq. (10.37), by summing the two vectors, multiplying by $1/2$, and adding $1/2$, we have the target vector `nboson` containing the number of photons in each mode.

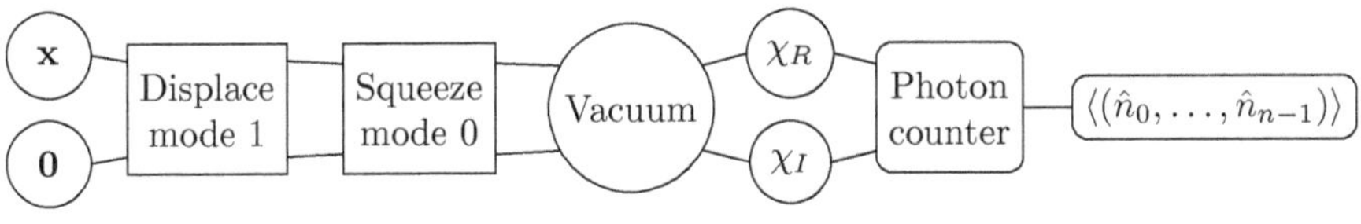

Fig. 10.3 Bug train representation for the model with two modes, one in a squeezed vacuum and one in a coherent state. A photon counting layer is used for the expected values of the number of bosons in each mode. The state is obtained by pulling back from vacuum a single-mode squeezing layer for mode 0 and a Glauber layer for mode 1

We test the `PhotonCountingLayer` by considering the model in Fig. 10.3, which contains one squeezed vacuum mode 0 and one coherent state in mode 1. The model is built by the following code.[14]

```
# parameters for the squeezer
r_np=0.1
phia_np=np.pi/2
# define the squeezing layer
squeezer=
SingleModeSqueezerLayer(N, r_np=r_np, theta_np=phia_np,
n_squeezed=0, trainable=False)
# parameters for the coherent state
A_np=1000
lambda_np=n.pi/2
dinput=np.zeros((N,1))
dinput[2]=np.sqrt(2)*A_np*np.cos(lambda_np)
dinput[3]=np.sqrt(2)*A_np*np.sin(lambda_np)
displacer = DisplacementLayerConstant(dinput)
# connect layers
xin = tf.keras.layers.Input(N)
x1, a1 = squeezer(xin)
x0, a0 = displacer(x1,a1)
chir, chii=vacuum(x0,a0)
modelPC=tf.keras.Model(inputs = xin, outputs=[chir,chii])
# add the photon counter layer
photon_current=PhotonCountingLayer(N) # define the layer
n_out = photon_current(chir, chii, modelPC) # output tensor
# model with photon counter
NphotonPC=tf.keras.model(inputs=xin, outputs=n_out)
```

The displacement vector `dinput` has components

$$d_0 = 0$$

[14] The code is in `jupyternotebooks/BellBS.ipynb`; other examples for the `PhotonCountingLayer` are in `jupyternotebooks/phasespace/photoncounting.ipynb`.

$$d_1 = 0$$
$$d_2 = \sqrt{2}A\cos(\lambda)$$
$$d_3 = \sqrt{2}A\sin(\lambda)\ .$$

$d_{0,1}$ refer to mode zero, being $\langle \hat{a}_0 \rangle = 0$ and $d_{2,3}$ to mode 1, with

$$\langle \hat{a}_1 \rangle = A\, e^{i\lambda}\ . \tag{10.41}$$

We run the model by using any input for $\boldsymbol{x}$, corresponding to one of the $(1, N)$ vectors `xtrain`, as the number of particles does not depend on $\boldsymbol{x}$.

```
print(NphotonPC(xtrain))
# output is
# tf.tensor([[1.e-02 1.00e+06]], shape=(1,2),dtype=float32)
```

If one applies the photon counting layer, this returns 0 (within numerical truncation) for mode 0, and A^2 for mode 1, as expected for a coherent state.[15]

10.11 Homodyne Detection

Beam splitters are widely used for quantum optical computing and for Bell measurements. Figure 10.4 shows an example of homodyne detection on a squeezed vacuum.

We consider a two-mode system, in which the mode with index 0 is a squeezed vacuum and the mode with index 1 is a coherent state. They correspond to the two modes that enter the beam splitter in Fig. 10.4. After the beam splitter, we have mixed modes with amplitudes $\hat{\tilde{a}}_0$ and $\hat{\tilde{a}}_1$.

Considering a 50 : 50 beam splitter (parameters in the beam splitter layer $\theta = \pi/4$, $\phi_0 = 0$ and $\phi_1 = 0$), we have

$$\hat{\tilde{a}}_0 = \frac{1}{\sqrt{2}}(\hat{a}_0 - \hat{a}_1) \tag{10.42}$$

$$\hat{\tilde{a}}_1 = \frac{1}{\sqrt{2}}(\hat{a}_1 + \hat{a}_0)\ . \tag{10.43}$$

We measure the expected number of photons for the two modes by the photocounters:

[15] The full code is in the file `jupyternotebooks/phasespace/BellBS.ipynb`.

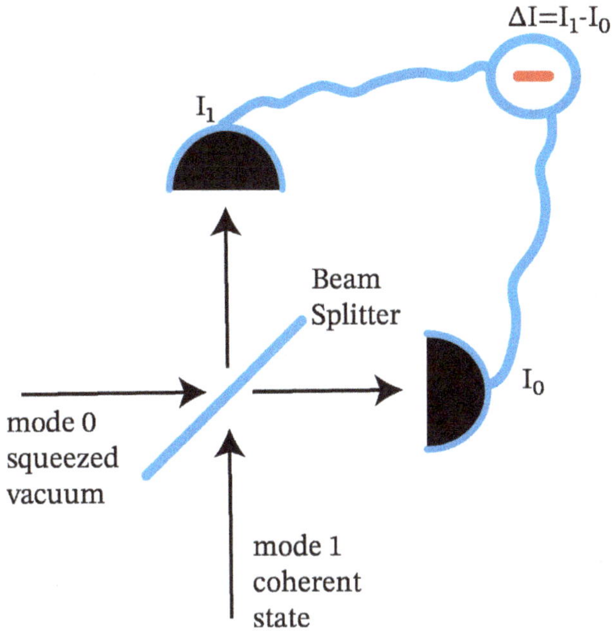

Fig. 10.4 In homodyne detection, a mode 0 beats with a local oscillator in a coherent state (mode 1), and the two are mixed by a 50 : 50 beam splitter. The differential current ΔI enables to measure the quadrature operator of mode 0, as detailed in the text

$$\langle I_0 \rangle = \langle \hat{\tilde{n}}_0 \rangle = \langle \hat{\tilde{a}}_0^{\dagger} \hat{\tilde{a}}_0 \rangle \tag{10.44}$$

$$\langle I_1 \rangle = \langle \hat{\tilde{n}}_1 \rangle = \langle \hat{\tilde{a}}_1^{\dagger} \hat{\tilde{a}}_1 \rangle \tag{10.45}$$

and subtract the two photocurrents. We have the resulting signal:

$$\langle \Delta I \rangle = \langle \hat{\tilde{a}}_1^{\dagger} \hat{\tilde{a}}_1 - \hat{\tilde{a}}_0^{\dagger} \hat{\tilde{a}}_0 \rangle = \langle \hat{a}_1^{\dagger} \hat{a}_0 + \hat{a}_0^{\dagger} \hat{a}_1 \rangle \tag{10.46}$$

Mode 1 is the "local oscillator," i.e., a coherent intense laser beam. If mode 1 is in a coherent state with a large number of photons, the amplitude fluctuation can be neglected, and we replace the operator $\hat{a}_1$ with its mean value α_1. For a coherent state, we have

$$\hat{a}_1 \simeq \alpha_1 = A\, e^{i\lambda}\ , \tag{10.47}$$

with λ the phase and $A = |\alpha_1|$ the modulus of α_1.

If we let $\hat{a}_1 \simeq Ae^{i\lambda}$ in Eq. (10.46), we have [2]

$$\langle \Delta I \rangle = A\langle \hat{a}_0 e^{-i\lambda} + \hat{a}_0^{\dagger} e^{i\lambda} \rangle = \sqrt{2} A \langle \hat{x}_0(\lambda) \rangle \tag{10.48}$$

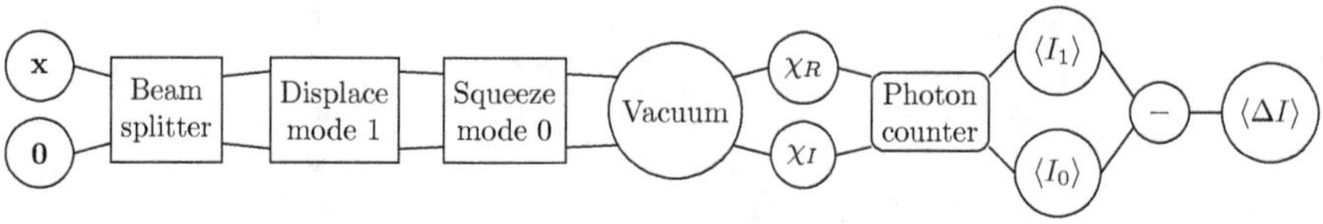

Fig. 10.5 Model for homodyne detection on $n = 2$ modes, with mode 1 representing the local oscillator

where $\hat{x}_0(\lambda)$ is the quadrature operator [1]:

$$\hat{x}_0(\lambda) = \frac{\hat{a}_0 e^{-i\lambda} + \hat{a}_0^\dagger e^{i\lambda}}{\sqrt{2}}, \tag{10.49}$$

such that, if $\lambda = 0$, we have $\hat{x}_0(\lambda) = \hat{q}_0$ and, if $\lambda = \pi/2$, $\hat{x}_0(\lambda) = \hat{p}_0$.

The homodyne detection scheme enables to measure any canonical observable by changing the phase of the coherent oscillator, and it is used for Bell measurements.

Figure 10.5 shows the neural network model for the homodyne detection in the particular case mode 0 is a squeezed vacuum. By pulling back a displacement layer for mode 1, and a single-mode squeezing layer for mode 0, we create the input 2-mode state, which then passes through a beam splitter as a further pullback. Note that the order of the squeezer and the displacer is not relevant as the they act on different modes (the two corresponding operators commute). This is different from the case discussed above (Fig. 10.2), where the layers affect the same mode. The model is built by the following code.

```
# parameters for the squeezer
r_np=1.0
phia_np=np.pi/2
# define the squeezing layer
squeezer=
SingleModeSqueezerLayer(N, r_np=r_np, theta_np=phia_np,
                         n_squeezed=0, trainable=False)
# parameters for the coherent state
A_np=10
lambda_np=n.pi/2
dinput=np.zeros((N,1))
dinput[2]=np.sqrt(2)*A_np*np.cos(lambda_np)
dinput[3]=np.sqrt(2)*A_np*np.sin(lambda_np)
displacer = DisplacementLayer(dinput)
# beam splitter
bs0 =
BeamSplitterLayer(N, theta=np.pi/4, phi0=0,
                  phi1=0, n_0=0, n_1=1, trainable_theta = True)
# connect layers
xin = tf.keras.layers.Input(N)
x2, a2 = bs0(xin)
x1, a1 = squeezer(x2, a2)
```

```
x0, a0 = displacer(x1,a1)
chir, chii=vacuum(x0,a0)
BellBs=tf.keras.Model(inputs = xin, outputs=[chir,chii])
```

In the neural network implementation, A and λ are determined by the displacement vector $\boldsymbol{d}_{\text{input}}$ of the Glauber layer, which creates the coherent state from the vacuum.

If mode 1 is displaced and mode 0 is left in a vacuum state, we have for the components of $\boldsymbol{d}_{\text{input}}$, $d_0 = d_1 = 0$, $d_2 = \sqrt{2}A\cos(\lambda)$, and $d_3 = \sqrt{2}A\sin(\lambda)$. In defining the displacement layer, we use the $N \times 1$ vector ($N = 4$ as we have two modes):

$$\boldsymbol{d}_{\text{input}} = \begin{pmatrix} 0 \\ 0 \\ \sqrt{2}A\cos(\lambda) \\ \sqrt{2}A\sin(\lambda) \end{pmatrix} \tag{10.50}$$

The components d_0 and d_1 of mode 0 are left equal to zero, so that the displacement operator does not affect mode 0. The single-mode squeezing layer only acts on mode 0 and leaves unchanged mode 1.

10.12 Measuring the Expected Value of the Quadrature Operator

We need to measure the photon number in the model; for this purpose, we will use a `PhotonCountingLayer` as follows.

```
# add the photon counting layer
photon_current=PhotonCountingLayer(N) # define the layer
n_out = photon_current(chir, chii, modelPC) # output tensor
# model with photon counter
NphotonPC=tf.keras.model(inputs=xin, outputs=n_out)
print(NphotonPC(xtrain))
# output is
# tf.tensor([[50.69 50.69]], shape=(1,2),dtype=float32)
```

The call to the photon counter returns the same value for the average currents, which are (within numerical precision) half of the input expected number of photons in the coherent state and in the squeezed state, which is $A^2 + \sinh(r)^2 = 101.38$ in the example, with r corresponding to `r_np=1.0`.

We obtain the difference $\langle \Delta I \rangle$ between the two currents by using a dedicated `TensorFlow` function, which returns the difference, as follows.

```
@tf.function
def differentia_detector(nphoton, A=1.0)
""" takes as input the tensor of nboson, returns \varDelta I"""
    n0=tf.slice(nphoton, [0,0],[1,1])
    n1=tf.slice(nphoton, [0,1],[1,1])
    return (n1-n0)/(np.sqrt(2)*A)
```

The method `tf.slice` returns the two elements of the tensor, the second argument is the starting element, and the third argument is the number of elements to extract. Note that we put as input the amplitude of the coherent state to scale out in the result. The function is evaluated as follows and returns 0 as expected, being the average value of the momentum operator $\hat{p}_0 = \hat{x}_0(\pi/2)$ on the vacuum state ($\lambda = \pi/2$ in the example).

```
differential_detector(Nphoton(xtrain),A_np)
# output is
# tf.tensor([[0.]], shape=(1,1), dtype=float32)
```

References

1. S.M. Barnett, P.M. Radmore, *Methods in Theoretical Quantum Optics* (Oxford University Press, New York, 1997)
2. X. Wang, T. Hiroshima, A. Tomita, M. Hayashi, Phys. Rep. **448**, 1 (2007). https://doi.org/10.1016/j.physrep.2007.04.005

Chapter 11
Uncertainties and Entanglement

One Index to rule them all

Abstract We study advanced layers for computing uncertainties and entanglement. These layers require higher-order derivatives of the characteristic functions by the `tf.GradientTape` method. We show the use of machine learning for maximizing entanglement with beam splitters.

11.1 Introduction

For many applications, we are interested not only in expected values as $\langle \hat{n}_0 \rangle$, but also in higher-order moments as $\langle \hat{n}_0^2 \rangle$. This is the case, for example, when we want to measure uncertainties as

$$\langle \Delta \hat{n}_0{}^2 \rangle = \langle \left(\hat{n}_0 - \langle \hat{n}_0 \rangle \right)^2 \rangle = \langle \hat{n}_0^2 \rangle - \langle \hat{n}_0 \rangle^2 \,. \tag{11.1}$$

In the general case, i.e., not limited to Gaussian states, computing $\langle \hat{n}_0^2 \rangle$ requires higher derivatives of the characteristic function.

We start considering the general case. We have, for a mode with index $j = 0, 1, \ldots, n-1$, the normal ordering product

$$\langle n_j^2 \rangle = \langle \hat{a}_j^\dagger \hat{a}_j \hat{a}_j^\dagger \hat{a}_j \rangle = \langle \hat{a}_j^\dagger \hat{a}_j^\dagger \hat{a}_j \hat{a}_j \rangle + \langle \hat{a}_j^\dagger \hat{a}_j \rangle \tag{11.2}$$

where we used $[\hat{a}_j, \hat{a}_j^\dagger] = \hat{a}_j \hat{a}_j^\dagger - \hat{a}_j^\dagger \hat{a}_j = 1$.

We know from Sect. 10.10

$$\langle \hat{a}_j^\dagger \hat{a}_j \rangle = \langle \hat{n}_j \rangle = -\left. \frac{\partial^2 \chi}{\partial z_j \, \partial z_j^*} \right|_{z=0} = -\frac{1}{2} \left(\partial_{q_j}^2 + \partial_{p_j}^2 \right) \chi_R \Big|_{x=0} - \frac{1}{2} \,. \tag{11.3}$$

C. Conti, *Quantum Machine Learning*, Quantum Science and Technology,
https://doi.org/10.1007/978-3-031-44226-1_11

We also have

$$\langle \hat{a}_j^\dagger \hat{a}_j^\dagger \hat{a}_j \hat{a}_j \rangle = \left(\frac{\partial}{\partial z_j}\right)^2 \left(-\frac{\partial}{\partial z_j^*}\right)^2 \chi \Big|_{x=0} - 2\left(\frac{\partial}{\partial z_j}\right)\left(-\frac{\partial}{\partial z_j^*}\right)\chi\Big|_{x=0} + \frac{1}{2}. \tag{11.4}$$

Equation (11.4) is expressed in terms of the derivatives w.r.t. q_j and p_j as follows:

$$\langle \hat{a}_j^\dagger \hat{a}_j^\dagger \hat{a}_j \hat{a}_j \rangle = \frac{1}{4}\left(\partial_{q_j}^2 + \partial_{p_j}^2\right)^2 \chi_R \Big|_{x=0} + \left(\partial_{q_j}^2 + \partial_{p_j}^2\right)\chi_R\Big|_{x=0} + \frac{1}{2} \tag{11.5}$$

where, as above, $q_j = x_{2j}$ and $p_j = x_{2j+1}$ with $j = 0, 1, ..., N/2 - 1$.

Only the real part of χ_R enters in Eq. (11.5) as $\langle n_j^2 \rangle$ and Eq. (11.2) is real-valued. The proof of Eqs. (11.4) and (11.5) is given in Sect. 11.5.

Hence, we have the following relations, which are helpful for designing the corresponding layers:

$$\begin{aligned}
\langle \hat{n}_j \rangle &= -\frac{1}{2}\nabla_j^2 \chi_R\Big|_{x=0} - \frac{1}{2} \\
\langle \hat{n}_j^2 \rangle &= \frac{1}{4}\nabla_j^4 \chi_R\Big|_{x=0} + \frac{1}{2}\nabla_j^2 \chi_R\Big|_{x=0} \\
\langle \Delta \hat{n}_j^2 \rangle &= \frac{1}{4}\nabla_j^4 \chi_R\Big|_{x=0} - \frac{1}{4}\left(\nabla_j^2 \chi_R\Big|_{x=0}\right)^2 - \frac{1}{4}
\end{aligned} \tag{11.6}$$

where $\nabla_j^2 = \partial_{q_j}^2 + \partial_{p_j}^2$ and $\nabla_j^4 = \left(\partial_{q_j}^2 + \partial_{p_j}^2\right)^2$.

For a general model, we need to evaluate the fourth-order derivatives: $\frac{\partial^4 \chi}{\partial x_s \partial x_p \partial x_q \partial x_r}$, at $\boldsymbol{x} = 0$, and combine some of them to compute the uncertainties. The computation of the fourth-order derivatives may be resource-demanding as it grows with N^4. For the specific case of Gaussian states, it can be simplified as the fourth-order derivatives are expressed in terms of the covariance matrix $\boldsymbol{g}$.

We first implement a general layer by using the `tf.jacobian` method in Sect. 11.2, and then we consider the special case of Gaussian states in Sect. 11.3.

11.2 The `HeisenbergLayer` for General States

For higher-order derivatives of χ, we write dedicated layers, which we combine to have the variances. We need the Laplacian ∇_j^2 and the square Laplacian ∇_j^4, also known as the "biharmonic" operator. Hence, we detail below the implementation of the corresponding `LaplacianLayer` and `BiharmonicLayer`.

11.2.1 The LaplacianLayer

We have discussed the PhotonCountingLayer in Sect. 10.10 that returns $\langle \hat{n}_j \rangle$. Here we start considering the LaplacianLayer, which returns $\nabla_j^2 \chi_R$ at $\boldsymbol{x} = 0$ of a tensor with shape (1,n), where the first index is 0 (derivatives of a scalar) and the other index corresponds to j. A similar layer is the CovarianceLayer in Sect. 9.7.

Here we will repeat the details on computing derivatives for the sake of completeness. The code for the LaplacianLayer is as follows.[1]

```
class LaplacianLayer(layers.Layer):
    """ Laplacian per mode at x=0

        The output is the vector

        laplacian_j = (d_{q_j}^2+d_{p_j}^2) chir at x=0

        Given
        x as 1xN vector
        d as Nx1 vector
        g as NxN symmetric matrix

       In the constructor:
        Parameters
        ----------
        N: size of the of vector

        Returns
        -------
        Laplacian : vector of laplacian (n,1)
    """
    def __init__(self, N,
                 **kwargs):
        super(LaplacianLayer, self).__init__(**kwargs)
        # vector size
        self.N = N
        # constant matrix to extract the modes
        Rq, Rp, _ = RQRP(N)
        self.Rq = tf.constant(Rq)
        self.Rp = tf.constant(Rp)
        # constant tensor of indices
        # for the diag derivatives
        list1 =[]
        for i in range(N):
            list1.append([0,i]*2);
        # list1 is
        # [[0,0,0,0],[0,1,0,1],[0,2,0,2],[0,3,0,3]] for N=4
```

[1] The code is in the file phasespace.py.

```
        self.indices_diag=tf.constant(list1)
        # zero constant for derivatives
        self.x0 = tf.constant(np.zeros((1,self.N)))

    def call(self, c1, c2, chi):
        # c1 and c2 are dummy needed for a model
        # chi is the model
        x = self.x0
        with tf.GradientTape() as t1:
            t1.watch(x)
            with tf.GradientTape() as t2:
            t2.watch(x)
                cr, _ = chi(x)
            # first the gradient shape [1,N]
            cr_x = t2.gradient(cr, x)
        # then the jacobian (shape [1,N,1,N])
        cr_xy=t1.jacobian(cr_x,x)
        # gather the diagonal part
        cr_xx1=tf.gather_nd(cr_xy,self.indices_diag)
        # reshape as [1,N]
        cr_xx=tf.reshape(cr_xx1,[1,self.N])
        # sum the derivatives to have the Laplacian
        dqq = tf.matmul(cr_xx,self.Rq)
        dpp = tf.matmul(cr_xx,self.Rp)
        lapl = dqq+dpp
        return lapl
```

This layer evaluates the Hessian, i.e., the matrix of derivatives $\partial^2\chi_R/\partial x_j\partial x_k$, and then extracts the diagonal part and combines the elements to have the Laplacian for each mode. To gather the diagonal part, we use the method `tf.gather_nd`. The Hessian matrix in the Variable `cr_xy` has shape `[1,N,1,N]`, the first two indices corresponding to j and the second two indices corresponding to k. For example, when $N = 2$, the diagonal elements for which $j = k$ correspond to elements $[[0, 0, 0, 0], [0, 1, 0, 1]]$.

To collect these components, we define a list of indices in `self.indices_diag` and then we extract the diagonal elements in the Variable `cr_xx` with shape $[1, N]$. As done for the `PhotonCountingLayer`, we build the Laplacian by using the matrices $\boldsymbol{R}_q$ and $\boldsymbol{R}_p$.[2]

11.2.2 The `BiharmonicLayer`

The biharmonic layer returns a tensor with shape $[1, n]$, which contains $\nabla_j^4\chi_R$ at $\boldsymbol{x} = 0$ with $j = 0, 1, \ldots, n-1$. We use a double call to `tf.jacobian` to evaluate

[2] All these operations and those below are detailed in the notebook `jupyternotebooks/phasespeace/uncertainty.ipynb`.

all the fourth-order derivatives with respect to the component of $\boldsymbol{x}$. Then we use tf.gather_nd to return the diagonal fourth-order derivatives and compose them as in the LaplacianLayer. The code is as follows.[3]

```
class BiharmonicLayer(layers.Layer):
    """ Returns the biharmonic per mode at x=0

        The output is the vector

        biharmonic_j = ((d_{q_j}^2+d_{p_j}^2))^2 chir at x=0

        and the

        laplacian_j = ((d_{q_j}^2+d_{p_j}^2)) chir at x=0

       In the constructor

       Parameters
       ----------
        N: size of the of vector

        In the call

        Parameters
        ----------
        c1: real part of chi
        c2: imag part of chi
        pullback: model to be derived wrt x

        Returns
        -------
        biharmonic : vector of laplacian squared (1,n)
        laplacian : vector of laplacian (1,n)
    """
    def __init__(self, N,
                 **kwargs):
        super(BiharmonicLayer, self).__init__(**kwargs)
        # vector size
        self.N = N
        # constant matrix to extract the modes
        Rq, Rp, _ = RQRP(N)
        self.Rq = tf.constant(Rq)
        self.Rp = tf.constant(Rp)
        # indices diagonal biharmonic and Laplacian
        list_2x =[]
        for i in range(N):
            list_2x.append([0,i]*2);
        # list_2x for N=4 is
        # [[0,0,0,0],[0,1,0,1],[0,2,0,2],[0,3,0,3]]
```

[3] The code is in the file phasespace.py.

```
            self.indices_2x=tf.constant(list_2x)
            # zero constant for derivatives
            self.x0 = tf.constant(np.zeros((1,self.N)))


    def call(self, c1, c2, chi):
        # c1 and c2 dummy needed to make a model
        # chi is the model
        x = self.x0
        with tf.GradientTape() as t4:
            t4.watch(x)
            with tf.GradientTape() as t3:
                t3.watch(x)
                with tf.GradientTape() as t2:
                    t2.watch(x)
                    with tf.GradientTape() as t1:
                        t1.watch(x)
                        cr, _ = chi(x) # shape [1,1]
                    cr_x = t1.gradient(cr, x) # shape [1,N]
                cr_xy=t2.jacobian(cr_x,x) # shape [1,N,1,N]
            cr_xyz=t3.jacobian(cr_xy,x) # shape [1,N,1,N,1,N]
        cr_xyzw=t4.jacobian(cr_xyz,x) # shape [1,N,1,N,1,N,1,N]
        # reshape as [N,N,N,N]
        cr_xyzw=tf.reshape(cr_xyzw,[self.N,self.N,self.N,self.N])
        # extract the diagonal part by gather and shape as [1,N]
        cr_2x=tf.reshape(tf.gather_nd(cr_xy,self.indices_2x),
                            [1, self.N])

        # sum the derivatives to have the squared laplacian
        dqq = tf.matmul(cr_2x,self.Rq)
        dpp = tf.matmul(cr_2x,self.Rp)
        laplacian = dqq+dpp

        # build the biharmonic
        biharmonic = tf.zeros_like(laplacian) #[1,N]
        for j in tf.range(self.n):
            qj = 2*j
            pj = 2*j+1
            dqqpp = tf.gather_nd(cr_xyzw, [[qj, qj, pj, pj]])
            dqqqq = tf.gather_nd(cr_xyzw, [[qj, qj, qj, qj]])
            dpppp = tf.gather_nd(cr_xyzw, [[pj, pj, pj, pj]])
            biharmonic = tf.tensor_scatter_nd_add(
                biharmonic,[[0,j]], dqqqq+dpppp+2*dqqpp)

        return biharmonic, laplacian
```

We use four `tf.GradientTape`, to store all the coefficients needed for the variables. After evaluating the gradient of `cr` by `tf.gradient`, we evaluate the fourth-order derivatives, by calling three times the `tf.jacobian`. Note the tensor `cr_xyzw` with the fourth-order derivatives has rank-8 shape `[1,N,1,N,1,N,1,N]`, as we have *four* variables, so that—for example—the derivatives with respect to x_0, x_2, x_3, and x_3 when $N = 4$ will correspond to the element [0,0,0,2,0,3,0,3].

To simplify the notation and the extraction of the diagonal part, we first reshape the tensor cr_xyzw to shape [N,N,N,N] to eliminate the redundant dimensions. Then we use tf.gather_nd to extract a list of elements corresponding, e.g., to the indices [[qj,qj,pj,pj]] for the term $\partial^4\chi_R/\partial q_j^2 \partial p_j^2$. Note that we use a for loop for computing the biharmonic at any mode, the for spans the mode index j, and we build the indices according to $q_j = 2j$ and $p_j = 2j + 1$.

We obtain the tensor dqq for the vector of elements $\partial^2_{q_j}\chi_R$ at $\boldsymbol{x} = 0$, the tensor dpp for the vector of elements $\partial^2_{p_j}\chi_R$ at $\boldsymbol{x} = 0$, the tensor dqqqq for the vector of elements $\partial^4_{q_j}\chi_R$ at $\boldsymbol{x} = 0$, the tensor dpppp for $\partial^4_{q_j}\chi_R$ at $\boldsymbol{x} = 0$, and the tensor dqqpp for $\partial^2_{q_j}\partial^2_{p_j}\chi_R$ at $\boldsymbol{x} = 0$.

Linear combinations of these tensors furnish the output tensors Laplacian and biharmonic.[4]

Note that for computational efficiency, we let the BiharmonicLayer return both the biharmonic and the Laplacian tensor, as during the computation of the biharmonic operator, one also computes the Laplacian. The BiharmonicLayer is helpful in the HeisenbergLayer below.

We also note that the computation of the biharmonic operator is demanding for large N, as it scales with N^4.

11.2.3 The HeisenbergLayer

The HeisenbergLayer returns the quantities $\langle\hat{n}\rangle$, $\langle\hat{n}^2\rangle$, and $\langle\Delta n\rangle$, by using the BiharmonicLayer.

The HeisenberLayer contains a BiharmonicLayer among its properties, and call it to evaluate the Laplacian and the biharmonic operator to compute the quantities in Eq. (11.6).

The code is as follows.[5]

```
class HeisenbergLayer(layers.Layer):
""" Compute mean values and uncertainties

    Returns
    -------
    <n>
    <n^2>
    <Dn^2>
        for each mode

        Use the BiharmonicLayer
```

[4] All this operations and those below are detailed in the notebook jupyternotebooks/phasespace/uncertainty.ipynb.

[5] The code is in the file phasespace.py.

```
        Given
        x as 1xN vector
        d as Nx1 vector
        g as NxN symmetric matrix

       In the constructor,

       Parameters
       ----------
       N: size of the of vector

        In the call,

       Parameters
       ----------
        c1: real part of chi (dummy)
        c2: imag part of chi (dummy)
        pullback: model to be derived wrt x

        Returns
        -------
        nboson: expected number of bosons per mode (1,n)
        n2: expected squared nboson  per mode (1,n)
        Dn2: expected values of squared uncertainties (1,n)
        """

    def __init__(self, N,
                 **kwargs):
        super(HeisenbergLayer, self).__init__(**kwargs)
        # vector size
            self.N = N
            # BiharmonicLayer
            self.Biharmonic=BiharmonicLayer(N)

    def call(self, c1, c2, chi):
        # c1 and c2 are dummy as in LaplacianLayer
        # chi is the model

        # call the biharmonic layer
        lapl2, lapl = self.Biharmonic(c1,c2,chi)

        # evaluate the number of bosons
        nboson = -0.5*lapl-0.5

        # evaluate the n2
        n2 = 0.25*lapl2+0.5*lapl

        # evaluate the uncertainties
        Dn2=0.25*lapl2 -0.25*tf.square(lapl)-0.25

        return nboson, n2, Dn2
```

We test the layer by creating a state with two modes, one vacuum mode (mode 0) and one mode as coherent state with complex displacement α (mode 1). We verify that for the vacuum, we have

$$\begin{aligned}\langle\hat{n}_0\rangle &= 0\\ \langle\hat{n}_0^2\rangle &= 0\\ \langle\Delta\hat{n}_0\rangle &= 0,\end{aligned} \tag{11.7}$$

and for the coherent state, we have[6]

$$\begin{aligned}\langle\hat{n}_1\rangle &= |\alpha|^2\\ \langle\hat{n}_1^2\rangle &= |\alpha|^2 + |\alpha|^4\\ \langle\Delta\hat{n}_1\rangle &= |\alpha|^2.\end{aligned} \tag{11.8}$$

11.3 Heisenberg Layer for Gaussian States

The `BiharmonicLayer` is a general-purpose layer for the uncertainties of any model. However, computing the biharmonic operator requires three calls to the `tf.Jacobian`, which is computationally demanding.

Gaussian states are determined by the covariance matrices $\boldsymbol{g}$ and the displacement $\boldsymbol{d}$, and we can use them to speed up computing the biharmonic operator.

For a Gaussian state, we have

$$\begin{aligned}\left.\frac{\partial^4\chi}{\partial x_s\partial x_p\partial x_q\partial x_r}\right|_{x=0} \\ &= \frac{1}{4}g_{sp}g_{qr} + \frac{1}{4}g_{sq}g_{pr} + \frac{1}{4}g_{sr}g_{pq}\\ &+ \frac{1}{2}g_{sp}d_q\,d_r + \frac{1}{2}g_{sq}d_p\,d_r + \frac{1}{2}g_{sr}d_p\,d_q\\ &+ \frac{1}{2}g_{pq}d_r\,d_s + \frac{1}{2}g_{pr}d_q\,d_s + \frac{1}{2}g_{qr}d_p\,d_s\\ &+ d_s d_p d_q d_r\ ,\end{aligned} \tag{11.9}$$

whose demonstration is given in Sect. 11.5.

[6] Details in the notebook `jupyternotebooks/phasespace/uncertainty.ipynb`.

When considering the diagonal terms, we have (no implicit sum is present in the following formula)

$$\left.\frac{\partial^4 \chi}{\partial x_q^4}\right|_{x=0} = \frac{3}{4} g_{qq}^2 + 3\, g_{qq} d_q^2 + d_q^4 \,, \tag{11.10}$$

Seemingly, we have

$$\left.\frac{\partial^4 \chi}{\partial x_q^2 \partial x_p^2}\right|_{x=0} = \frac{1}{4} g_{qq} g_{pp} + \frac{1}{2} g_{pq}^2 + \frac{1}{2} g_{pp} d_q^2 + \frac{1}{2} g_{qq} d_p^2 + 2 g_{pq} d_p \, d_q + d_p^2 d_q^2 \,. \tag{11.11}$$

Hence, for Gaussian states, we do not need to evaluate explicitly the fourth-order derivatives, but we can combine the components of $\boldsymbol{g}$ and $\boldsymbol{d}$.

For a model with many layers, we do not know $\boldsymbol{g}$ and $\boldsymbol{d}$, but we can compute them by the covariance layer (see Sect. 9.7). After that, we can use the `TensorFlow` methods for the outer product of tensors to build the derivatives in Eq. (11.9).

We define a dedicated biharmonic layer for Gaussian states as follows.

```
class BiharmonicGaussianLayer(layers.Layer):
    """ Biharmonic at x=0 for Gaussian state
        Returns
        -------
        biharmonic : vector of laplacian squared (n,1)
        laplacian : vector of laplacian (n,1)
    """
    def __init__(self, N,
                 **kwargs):
            super(BiharmonicGaussianLayer,
            ↪   self).__init__(**kwargs)
            # vector size
            self.N = N
            # covariance layer
            self.covariance = CovarianceLayer(N)
            # constant matrix to extract the modes
            Rq, Rp, _ = RQRP(N)
            self.Rq = tf.constant(Rq)
            self.Rp = tf.constant(Rp)


    def call(self, c1, c2, chi):

        # evaluate the covariace and the displacement
        g, d, hessian = self.covariance(c1, c2, chi)

        # evaluate the Laplacian by diagonal elements of Hessian
        cr_2x =
        ↪   tf.reshape(tf.linalg.diag_part(hessian),[1,self.N])
```

```
        # sum the derivatives to have the squared Laplacian
        dqq = tf.matmul(cr_2x,self.Rq)
        dpp = tf.matmul(cr_2x,self.Rp)
        laplacian = dqq+dpp

        # evaluate the biharmonic by diag g and d
        d = tf.squeeze(d) #[N,] dj
        gd =tf.linalg.diag_part(g) #[N,] gjj
        d2 = tf.square(d) #[N,] dj^2
        d4 = tf.tensordot(d2,d2,axes=0) #[N,N] dj^2 dk^2
        dd = tf.tensordot(d,d, axes=0) #[N,N] dj dk

        djjkk= 0.25*tf.tensordot(gd,gd,axes=0)+\
            0.5*tf.square(g)+\
                0.5*tf.tensordot(gd, d2,axes=0)+\
                    0.5*tf.tensordot(d2, gd,axes=0)+\
                        2.0*tf.multiply(g,dd)+d4

        biharmonic = tf.zeros_like(laplacian) #[1,N]
        for j in tf.range(self.n):
            qj = 2*j
            pj = 2*j+1
            dqqpp = tf.gather_nd(djjkk, [[qj, pj]])
            dqqqq = tf.gather_nd(djjkk, [[qj, qj]])
            dpppp = tf.gather_nd(djjkk, [[pj, pj]])
            biharmonic = tf.tensor_scatter_nd_add(
                biharmonic,[[0,j]],
                dqqqq+dpppp+2*dqqpp)

        return biharmonic, laplacian
```

The `BiharmonicGaussianLayer` contains a `CovarianceLayer`, which is used to retrieve $\boldsymbol{g}$ and $\boldsymbol{d}$ and used to compute the square Laplacian by Eq. (11.10) and also the Laplacian. Various external products of tensors are used to build the terms in Eq. (11.11), and then we use a `for` loop to gather the elements to compose the `biharmonic` tensor. We use the `BiharmonicGaussianLayer` to define a specialized `HeisenbergGaussianLayer` as follows.

```
class HeisenbergGaussianLayer(layers.Layer):

    def __init__(self, N,**kwargs):
        super(HeisenbergGaussianLayer, self).__init__(**kwargs)
        # vector size
        self.N = N
        # BiharmonicLayer
        self.Biharmonic=BiharmonicGaussianLayer(N)

    def call(self, c1, c2, chi):

        # call the biharmonic layer
```

```
lapl2, lapl = self.Biharmonic(c1,c2,chi)

# evaluate the number of bosons
nboson = -0.5*lapl-0.5

# evaluate the n2
nboson2 = 0.25*lapl2+0.5*lapl

# evaluate the uncertainties
Dnboson2=0.25*lapl2 -0.25*tf.square(lapl)-0.25

return nboson, nboson2, Dnboson2
```

The use of these specialized Gaussian layers speeds up evaluating the uncertainties for Gaussian states.[7]

11.4 Testing the `HeinsenbergLayer` with a Squeezed State

We test the former layer by considering a two-mode state with mode 0 squeezed and mode 1 vacuum. For a squeezed state, with squeezing parameter $\zeta = r\,e^{\mathrm{i}\theta}$, we have [1]

$$\begin{aligned} \langle \hat{n}_0 \rangle &= n_0 = \sinh(r)^2 \\ \langle \hat{n}_0^2 \rangle &= 3n_0^2 + 2n_0 \\ \langle \Delta \hat{n}_0^2 \rangle &= 2n_0^2 + 2n_0 \,. \end{aligned} \tag{11.12}$$

In the notebook `jupyternotebooks/phasespace/singlemodesqueezer.ipynb`, we verify that the output of the model is in agreement with Eqs. (11.12).

11.5 Proof of Eqs. (11.4) and (11.5)* and (11.9)*

To prove Eq. (11.4), we use the derivatives of the characteristic function.

Proof The expectation value

$$\langle \hat{a}_j^\dagger \hat{a}_j^\dagger \hat{a}_j \hat{a}_j \rangle = \mathrm{Tr}\left[\hat{\rho}\hat{a}_j^\dagger \hat{a}_j^\dagger \hat{a}_j \hat{a}_j\right] \tag{11.13}$$

is written as

[7] These layers are tested and compared in the notebook `jupyternotebooks/phasespace/uncertainty.ipynb`.

$$\langle \hat{a}_j^\dagger \hat{a}_j^\dagger \hat{a}_j \hat{a}_j \rangle = \mathrm{Tr} \left\{ \hat{\rho} \left[\left(\frac{\partial}{\partial z_j} \right)^2 e^{\sum_k z_k a_k^\dagger} \right] \right.$$
$$\left. \times \left[\left(-\frac{\partial}{\partial z_j^*} \right)^2 e^{-\sum_k z_k^* a_k} \right] e^{-\sum_k \frac{|z_k|^2}{2}} \right\} \Bigg|_{z=0} . \tag{11.14}$$

We consider the characteristic function:

$$\chi(z, z^*) = \mathrm{Tr} \left[\hat{\rho} \exp \left(\sum_k z_k \hat{a}_k^\dagger - z_k^* a_k \right) \right] = \mathrm{Tr} \left[\hat{\rho} e^{\sum_k z_k \hat{a}_k^\dagger} e^{-\sum_k z_k^* a_j} e^{-\frac{1}{2} \sum_k z_k z_k^*} \right], \tag{11.15}$$

and its derivatives with respect to z_j and z_j^*:

$$\frac{\partial \chi}{\partial z_j}, \quad \frac{\partial \chi}{\partial (-z_j^*)}, \quad \frac{\partial^2 \chi}{\partial z_j^2}, \quad , \ldots \frac{\partial^2 \chi}{\partial (-z_j^*)^2}, \quad \frac{\partial \chi}{\partial z_j^2 \partial (-z_j^*)^2} . \tag{11.16}$$

As all the variables different from j are traced out, we can simplify the notation by considering the following operator:

$$\hat{\chi}_j = e^{z_j \hat{a}_j^\dagger} e^{-z_j^* \hat{a}_j} e^{-\frac{1}{2} z_j z_j^*}, \tag{11.17}$$

such that

$$\chi = \mathrm{Tr} \left(\hat{\rho} \prod_k \hat{\chi}_k \right) \tag{11.18}$$

and

$$\langle \hat{a}_j^\dagger \hat{a}_j^\dagger \hat{a}_j \hat{a}_j \rangle = \mathrm{Tr} \left(\hat{\rho} \hat{a}_j^\dagger \hat{a}_j^\dagger \hat{a}_j \hat{a}_j \prod_k \hat{\chi}_k \right) \Bigg|_{z=0} . \tag{11.19}$$

Note that $\hat{\chi}_j$ and $\hat{\chi}_k$ commute as $j \neq k$.

Also, we define the symbols

$$\partial_j \equiv \frac{\partial}{\partial z_j}$$
$$\bar{\partial}_j \equiv \frac{\partial}{\partial (-z_j)^*} \tag{11.20}$$

and

$$\begin{aligned}\hat{E}_j &= e^{z_j \hat{a}_j^\dagger}\\ \hat{\hat{E}}_j &= e^{-z_j^* \hat{a}_j}\\ e_j &= e^{-|z_j|^2/2}\end{aligned} \tag{11.21}$$

such that $\hat{\chi}_j = \hat{E}_j \hat{\hat{E}}_j e_j$.

We have

$$\begin{aligned}\partial_j \hat{E}_j &= \hat{a}_j^\dagger \hat{E}_j\\ \bar{\partial}_j \hat{\hat{E}}_j &= \hat{a}_j \hat{\hat{E}}_j\\ \partial_j e_j &= -\frac{z_j^*}{2} e_j\\ \bar{\partial}_j e_j &= \frac{z_j}{2} e_j \ .\end{aligned} \tag{11.22}$$

We write

$$\begin{aligned}\partial \hat{\chi} &= \hat{a}^\dagger \hat{E} \hat{\hat{E}} e - \frac{z^*}{2} \hat{\chi}\\ \bar{\partial} \hat{\chi} &= \hat{E} \hat{a}^\dagger \hat{\hat{E}} e + \frac{z}{2} \hat{\chi}\\ \partial \bar{\partial} \hat{\chi} &= \hat{a}^\dagger \hat{E} \hat{a} \hat{\hat{E}} e - \frac{z^*}{2} \hat{E} \hat{a} \hat{\hat{E}} e + \frac{z}{2} \hat{a}^\dagger \hat{E} \hat{\hat{E}} e - \frac{|z|^2}{4} \hat{E} \hat{\hat{E}} e + \frac{1}{2} \hat{\chi}\\ \bar{\partial}^2 \hat{\chi} &= \hat{E} (\hat{a})^2 \hat{\hat{E}} e + z \hat{E} \hat{a} \hat{\hat{E}} e + \frac{(z)^2}{4} \hat{\chi}\\ \partial \bar{\partial}^2 \hat{\chi} &= \hat{a}^\dagger \hat{E} (\hat{a})^2 \hat{\hat{E}} e - \frac{z^*}{2} \hat{E} (\hat{a})^2 \hat{\hat{E}} e + \hat{E} \hat{a} \hat{\hat{E}} e + z \hat{a}^\dagger \hat{E} \hat{a} \hat{\hat{E}} e - \frac{|z^2|}{2} \hat{E} \hat{a} \hat{\hat{E}}\\ &\quad + \frac{z}{2} \hat{\chi} + \frac{z^2}{4} \partial \hat{\chi}\end{aligned} \tag{11.23}$$

where we omitted the index j to simplify the notation.
As $\hat{E}_j \to 1$, $\hat{\hat{E}}_j \to 1$ and $\hat{\chi} \to 1$ when $z \to 0$, we have

$$\partial_j \bar{\partial}_j \hat{\chi}_j = \hat{a}_j^\dagger \hat{a}_j + \frac{1}{2} + o(z) \tag{11.24}$$

where $o(z)$ is after the Bachmann-Landau notation, i.e., $o(z) \to 0$ as $z \to 0$. By Eqs. (11.23), we have

$$\partial^2\bar{\partial}^2\hat{\chi} = (\hat{a}^\dagger)^2\hat{E}(\hat{a})^2\hat{\hat{E}}e + 2\hat{a}^\dagger\hat{E}\hat{a}\hat{\hat{E}} + \frac{1}{2}\hat{\chi} + o(z), \tag{11.25}$$

which is equivalent to

$$\partial_j^2\bar{\partial}_j^2\hat{\chi}_j = (\hat{a}_j^\dagger)^2(\hat{a}_j)^2 + 2\hat{a}_j^\dagger\hat{a} + \frac{1}{2} + o(z)\,. \tag{11.26}$$

From Eq. (11.23) we have

$$\begin{aligned} \hat{a}_j^\dagger\hat{a}_j\hat{\chi}_j &= \partial_j\bar{\partial}_j\hat{\chi}_j - \frac{1}{2}\hat{\chi}_j + o(z) \\ \hat{a}_j^\dagger\hat{a}_j^\dagger\hat{a}_j\hat{a}_j\hat{\chi}_j &= \partial_j^2\bar{\partial}_j^2\hat{\chi}_j - 2\partial_j\bar{\partial}_j\hat{\chi}_j + \frac{1}{2}\hat{\chi}_j + o(z)\,. \end{aligned} \tag{11.27}$$

Using Eq. (11.19) we have

$$\begin{aligned} \langle\hat{a}_j^\dagger\hat{a}_j^\dagger\hat{a}_j\hat{a}_j\rangle &= \mathrm{Tr}[\hat{\rho}\left(\partial_j^2\bar{\partial}_j^2\hat{\chi}_j - 2\partial_j\bar{\partial}_j\hat{\chi}_j + \frac{1}{2}\hat{\chi}_j\right)\prod_{k\neq j}\hat{\chi}_k] = \\ &\mathrm{Tr}[\hat{\rho}\left(\partial_j^2\bar{\partial}_j^2 - 2\partial_j\bar{\partial}_j + \frac{1}{2}\right)\hat{\chi}_j\prod_{k\neq j}\hat{\chi}_k] + o(z) = \\ &\left(\partial_j^2\bar{\partial}_j^2 - 2\partial_j\bar{\partial}_j + \frac{1}{2}\right)\mathrm{Tr}[\hat{\rho}\prod_j\hat{\chi}_k] + o(z), \end{aligned} \tag{11.28}$$

which gives the proof.[8]

Equation (11.5) is obtained by changing from the variables to z_j and z_j^* to q_j and p_j, using

$$\begin{aligned} \partial_j &= -\frac{\mathrm{i}}{\sqrt{2}}\frac{\partial}{\partial q_j} + \frac{1}{\sqrt{2}}\frac{\partial}{\partial p_j} \\ \bar{\partial}_j &= -\frac{\mathrm{i}}{\sqrt{2}}\frac{\partial}{\partial q_j} - \frac{1}{\sqrt{2}}\frac{\partial}{\partial p_j}\,. \end{aligned} \tag{11.29}$$

For Eq. (11.9), we consider a Gaussian characteristic function by using the Einstein convention, i.e., omitting the summation over repeated indices

$$\chi = \exp\left(-\frac{1}{4}g_{jk}x_jx_k + \mathrm{i}d_qx_q\right), \tag{11.30}$$

[8] One can test this proof with symbolic computation as with `Mathematica` and non-commuting derivatives; see the file `mathematica/noncommutative.nb`.

we obtain Eq. (11.9) by direct derivation as follows.

Proof Using the notation $\partial_q = \frac{\partial}{\partial x_q}$, with $\partial_q x_k = \delta_{qk}$, and $g_{jk} = g_{kj}$, we have

$$
\begin{aligned}
\partial_r \chi &= \left(-\frac{1}{2}g_{rk}x_k + \mathrm{i}d_r\right)\chi \\
\partial_q\partial_r \chi &= \left(-\frac{1}{2}g_{qk}x_k + \mathrm{i}d_q\right)\left(-\frac{1}{2}g_{rk}x_k + \mathrm{i}d_r\right)\chi - \frac{1}{2}g_{qr}\chi \\
&= A_{qrlm}x_l x_m + B_{qrn}x_n + C_{qr}
\end{aligned}
\tag{11.31}
$$

with all the Latin indices in the range $0, 1, \ldots, N-1$ and being

$$
\begin{aligned}
A_{qrlm} &= \frac{1}{4}g_{lr}g_{mq} \\
B_{qrl} &= -\frac{\mathrm{i}}{2}g_{nr}d_q - \frac{\mathrm{i}}{2}g_{nq}d_r \\
C_{qr} &= -\frac{1}{2}g_{qr} - d_q d_r
\end{aligned}
\tag{11.32}
$$

For the third-order derivative, we have a third polynomial times χ:

$$
\partial_p\partial_q\partial_r\chi = D_{pqrjkl}x_j x_k x_l \chi + E_{pqrkl}x_k x_l \chi + F_{pqrk}x_k \chi + G_{pqr}\chi \tag{11.33}
$$

and the fourth order reads

$$
\partial_s\partial_p\partial_q\partial_r\chi = F_{pqrs} + \mathrm{i}G_{pqr}d_s + o(\boldsymbol{x}) \tag{11.34}
$$

with

$$
\begin{aligned}
F_{pqrs} &= \frac{1}{4}g_{rs}g_{pq} + \frac{1}{4}g_{rp}g_{qs} + \frac{1}{4}g_{sp}g_{qr} \\
&\quad + \frac{1}{2}g_{sr}d_p d_q + \frac{1}{2}g_{sq}d_r d_p + \frac{1}{2}g_{sp}d_q d_r \\
\mathrm{i}G_{pqr}d_s &= \frac{1}{2}g_{pr}d_q d_s + \frac{1}{2}g_{pq}d_r d_s + \frac{1}{2}g_{qr}d_p d_s + d_s d_p d_q d_r \,.
\end{aligned}
\tag{11.35}
$$

We do not report the expression for the D_{pqrjkl} and E_{pqrkl} as they are cumbersome and not needed. Equation (11.35) used in (11.34) gives the proof.[9]

[9] This proof can also be tested by `Mathematica` as in the file `mathematica/tensorgaussian.nb`.

11.6 `DifferentialGaussianLayer`

Here we reconsider the example in Sect. 10.11 and add a layer to measure the quadratures in the uncertainty.

We are interested in measuring the uncertainty of $\hat{x}_0(\lambda)$, that is, the uncertainty of the differential current. We have, following Eq. (10.46) and omitting the tilde to simplify the notation,

$$\langle \Delta I^2 \rangle = \left\langle \left(\hat{a}_1^\dagger \hat{a}_1 - \hat{a}_0^\dagger \hat{a}_0 \right)^2 \right\rangle \tag{11.36}$$

As the number operators for different modes commute, we have

$$\langle \Delta I^2 \rangle = \langle \hat{a}_1^\dagger \hat{a}_1 \hat{a}_1^\dagger \hat{a}_1 \rangle + \langle \hat{a}_0^\dagger \hat{a}_0 \hat{a}_0^\dagger \hat{a}_0 \rangle - 2 \langle \hat{a}_1^\dagger \hat{a}_1 \hat{a}_0^\dagger \hat{a}_0 \rangle \,. \tag{11.37}$$

In this chapter, we develop a dedicated layer for applying Eq. (11.37).

11.7 Uncertainties in Homodyne Detection

As for the `HeisenbergLayer` above, as in Eq. (11.6), we use the fourth-order derivatives. For the mixed term $\langle \hat{a}_1^\dagger \hat{a}_1 \hat{a}_0^\dagger \hat{a}_0 \rangle$ in (11.80), we have

$$\begin{aligned} \langle \hat{a}_1^\dagger \hat{a}_1 \hat{a}_0^\dagger \hat{a}_0 \rangle = & \left(\frac{\partial}{\partial z_1} \right) \left(-\frac{\partial}{\partial z_1^*} \right) \left(\frac{\partial}{\partial z_0} \right) \left(-\frac{\partial}{\partial z_0^*} \right) \chi \bigg|_{z=0} \\ & - \frac{1}{2} \left(\frac{\partial}{\partial z_0} \right) \left(-\frac{\partial}{\partial z_0^*} \right) \chi \bigg|_{z=0} - \frac{1}{2} \left(\frac{\partial}{\partial z_1} \right) \left(-\frac{\partial}{\partial z_1^*} \right) \chi \bigg|_{z=0} + \frac{1}{4} . \end{aligned} \tag{11.38}$$

In terms of the $\boldsymbol{x}-$operators ∇_j^2, Eq. (11.38) reads

$$\langle \hat{a}_1^\dagger \hat{a}_1 \hat{a}_0^\dagger \hat{a}_0 \rangle = \frac{1}{4} \nabla_1^2 \nabla_0^2 \chi_R + \frac{1}{4} \nabla_0^2 \chi_R + \frac{1}{4} \nabla_1^2 \chi_R + \frac{1}{4} \bigg|_{\boldsymbol{x}=0} \,. \tag{11.39}$$

Equation (11.39) holds true for any many-body state. Considering two modes with indices j and $k \neq j$, we have

$$\langle \hat{a}_j^\dagger \hat{a}_j \hat{a}_k^\dagger \hat{a}_k \rangle = \langle \hat{n}_j \hat{n}_k \rangle = \frac{1}{4} \nabla_j^2 \nabla_k^2 \chi_R + \frac{1}{4} \nabla_j^2 \chi_R + \frac{1}{4} \nabla_k^2 \chi_R + \frac{1}{4} \bigg|_{\boldsymbol{x}=0} \,. \tag{11.40}$$

The proof of Eqs. (11.39) and (11.40) is given in Sect. 11.7.1.

By using Eq. (11.6), we have

$$
\begin{aligned}
\langle \Delta I^2 \rangle &= \langle (\hat{n}_1 - \hat{n}_0)^2 \rangle \\
&= \frac{1}{4} \left(\nabla_1^2 - \nabla_0^2\right)^2 \chi_R \Big|_{x=0} - \frac{1}{2} \\
&= \frac{1}{4} \nabla_1^4 \chi_R + \frac{1}{4} \nabla_0^4 \chi_R - \frac{1}{2} \nabla_0^2 \nabla_1^2 \chi_R - \frac{1}{2} \Big|_{x=0} .
\end{aligned}
\tag{11.41}
$$

For two indices j and $k \neq j$, we have

$$
\begin{aligned}
\langle (\hat{n}_j - \hat{n}_k)^2 \rangle &= \frac{1}{4} \left(\nabla_j^2 - \nabla_k^2\right)^2 \chi_R \Big|_{x=0} - \frac{1}{2} \\
&= \frac{1}{4} \nabla_j^4 \chi_R + \frac{1}{4} \nabla_k^4 \chi_R - \frac{1}{2} \nabla_j^2 \nabla_k^2 \chi_R - \frac{1}{2} \Big|_{x=0} .
\end{aligned}
\tag{11.42}
$$

For a general state, one can use multiple calls to `tf.jacobian` for Eq. (11.42), as in the Heisenberg layer in Sect. 11.2.

11.7.1 Proof of Eqs. (11.39) and (11.40)*

Proof We have

$$
\langle (\hat{n}_1 - \hat{n}_0)^2 \rangle = \langle \hat{n}_0^2 \rangle + \langle \hat{n}_1^2 \rangle - 2 \langle \hat{n}_0 \hat{n}_1 \rangle .
\tag{11.43}
$$

We consider the characteristic function for the two modes with indices 0 and 1:

$$
\chi = \mathrm{Tr}\left(\rho e^{z_0 \hat{a}_0^\dagger} e^{-z_0^* \hat{a}_0} e^{z_1 \hat{a}_1^\dagger} e^{-z_1^* \hat{a}_1} e^{-|z_0|^2/2} e^{-|z_1|^2/2} \right)
\tag{11.44}
$$

and find the derivatives with the notation in Sect. 11.5:

$$
\partial_0 \chi = \mathrm{Tr}\left[\rho \left(\partial_0 e^{z_0 \hat{a}_0^\dagger} \right) \ldots \right] - \frac{z_0^*}{2} \chi
\tag{11.45}
$$

$$
\begin{aligned}
\bar{\partial}_0 \partial_0 \chi &= \mathrm{Tr}\left[\rho \left(\partial_0 e^{z_0 \hat{a}_0^\dagger} \right) \left(\bar{\partial}_0 e^{-z_0^* \hat{a}_0} \right) \ldots \right] \\
&\quad + \frac{z_0}{2} \mathrm{Tr}\left[\left(\partial_0 e^{z_0 \hat{a}_0^\dagger} \right) \right] + \frac{1}{2} \chi
\end{aligned}
\tag{11.46}
$$

$$
\begin{aligned}
\partial_1 \bar{\partial}_0 \partial_0 \chi &= \mathrm{Tr}\left[\rho \left(\partial_0 e^{z_0 \hat{a}_0^\dagger} \right) \left(\bar{\partial}_0 e^{-z_0^* \hat{a}_0} \right) \left(\partial_1 e^{z_1 \hat{a}_1^\dagger} \right) \ldots \right] \\
&\quad - \frac{z_1^*}{2} \mathrm{Tr}\left[\left(\partial_0 e^{z_0 \hat{a}_0^\dagger} \right) \left(\bar{\partial}_0 e^{-z_0^* \hat{a}_0} \right) \ldots \right] + \frac{1}{2} \partial_1 \chi + o(z_0)
\end{aligned}
\tag{11.47}
$$

$$\bar{\partial}_1\partial_1\bar{\partial}_0\partial_0\chi = \mathrm{Tr}\left[\rho\left(\partial_0 e^{z_0\hat{a}_0^\dagger}\right)\left(\bar{\partial}_0 e^{-z_0^*\hat{a}_0}\right)\left(\partial_1 e^{z_1\hat{a}_1^\dagger}\right)\left(\bar{\partial}_1 e^{-z_1^*\hat{a}_1}\right)\ldots\right]$$
$$+\frac{1}{2}\mathrm{Tr}\left[\rho\left(\partial_0 e^{z_0\hat{a}_0^\dagger}\right)\left(\bar{\partial}_0 e^{-z_0^*\hat{a}_0}\right)\ldots\right]$$
$$+\frac{1}{2}\bar{\partial}_1\partial_1\chi + o(z_0, z_1)\,. \tag{11.48}$$

From the previous equations, we get

$$\langle\hat{n}_0\hat{n}_1\rangle = \bar{\partial}_1\partial_1\bar{\partial}_0\partial_0\chi - \frac{1}{2}\bar{\partial}_0\partial_0\chi - \frac{1}{2}\bar{\partial}_1\partial_1\chi + \frac{1}{4}\chi\bigg|_{z=0}\,. \tag{11.49}$$

Seemingly, we have

$$\langle\hat{n}_0^2\rangle = \bar{\partial}_0^2\partial_0^2\chi - \partial_0\bar{\partial}_0\chi\bigg|_{z=0} \tag{11.50}$$

and analogous for $\langle\hat{n}_1^2\rangle$.

Finally by using

$$\bar{\partial}_0\partial_0 = -\frac{1}{2}\nabla_0^2 \tag{11.51}$$

$$\bar{\partial}_1\partial_1 = -\frac{1}{2}\nabla_1^2\,, \tag{11.52}$$

we get Eqs. (11.39) and, equivalently, (11.40). □

11.8 Uncertainties for Gaussian States

We consider explicitly the case of Gaussian states, use the general results in Eq. (11.9), and express the uncertainties in terms of the covariance matrix g_{jk} and the displacement d_j. We have

$$\frac{\partial^2\chi_R}{\partial x_j^2\partial x_k^2} = \frac{1}{4}g_{jj}g_{kk} + \frac{1}{2}g_{jk}^2 + \frac{1}{2}g_{jj}d_k^2 + 2g_{jk}d_jd_k + \frac{1}{2}g_{kk}d_j^2 + d_j^2d_k^2, \tag{11.53}$$

which is used in

$$\nabla_j^2\nabla_k^2\chi_R = \frac{\partial^4\chi_R}{\partial q_j^2\partial q_k^2} + \frac{\partial^4\chi_R}{\partial p_j^2\partial p_k^2} + \frac{\partial^4\chi_R}{\partial q_j^2\partial p_k^2} + \frac{\partial^4\chi_R}{\partial p_j^2\partial q_k^2} \tag{11.54}$$

being, as above, $q_k = x_{2k}$ and $p_k = x_{2k+1}$

The previous equations for the fourth-order derivatives in terms of the components of $\boldsymbol{g}$ and $\boldsymbol{d}$ allow introducing a `DifferentialGaussianLayer`, which returns the following quantities for any couple of modes j, k in a Gaussian model:

$$\begin{aligned} &\langle n_j \rangle \\ &\langle \Delta n_{jk} \rangle = \langle \hat{n}_j - \hat{n}_k \rangle \\ &\langle \Delta n_{jk}^2 \rangle = \langle (\hat{n}_j - \hat{n}_k)^2 \rangle \, . \end{aligned} \tag{11.55}$$

The code for the `DifferentialGaussianLayer` is as follows.[10]

```
class DifferentialGaussianLayer(layers.Layer):
    """ Returns mean nj-nk and (nk-nk)^2 for a Gaussian model

        Use the biharmonic Gaussian layer

       In the constructor,

        Parameters
        ----------
        N: size of the of vector

        In the call,

        Parameters
        ----------
        c1: real part of chi
        c2: imag part of chi
        pullback: model to be derived wrt x

        Returns
        -------
        nboson : (1,n) tensor with <nj>
        dn : (n,n) tensor with <nj-nk>
        dn2 : (n,n) tensor with <(nj-nk)^2>

    """

    def __init__(self, N,
                 **kwargs):
            super(DifferentialGaussianLayer,
                  self).__init__(**kwargs)
            # vector size
            self.N = tf.constant(N)
            self.n = tf.constant(np.floor_divide(N,2))
            # covariance layer
            self.covariance = CovarianceLayer(N)
```

[10] The code is in the file `phasespace.py`.

```
            # constant matrix to extract the modes
            Rq, Rp, _ = RQRP(N)
            self.Rq = tf.constant(Rq)
            self.Rp = tf.constant(Rp)
            # tensors of mixed derivatives
            self.cxx = tf.zeros([1,N])
            self.cxxyy = tf.zeros([N,N])

    @tf.function
    def build_differences(self, nb1):
        # build a tensor with the differences
        #
        # Parameters
        # ----------
        # nb1: tensor with shape (1,n)
        #
        # Returns
        # -------
        # dn: tensor nj-nk with shape (n,n)

        dn = tf.zeros([self.n, self.n])
        for j in tf.range(self.n):
            nj = tf.gather_nd(nb1, [[0,j]])
            for k in tf.range(j+1, self.n):
                nk = tf.gather_nd(nb1, [[0,k]])
                indices = [[j,k]]
                dn = tf.tensor_scatter_nd_add(dn,
                                              indices, nj-nk)
                indices = [[k,j]]
                dn = tf.tensor_scatter_nd_add(dn,
                                              indices, -nj+nk)

        return dn

    def call(self, c1, c2, chi):
        # c1 and c2 are dummy needed to model
        # chi is the model

        # evaluate the covariace and the displacement
        g, d, hessian = self.covariance(c1, c2, chi)

        # evaluate the laplacian
        # by gathering the diagonal elements of gaussian
        cr_2x = tf.reshape(tf.linalg.diag_part(hessian),
                           [1, self.N])

        # sum the derivatives to squared laplacian
        dqq = tf.matmul(cr_2x, self.Rq)
        dpp = tf.matmul(cr_2x, self.Rp)
        lapl = dqq+dpp

        # evaluate the number of bosons
        nboson = -0.5*lapl-0.5
```

```
# tensor with differences
dn = self.build_differences(nboson)

# evaluate biharmonic
gdiag = tf.linalg.diag_part(g)
d2 = tf.square(d)
d4 = tf.square(d2)

# build nabla2 nabla2
d2 = tf.squeeze(d2)
d4 = tf.squeeze(d4)

# build tensor gjk^2 [N,N]
gjkjk = tf.square(g)

# build tensor gjj gkk [N,N]
gjjkk = tf.tensordot(gdiag,gdiag, axes=0)

# build tensor gjj dk dk [N,N]
gjjdkdk = tf.tensordot(gdiag, d2, axes =0)

# build tensor gkk dj dj [N,N]
gkkdjdj = tf.tensordot(d2, gdiag, axes =0)

# build tensor gjk dj dk [N,N]
dd = tf.squeeze(tf.tensordot(d,d,axes=0)) #[N,N]
gjkdjdk = tf.multiply(g, dd)

# build tensor dj dj dk dk [N,N]
djdjdkdk = tf.tensordot(d2, d2, axes = 0)

# this is a NxN tensor [N,N]
cr_xxyy= 0.5*gjkjk \
         +0.25*gjjkk+0.5*gjjdkdk \
         +2.0*gjkdjdk+0.5*gkkdjdj+djdjdkdk

# make a double loop to build the final tensor
dn2 = tf.zeros([self.n,self.n])
for j in tf.range(self.n):
    qj = 2*j
    pj = 2*j+1
    # nabla^4 j
    # d_qjqjqjqj
    d_qj4 = tf.gather_nd(cr_xxyy,  [[qj,qj]])
    # d_qjqjqjqj
    d_pj4 = tf.gather_nd(cr_xxyy,  [[pj,pj]])
    # d_qjqj
    d_qj2 = tf.gather_nd(cr_2x,  [[0,qj]])
    # d_pjpj
    d_pj2 = tf.gather_nd(cr_2x,  [[0,pj]])
    # nabla^4_j
    n4j = d_qj4+d_pj4+2*d_qj2*d_pj2
    for k in tf.range(j+1,self.n):
        qk = 2*k
```

```
            pk = 2*k+1
            # nabla^4 k
            # d_qjqjqjqj
            d_qk4 = tf.gather_nd(cr_xxyy,  [[qk,qk]])
             # d_qjqjqjqj
            d_pk4 = tf.gather_nd(cr_xxyy,  [[pk,pk]])
            # d_qjqj
            d_qk2 = tf.gather_nd(cr_2x,  [[0,qk]])
            # d_pjpj
            d_pk2 = tf.gather_nd(cr_2x,  [[0,pk]])
            n4k = d_qk4+d_pk4+2*d_qk2*d_pk2 #nabla^4 k

            # nabla^2_j nabla^2_k
            n2j2k = tf.gather_nd(cr_xxyy, [[qj,qk]])+\
                tf.gather_nd(cr_xxyy, [[qj,pk]])+\
                tf.gather_nd(cr_xxyy, [[qk,pj]])+\
                tf.gather_nd(cr_xxyy, [[pk,pj]])
            # store the final value
            njnk = 0.25*(n4j+n4k)-0.5*n2j2k-0.5
            indices =[[j,k]]
            dn2 = tf.tensor_scatter_nd_add(dn2,
                             indices, njnk)
            indices =[[k,j]]
            dn2 = tf.tensor_scatter_nd_add(dn2,
                             indices, njnk)

    return nboson, dn, dn2
```

To build the tensor elements, we need to extract elements from the various tensor, and we use the method `tf.gather_nd`; seemingly to assign the elements of the target tensors dn and dn2, we use the method `tf.tensor_scatter_nd_add`.

By the differential layer, when $n = 2$ and for two modes 0 and 1, we have $\langle \Delta I \rangle = \langle \Delta n_{01} \rangle$ and $\langle \Delta I^2 \rangle = \langle \Delta n_{01}^2 \rangle$.

11.9 DifferentialGaussianLayer on Coherent States

Before homodyne detection, we test the `DifferentialLayer` for a two-mode coherent state with parameters α_0 and α_1, built with the following code.[11]

```
A_np = 10.0
lambda_np = np.pi/2
dinput=np.zeros((N,1));
```

[11] This example in the notebook `jupyternotebooks/phasespace/differentialla yer.ipynb`.

```
dinput[2]=np.sqrt(2)*A_np*np.cos(lambda_np);
dinput[3]=np.sqrt(2)*A_np*np.sin(lambda_np);
displacer=ps.DisplacementLayerConstant(dinput)
xin = tf.keras.layers.Input(N)
x0, a0 = displacer(xin)
chir, chii = vacuum(x0, a0)
model = tf.keras.Model(inputs = xin, outputs=[chir, chii])
```

Here the vector `dinput` corresponds to the vector

$$\boldsymbol{d}_{input} = \begin{pmatrix} \sqrt{2}\Re(\alpha_0) \\ \sqrt{2}\Im(\alpha_0) \\ \sqrt{2}\Re(\alpha_1) \\ \sqrt{2}\Im(\alpha_1) \end{pmatrix} . \tag{11.56}$$

For this state, we expect

$$\begin{aligned} \langle \hat{n}_0 \rangle &= |\alpha_0|^2 \\ \langle \hat{n}_1 \rangle &= |\alpha_1|^2 \\ \langle \hat{n}_0 - \hat{n}_1 \rangle &= |\alpha_0|^2 - |\alpha_1|^2 \\ \langle (\hat{n}_1 - \hat{n}_0)^2 \rangle &= \langle \hat{n}_0^2 \rangle + \langle \hat{n}_1^2 \rangle - 2\langle \hat{n}_0 \rangle \langle \hat{n}_1 \rangle = (|\alpha_1|^2 - |\alpha_0|^2)^2 + |\alpha_0|^2 + |\alpha_1|^2 , \end{aligned} \tag{11.57}$$

where we used $\langle \hat{n}_0 \hat{n}_1 \rangle = \langle \hat{n}_1 \rangle \langle \hat{n}_0 \rangle$, which holds for a two-mode coherent state that has a factorizable characteristic function. One can verify these values, by introducing a `DifferentialGaussianLayer`, adding to the model as in Fig. 11.1, and evaluating on training points (whose value is not affecting the final result), as in

```
Diff=ps.DifferentialGaussianLayer(N)
nboson, Dn, Dn2 = Diff(chir,chii, model)
HModel = tf.keras.Model(inputs = xin, outputs=[nboson, Dn, Dn2])
# run the model
nboson1, Dn, Dn2=HModel(xtrain)
# print the differential matrix
tf.print(Dn)
# output is
# [[0 10100]
# [10100 0]
```

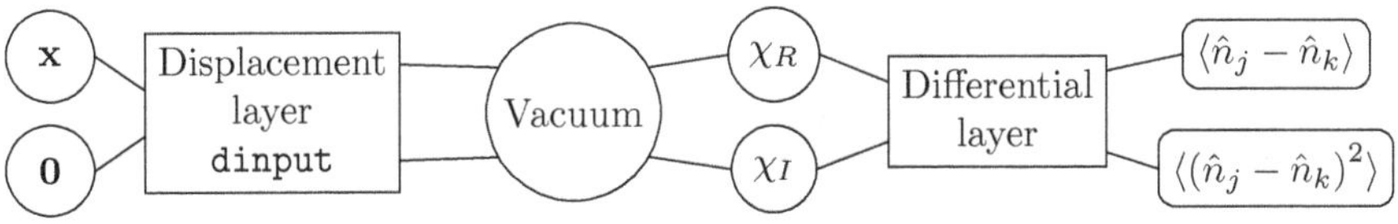

Fig. 11.1 Model for testing the differential Gaussian layer on a coherent state

11.10 DifferentialGaussianLayer in Homodyne Detection

The model considered in Sect. 10.11 includes a squeezed vacuum mode with squeezing parameter $\zeta = r\, e^{i\phi}$, a coherent state with $\alpha = A\, e^{i\lambda}$, and a 50 : 50 beam splitter, as follows.

```
xin = tf.keras.layers.Input(N)
x2, a2 = bs0(xin)
x1, a1 = squeezer(x2,a2)
x0, a0 = displacer(x1,a1)
chir, chii = vacuum(x0, a0)
BellBS = tf.keras.Model(inputs = xin, outputs=[chir, chii])
```

The coherent state has a phase θ that enables the measurement of the quadrature $\hat{x}_0(\lambda)$. We are interested in determining the following quantities:

$$\langle\, x_0(\lambda)\rangle \,, \tag{11.58}$$

$$\langle\, x_0^2(\lambda)\rangle \,, \tag{11.59}$$

$$\langle\, \Delta x_0^2(\lambda)\rangle = \langle x_0^2(\lambda)\rangle - \langle x_0(\lambda)\rangle^2 \,. \tag{11.60}$$

We add a differential layer to the model detailed in Sect. 10.11 as follows.[12]

```
Heisenberg=ps.DifferentialGaussianLayer(N)
nH_out, n2, Dn2 = Heisenberg(chir,chii, BellBS)
HModel = tf.keras.Model(inputs = xin, outputs=[nH_out, n2, Dn2])
```

The corresponding diagram is in Fig. 11.2. The expected uncertainty for the operator $\hat{x}_0(\lambda)$ is [1, 2]

[12] Details in the notebook jupyternotebooks/phasespace/BellBS.ipynb.

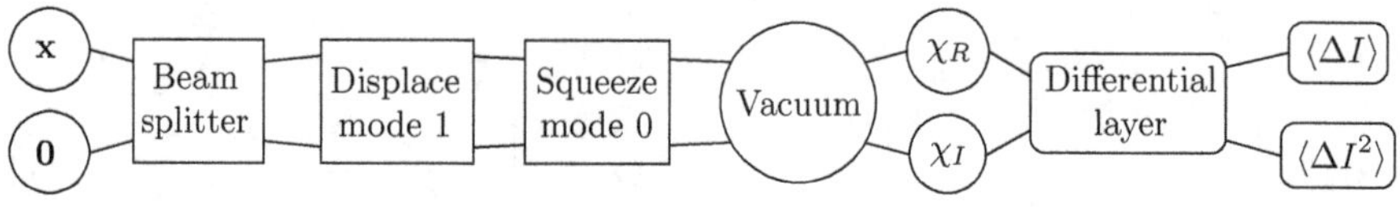

Fig. 11.2 Model for homodyne detection on $n = 2$ modes, with mode 1 representing the local oscillator, including the differential layer. Here $I_1 = \hat{n}_1$ and $I_0 = \hat{n}_0$

$$\langle \Delta \hat{x}_0(\lambda)^2 \rangle = \frac{1}{2} e^{-2r} \cos\left(\lambda - \frac{\phi}{2}\right)^2 + \frac{1}{2} e^{2r} \sin\left(\lambda - \frac{\phi}{2}\right)^2 \tag{11.61}$$

with θ the phase of the local oscillator, ϕ the squeezing parameter phase, and r the modulus of the squeezing parameter.

The output of the homodyne experiment, including the local oscillator, is

$$\langle \Delta \hat{I}^2 \rangle = |\alpha|^2 \left[e^{-2r} \cos\left(\lambda - \frac{\phi}{2}\right)^2 + e^{2r} \sin\left(\lambda - \frac{\phi}{2}\right)^2 \right]. \tag{11.62}$$

In our example, we choose $\phi = 0$, $\lambda = 0$ (phase locking between the local oscillator and the squeezing device pump). Letting $|\alpha| = 10$ and $r = 0.1$, we get

```
# call the model for lambda=0
print(HModel(xtrain))
# output:
#[<tf.Tensor: shape=(1, 2), dtype=float32,
#numpy=array([[50.01, 50.01]], dtype=float32)>,
#<tf.Tensor: shape=(2, 2), dtype=float32, numpy=
#array([[0., 0.],
#       [0., 0.]], dtype=float32)>,
#<tf.Tensor: shape=(2, 2), dtype=float32, numpy=
#array([[ 0.   , 81.88],
#       [81.88,  0.   ]], dtype=float32)>]
```

which corresponds to $\langle \hat{x}_0^2(0) \rangle = 2|\alpha|^2 \langle \hat{x}_0^2 \rangle \simeq 82$, or $\langle \hat{x}_0^2 \rangle \simeq 0.41$ (see Eq. 11.60).

We then consider $\lambda = \pi/2$ and left unchanged r, ϕ, and A, and we obtain $\langle \hat{x}_0^2(\frac{\pi}{2}) \rangle = 2|\alpha|^2 \langle \hat{p}_0^2 \rangle \simeq 122$, or $\langle \hat{p}_0^2 \rangle = 0.611$.

As one can verify, these results are consistent with the expected theoretical values in Eq. (11.61) and also with the Heisenberg principle as $\Delta x_0^2 \, \Delta p_0^2 = 0.250$ (we use units with $\hbar = 1$), being $\Delta x_0 < \Delta p_0$, as the state is squeezed.

11.11 Entanglement with Gaussian States

Coherent states remain classical states (i.e., are not squeezed or entangled) when propagating through beam splitters and interferometers. A coherent state has unitary eigenvalues of the covariance matrix, and this does not change when the state passes through beam splitters or other mode-mixers. The eigenvalues of the covariance matrix determine the degree of entanglement. Hence, one has to start with nonclassical Gaussian states, typically squeezed states, and then modes are mixed to create entanglement.

A beam splitter does not produce entangled states if the inputs are classical [3]. On the other hand, starting from a nonclassical state, as a squeezed state, one can use a beam splitter to maximize entanglement [4]. However, quantifying entanglement is not an easy task, and this difficulty holds for multimodal Gaussian states.

The simplest approach is considering a composite systems by two Hilbert spaces H_A and H_B, as the two input modes in Fig. 11.3. The state of the system is Ψ. Measurement performed on subsystem A corresponds to the reduced density operator:

$$\rho_A = \text{Tr}_B\left(\rho\right) . \tag{11.63}$$

The simplest measure for entanglement $E(\Psi)$ between A and B is the von Neumann entropy $S(\rho_A)$ of ρ_A, [5], and we have

$$E(\Psi) = -\text{Tr}\,\rho_A \log_2 \rho_A = S(\rho_A) = S(\rho_B) , \tag{11.64}$$

where ρ_B is the reduced density operator of ρ_B. For a pure state $S(\rho) = 0$, however, the partial trace unveils correlations between subsystems; hence, one can have $E(\Psi) \neq 0$. The larger is the entropy of the reduced state, the more is the information loss when we trace out one subsystem, corresponding to the fact that the two subsystems are strongly correlated. On the contrary, for a product state, the entropy of entanglement is zero (see Sect. 11.13).

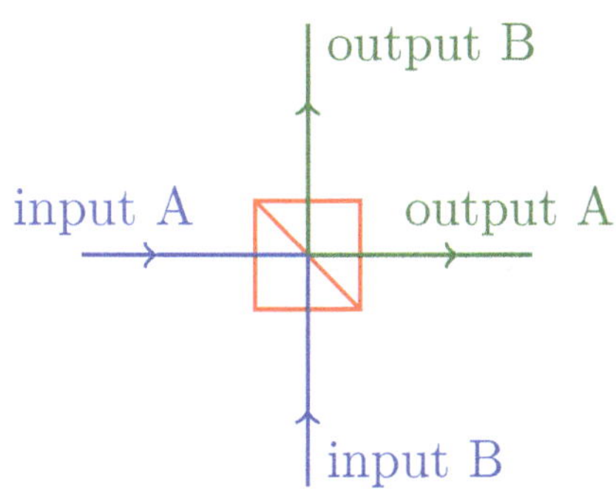

Fig. 11.3 Beam splitter for mixing different modes and controlling entanglement when the input is nonclassical

A beam splitter can create entanglement in the sense that it changes $E(\Psi)$, when varying parameters as the splitting ration. However, this occurs only if the input state is nonclassical, for example, if one of the modes in input is squeezed. We will study with detail this property and use the NN model to compute and maximize entanglement.

The more general case of Gaussian states in an arbitrary system will be addressed in a later chapter. The interested reader may consider extensive reviews as in [6–8].

11.12 Beam Splitters as Entanglers

A simple well-studied example is the case in which two single-mode squeezed Gaussian states interfere at a beam splitter (Fig. 11.3). Letting $\hat{\mathcal{B}}$ be the beam splitter operator, the two-mode state we are considering is the following [4]:

$$\hat{\mathcal{B}}\hat{S}_a(\zeta_a)\hat{S}_b(\zeta_b)|00\rangle \; . \tag{11.65}$$

The corresponding NN model is shown in Fig. 11.4. The code is as follows.[13]

```
# first single mode squeezer layer (not trainable)
squeezer_a=ps.SingleModeSqueezerLayer(N, r_np=ra,
theta_np=phia, n_squeezed=0, trainable=False)
# second single mode squeezer layer (not trainable)
squeezer_b=ps.SingleModeSqueezerLayer(N, r_np=rb,
theta_np=phib, n_squeezed=1, trainable=False)
# beam splitter (only phi0 and phi1 are trainable)
theta=0.5
phi0=0.0
phi1=0.0
bs = ps.BeamSplitterLayer(N, theta=theta,
                          phi0=phi0, phi1=phi1,
                          n_0=0, n_1=1,  trainable_theta=True)
# connect layer and input
xin= tf.keras.layers.Input(N)
x0,d0 = bs(xin)
x1, d1 = squeezer_b(x0,d0)
x2, d2 = squeezer_a(x1,d1)
chir, chii = vacuum(x2, d2)
squeezed1mBS = tf.keras.Model(inputs = xin, outputs=[chir, chii])
```

[13] The code for this example is in jupyternotebooks/phasespace/singlemodesqueez erBS.ipynb.

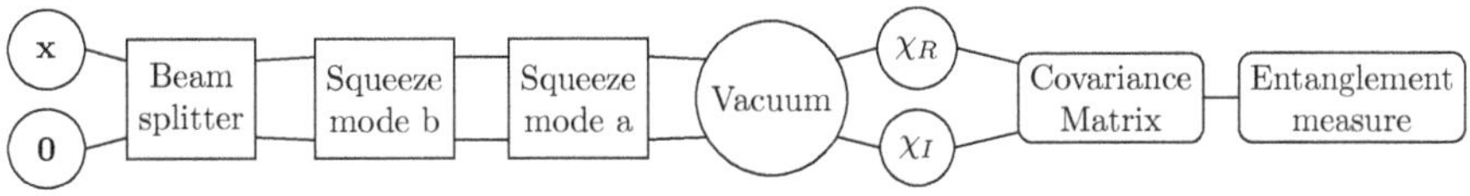

Fig. 11.4 Pullback model for two squeezed modes entering a beam splitter. To measure and maximize entanglement, we first determine the covariance matrix $\boldsymbol{g}$ and then we use an entanglement layer for the beam splitter

The parameters in the model are as follows:

- The squeezing parameters of the layers a and b: r_a,θ_a,r_b, and θ_b, corresponding to the complex degrees of squeezing $\zeta_{a,b} = r_{a,b} \exp(\mathrm{i}\theta_{a,b})$. These are determined by the variables `ra` and `phia` for layer a and `rb` and `phib` for layer b.
- The parameters of the beam splitter layer θ, ϕ_0, and ϕ_1, corresponding to the variables `theta`, `phi0`, and `phi1`.

11.13 Entanglement by the Reduced Characteristic Function

We review the treatment in [4]. One starts from the characteristic function of the bipartite system, with the two modes A and B, corresponding to `0` and `1` in our model. The characteristic function is

$$\chi(\boldsymbol{x}) = \chi(\boldsymbol{x}_A, \boldsymbol{x}_B) = \mathrm{Tr}\left(\rho e^{\mathrm{i}\boldsymbol{x}\hat{\boldsymbol{R}}}\right) = \mathrm{Tr}\left(\rho e^{\mathrm{i}\boldsymbol{x}_A\hat{\boldsymbol{R}}_A + \mathrm{i}\boldsymbol{x}_B\hat{\boldsymbol{R}}_B}\right) \tag{11.66}$$

where we made explicit the degrees of freedom of the two modes by defining the reduced observable vectors:

$$\hat{\boldsymbol{R}}_{A,B} = \begin{pmatrix} \hat{q}_{A,B} \\ \hat{p}_{A,B} \end{pmatrix} \tag{11.67}$$

and the corresponding row vectors:

$$\boldsymbol{x}_{A,B} = \begin{pmatrix} q_{A,B} & p_{A,B} \end{pmatrix} . \tag{11.68}$$

We are interested in the partial trace with respect to the degree of freedom of system B, which is obtained by computing χ at the points $\boldsymbol{x}_B = 0$, i.e.,

$$\chi_A(\boldsymbol{x}_A) = \chi(\boldsymbol{x}_A, \boldsymbol{0}) = \mathrm{Tr}\left(\rho e^{\mathrm{i}\boldsymbol{R}_A \cdot \boldsymbol{x}_A}\right) = \mathrm{Tr}_B\left(\rho e^{\mathrm{i}\boldsymbol{R}_A \cdot \boldsymbol{x}_A}\right) , \tag{11.69}$$

where we denote $\chi_A(\boldsymbol{x}_A)$ the reduced characteristic function for subsystem A. The partial trace is obtained by the general characteristic function letting $\boldsymbol{x}_B = 0$.

As we are dealing with Gaussian states, the characteristic function is by pullback, as detailed in the previous chapters. The covariance matrix $\boldsymbol{g}$ by the model in Fig. 11.4 is written as the 4×4 matrix:

$$\boldsymbol{g} = \begin{pmatrix} g_{00} & g_{01} & g_{02} & g_{03} \\ g_{10} & g_{11} & g_{12} & g_{13} \\ g_{20} & g_{21} & g_{22} & g_{23} \\ g_{30} & g_{31} & g_{32} & g_{33} \end{pmatrix} . \tag{11.70}$$

The overall model is a Gaussian state with no displacement $\boldsymbol{d} = 0$, as in Eq. (7.37):

$$\chi(\boldsymbol{x}) = \exp\left(-\frac{1}{4}\boldsymbol{x}\boldsymbol{g}\boldsymbol{x}^{\top}\right) . \tag{11.71}$$

Computing (11.71) at $\boldsymbol{x}_B = 0$, we get

$$\chi_A(\boldsymbol{x}_A) = \exp\left(-\frac{1}{4}\boldsymbol{x}_A\boldsymbol{g}_A\boldsymbol{x}_A^{\top}\right) , \tag{11.72}$$

with the reduced covariance matrix g_A given by

$$\boldsymbol{g}_A = \begin{pmatrix} g_{00} & g_{01} \\ g_{10} & g_{11} \end{pmatrix} . \tag{11.73}$$

By a proper rotation of the vector $\boldsymbol{x}_A$ into a new vector $\tilde{\boldsymbol{x}}_A$, which corresponds to an unitary transformation for system A, one can write the reduced density matrix in a diagonal form, such that

$$\chi_R = \exp\left(-\frac{1}{4}\tilde{\boldsymbol{x}}_A\tilde{\boldsymbol{g}}_A\tilde{\boldsymbol{x}}_A\right) , \tag{11.74}$$

being

$$\tilde{\boldsymbol{g}}_A = \begin{pmatrix} 2\delta & 0 \\ 0 & 2\delta \end{pmatrix} . \tag{11.75}$$

Equation (11.75) corresponds to the characteristic matrix of a thermal state with parameter β [4], as in Eq. (7.62) being

$$\delta = \frac{e^{\beta} + 1}{e^{\beta} - 1} = (2\bar{n} + 1) . \tag{11.76}$$

Hence, after tracing out subsystem B, subsystem A is equivalent to a thermal state with reduced density matrix:

$$\rho_A = (1 - e^{-\beta})e^{-\beta\hat{n}} = (1 - e^{-\beta})\sum_{n=0}^{\infty} e^{-\beta n}|n\rangle\langle n| \tag{11.77}$$

with

$$e^{\beta} = \frac{\delta + 1}{\delta - 1} . \tag{11.78}$$

The corresponding von Neumann entropy can be used readily to measure entanglement:

$$E(\Psi) = -\operatorname{Tr} \rho_A \ln \rho_A = -\sum_{n=0}^{\infty}(1 - e^{-\beta})e^{-\beta n} \ln(1 - e^{-\beta})e^{-\beta n} . \tag{11.79}$$

Note that we are using the natural logarithm ln in Eq. (11.79) at variance with Eq. (11.64) to simplify the notation; the resulting entropy will differ by a positive multiplicative factor.

After Eqs. (11.79) and (11.78), we have

$$E(\delta) = -\ln \frac{2}{\delta + 1} - \frac{\delta - 1}{2} \ln \frac{\delta - 1}{\delta + 1} . \tag{11.80}$$

Equation (11.80) is a growing function with respect to δ, which is proportional to the square root of the determinant $4\delta^2$ of $\boldsymbol{g}_A$. By varying the parameters of the beam splitter, we change δ, as detailed in the next section. This implies that, starting from a nonclassical state (e.g., a squeezed vacuum), we can use the beam splitter to maximize entanglement. Specifically, we need to maximize the determinant of $\boldsymbol{g}_A$.

11.14 Computing the Entanglement

The first step is done by using the covariance layer defined above. For the last steps, we define a dedicated layer that takes as input the covariance matrix, extracts the submatrix $\boldsymbol{g}_A$, and evaluates its determinant.[14]

```
class entanglementBSLayer(layers.Layer):
    """
    Return the entanglement of a mode evaluated as the
    the delta parameter and
    E = -ln (2/(delta+1))-((delta-1)/2)ln((nu-1)(nu+1))
    as for a beam splitter with non-classical states
```

[14] The code for the layer is in the file `phasespace.py`.

```
    Take as input the covariance matrix of a model

        Parameters
        ----------
        ne: index of the mode to be calculated
        cov: covariance matrix of the model
    """

    def __init__(self, N, ne=0,
                 **kwargs):
            super(entanglementBSLayer,
            ↪  self).__init__(autocast=False, **kwargs)
            assert ne < N//2, " mode index must be smaller than
            ↪  N/2"
            self.N = N
            self.ne = ne

    def Enuf(self,delta):
    """ Compute entanglement entropy for a given nu """
        E=-tf.math.log(2/(delta+1))-(nu-1)*0.5*tf.math.log
          ((delta-1)/(delta+1))
        return E

    def call(self, cov):
        """
            Return parameter delta, and E
        """
        N0=cov.shape[0]
        N1=cov.shape[1]
        # extract the reduced covariance matrix
        ca = tf.slice(cov, [2*self.ne, 2*self.ne],[2,2])
        # compute the determinant nu=4 delta^2
        nu = tf.linalg.det(ca)
        Enu = self.Enuf(0.5*tf.math.sqrt(nu))
        return nu, Enu, ca
```

The layer first extracts the submatrix $\mathbf{g}_A$ by the covariance matrix by using the method `tf.slice`, then evaluates the determinant by `tf.linalg.det`, and finally calls the `tf.function Enuf` that we define in the layer class.

Following [4],[15] we have

[15] Within a factor 2 in the definition of the covariance matrix in [4].

$$
\begin{aligned}
g_{00} &= 2\Sigma_a \cos^2(\theta) + 2\Sigma_b \sin^2(\theta) + 4x_a \cos^2(\theta)\cos(\Delta_a) + 4x_b \sin^2(\theta)\cos(\Delta_b) \\
g_{01} &= g_{10} = 4x_a \cos^2(\theta)\sin(\Delta_a) + 4x_b \sin^2(\theta)\sin(\Delta_b) \\
g_{11} &= 2\Sigma_a \cos^2(\theta) + 2\Sigma_b \sin^2(\theta) - 4x_a \cos^2(\theta)\cos(\Delta_a) - 4x_b \sin^2(\theta)\cos(\Delta_b) \\
\Delta_a &= 2\phi_0 - \phi_a \\
\Delta_b &= 2\phi_1 - \phi_b \\
\Sigma_a &= \cosh^2(r_a) + \sinh^2(r_a) \\
\Sigma_b &= \cosh^2(r_b) + \sinh^2(r_b) \\
x_a &= \sinh(r_a)\cosh(r_a) \\
x_b &= \sinh(r_b)\cosh(r_b) \\
\delta^2 &= 1 + \frac{1}{4}[1 - \cos(4\theta)]\,[\cosh(2r_a)\cosh(2r_b) \\
&\quad - \sinh(2r_a)\sinh(2r_b)\cos(\Delta_a - \Delta_b)]
\end{aligned}
\tag{11.81}
$$

From (11.81), $\delta^2 = 1$ and $E = 0$, if $\theta = 0$, which corresponds to the absence of a beam splitter. Even if mode A is squeezed, it is not entangled with the other squeezed B.

Maximum entanglement, after the beam splitter, is for $\theta = \pi/4$, corresponding to a 50 : 50 splitting ratio. Also, δ^2 is maximized if

$$
\Delta_b - \Delta_a = 2\phi_1 - 2\phi_0 - \phi_b + \phi_a = (2k+1)\pi \tag{11.82}
$$

with k as an integer.

We find the degree of the entanglement by adding one layer for the `covariance` and one `entanglementBS` layer, as follows.

```
# layer for evaluating the covariance
cov_layer = ps.covariance(N)
# layer for evaluating the entanglement
e_layer = ps.entanglementBS(N)
# connect the layers
cova, _ = cov_layer(chir,chii, squeezed1mBS)
delta, Edelta, _ = e_layer(cova)
entangler = tf.keras.Model(inputs = xin, outputs=delta)
entanglement_value = tf.keras.Model(inputs = xin, outputs=Edelta)
```

Note that we have defined two models:

- The model `entangler` returning δ
- The model `entanglement_value` returning $E(\delta)$

Running the models, for any `xtrain`, we obtain, for example, $\delta = 3.5$ and $E = -0.527$ letting $r_a = 0.2$, $r_b = 3.0$, $\chi_a = \pi/4$, $\chi_b = \pi/3$, $\theta = 0.5$, $\phi_0 = 0$, and $\phi_1 = 0$. This result agrees with (11.81).[16]

11.15 Training the Model to Maximize the Entanglement

We train the model to maximize the degree of entanglement. The training determines the parameters for the beam splitter and we check the agreement with (11.81).

To train the model, we define a loss function, which is minimal, when E and δ are maximized for any input data $\boldsymbol{x}$. We remark that the model `entangler` takes as input $\boldsymbol{x}$ and returns δ, which is independent of $\boldsymbol{x}$.

In `tensorflow`, we add a loss to the model by using a variable to be minimized. We have the value of the entanglement in the variable `Edelta`. When `Edelta` is maximum, the quantity `-tf.math.log(Edelta)` is minimum, and we use the latter as the loss function. The logarithm increases the accuracy in computing the loss.

We add the loss as follows, before compiling the model. Then we launch the `fit` method to perform the training for 1000 epochs.

```
# add loss
entangler.add_loss(-tf.math.log(Edelta))
# compile the model with a given learning_rate
entangler.compile(optimizer=tf.keras.optimizers.Adam(learning_rate
 =0.01))
#fit the model
history = entangler.fit(x=xtrain,y=ytrain,epochs=1000,verbose=1)
```

As an example, we choose the initial parameters as follows:

$$r_a = 0.2 \tag{11.83}$$

$$\phi_a = \pi/4 \tag{11.84}$$

$$r_b = 3.0 \tag{11.85}$$

$$\phi_b = \pi/3 \tag{11.86}$$

$$\theta = 0.5 \tag{11.87}$$

$$\phi_0 = 0.0 \tag{11.88}$$

$$\phi_1 = 0.0 \tag{11.89}$$

[16] In the file `singlemodesqueezerBS`, we define a python function to compute (11.81) from θ, ϕ_0, and ϕ_1.

corresponding to `theta=0.5`, `phi0=0`, `phi1=0`, `ra=0.2`, `phia=np.pi/4`, `rb=3.0`, and `phib=np.pi/3`. We use as trainable weights the parameters of the beam splitter θ, ϕ_0, and ϕ_1.

After the training with randomly generated `xtrains`, the resulting parameters of the model are `theta=0.785`, `phi0=0.720`, `phi1=-0.720`, which corresponds (within numerical precision) to a 50:50 beam splitter with $\theta = \pi/4$, and satisfying (11.82). We have $\delta = 6.143$, and $E = 2.118$.

This approach may be extended to more complicated situations, e.g., a multimode optical beam traveling through a cascade of trainable interferometers and random media. Machine learning technology enables the design of devices with highly nonclassical outputs for many applications.

In the following chapters, we will focus on training boson sampling and variational ansatzes for many-body Hamiltonians.

11.16 Further Reading

- Extensive investigations and details on Gaussian states and entanglement are in [6–8]

References

1. S.M. Barnett, P.M. Radmore, *Methods in Theoretical Quantum Optics* (Oxford University Press, New York, 1997)
2. X. Wang, T. Hiroshima, A. Tomita, M. Hayashi, Phys. Rep. **448**, 1 (2007). https://doi.org/10.1016/j.physrep.2007.04.005
3. W. Xiang-bin, Phys. Rev. A **66**, 024303 (2002). https://doi.org/10.1103/physreva.66.024303
4. W. Xiang-bin, Phys. Rev. A **66**, 064304 (2002). https://doi.org/10.1103/physreva.66.064304
5. C. Gardiner, P. Zoller, *The Quantum World of Ultra-Cold Atoms and Light - Book II - The Physics of Quantum-Optical Devices* (Imperial College Press, London, 2015)
6. A. Ferraro, S. Olivares, M.G.A. Paris, arXiv:0503237 (2005)
7. R. Horodecki, P. Horodecki, M. Horodecki, K. Horodecki, Rev. Mod. Phys. **81**, 865 (2009). https://doi.org/10.1103/RevModPhys.81.865
8. M.A. Nielsen, I.L. Chuang, *Quantum Computation and Quantum Information (10th Anniversary edition)* (Cambridge University Press, Cambridge, 2016).

Chapter 12
Gaussian Boson Sampling

Yes we can

Abstract We introduce the fundamental models for Gaussian boson sampling and the link with the computation of the Hafnian. We show how to compute and train boson sampling patterns by machine learning.

12.1 Introduction

Gaussian boson sampling refers to computing the probability of multiparticle patterns in Gaussian states. The most studied case is when Gaussian states propagate in a system made of squeezers and interferometers [1, 2]. This problem demonstrated the quantum advantage at an impressive scale [3], following earlier realizations [4–9] of the original proposal by Aharanson and Arkhipov [10]. The theory of Gaussian boson sampling (GBS) heavily relies on phase-space methods [11], making it an exciting NN testbed [12].

For special classes of states, GBS is computationally demanding for a classical computer. A device that returns the boson sampling pattern probability by direct measurements is a quantum computing system that has a computational advantage. The class of states for which the classical computation is hard are many-body squeezed vacuum states through a random interferometer.

In the following, we consider GBS to compute the particle number distribution by machine learning. The first goal is to obtain the probability of having a given particle pattern in the modes. We start considering the case of a single mode.

In comparison with previously described coding solutions, the main difference is the use of functions that return functions. This enables to implement operators on models to change the phase-space representation and computing higher-order derivatives in an efficient way.

C. Conti, *Quantum Machine Learning*, Quantum Science and Technology,
https://doi.org/10.1007/978-3-031-44226-1_12

12.2 Boson Sampling in a Single Mode

Given a single-mode density matrix $\rho = |\psi\rangle\langle\psi|$, we consider a number state $|\bar{n}_0\rangle$ and

$$\Pr(\bar{n}_0) = \mathrm{Tr}\left[\rho|\bar{n}_0\rangle\langle\bar{n}_0|\right] = \langle\bar{n}_0|\rho|\bar{n}_0\rangle = |\langle n_0|\psi\rangle|^2 \,, \tag{12.1}$$

namely, the probability of finding $\bar{n}_0$ particles or, equivalently, finding the system in the state $|n_0\rangle$.

The projector $|\bar{n}_0\rangle\langle\bar{n}_0|$ has the following representation in terms of coherent states [13]:

$$|\bar{n}_0\rangle\langle\bar{n}_0| = \left[\frac{1}{\bar{n}_0}\left(\frac{\partial^2}{\partial\alpha_0\partial\alpha_0^*}\right)^{\bar{n}_0} e^{|\alpha|_0^2}|\alpha_0\rangle\langle\alpha_0|\right]_{\alpha_0=0} \,, \tag{12.2}$$

with α_0 the complex displacement.

Equation (12.2) used in (12.1) gives

$$\Pr(\bar{n}_0) = \frac{\pi}{\bar{n}_0!}\left(\frac{\partial^2}{\partial\alpha_0\partial\alpha_0^*}\right)^{\bar{n}_0} e^{|\alpha_0|^2} Q_\rho(\alpha_0, \alpha_0^*)\Bigg|_{\alpha_0=0} \,, \tag{12.3}$$

where we have introduced the $Q-$representation of the density matrix [14]

$$Q_\rho(\alpha_0, \alpha_0^*) = \frac{1}{\pi}\langle\alpha_0|\rho|\alpha_0\rangle \,. \tag{12.4}$$

Thus, the problem is reduced to finding the derivatives of the $Q-$function Q_ρ.

If we have a NN representation of Q_ρ, we can use the automatic differentiation routines. So far, we considered neural network models of the characteristic function χ, but χ and Q_ρ are related as follows [14]:

$$Q_\rho(\alpha, \alpha^*) = \frac{1}{\pi^2}\int \chi(z, z^*) e^{-\frac{|z|^2}{2}} e^{\alpha z^* - \alpha^* z} \mathrm{d}^2 z \tag{12.5}$$

with $\mathrm{d}^2 z = \mathrm{d}\Re(z)\mathrm{d}\Im(z)$. For a Gaussian state, χ is a Gaussian function, and the integral in Eq. (12.5) is evaluated explicitly, as in the following.

We introduce the real variables

$$\begin{aligned} k_0 &= \frac{\alpha + \alpha^*}{\sqrt{2}} \\ k_1 &= \frac{\alpha^* - \alpha}{\sqrt{2}\mathrm{i}} \,, \end{aligned} \tag{12.6}$$

and

$$\begin{aligned} x_0 &= \frac{z - z^*}{\sqrt{2}\mathrm{i}} \\ x_1 &= \frac{z + z^*}{\sqrt{2}}\,, \end{aligned} \tag{12.7}$$

and we have

$$Q_\rho(\boldsymbol{k}) = \int \chi(\boldsymbol{x}) e^{\frac{x_0^2+x_1^2}{2}} e^{-\mathrm{i}x_0k_0 - \mathrm{i}x_1k_1} \mathrm{d}^2\boldsymbol{x} = \int \chi(\boldsymbol{x}) e^{\frac{|\boldsymbol{x}|^2}{2}} e^{-\mathrm{i}\boldsymbol{x}\cdot\boldsymbol{k}} \mathrm{d}\boldsymbol{x} \tag{12.8}$$

with $\boldsymbol{k} = (k_0, k_1)$, $\boldsymbol{x} = (x_0, x_1)$, and $\mathrm{d}\boldsymbol{x} = \mathrm{d}x_0\mathrm{d}x_1$.

We express the derivatives in (12.3) as derivatives with respect to real variables by using

$$\left(\frac{\partial^2}{\partial\alpha\partial\alpha_0^*}\right)^{\bar{n}_0} = \frac{1}{2^{\bar{n}_0}}\left(\tilde{\nabla}_0^2\right)^{\bar{n}_0} \tag{12.9}$$

with

$$\tilde{\nabla}_0^2 = \frac{\partial^2}{\partial k_0^2} + \frac{\partial^2}{\partial k_1^2} \tag{12.10}$$

and Eq. (12.3) is written as

$$\Pr(\bar{n}_0) = \frac{\pi}{2^{\bar{n}_0}\bar{n}_0!}\left(\tilde{\nabla}_0^2\right)^{\bar{n}_0} e^{\frac{k^2}{2}} Q_\rho(\boldsymbol{k})\bigg|_{\boldsymbol{k}=0}. \tag{12.11}$$

Equation (12.11) gives $\Pr(\bar{n}_0)$ as derivatives with respect to real variables, which is helpful when coding NNs.

12.3 Boson Sampling with Many Modes

For many modes, letting n be the number of modes, boson sampling corresponds to computing the probability of finding $\bar{n}_0$ particles in mode 0, $\bar{n}_1$ particles in mode 1, and so forth. In the following, we denote

$$\bar{\mathbf{n}} = (\bar{n}_0, \bar{n}_1, \ldots, \bar{n}_{n-1}) \tag{12.12}$$

as a given particle pattern.

By introducing the operator

$$|\bar{\mathbf{n}}\rangle\langle\bar{\mathbf{n}}| = \otimes_{j=0}^{n-1}|\bar{n}_j\rangle\langle\bar{n}_j|, \tag{12.13}$$

we have

$$\Pr(\bar{\mathbf{n}}) = \mathrm{Tr}\left[\rho|\bar{\mathbf{n}}\rangle\langle\bar{\mathbf{n}}|\right]. \tag{12.14}$$

Considering the complex displacements α_j, for many modes, Eq. (12.2) becomes

$$\begin{aligned}
|\bar{\mathbf{n}}\rangle\langle\bar{\mathbf{n}}| &= \frac{1}{\bar{n_0}!\bar{n_1}!\ldots\bar{n}_{n-1}!} \\
&\quad\times\left(\frac{\partial^2}{\partial\alpha_0\partial\alpha_0^*}\right)^{\bar{n}_0}\left(\frac{\partial^2}{\partial\alpha_1\partial\alpha_1^*}\right)^{\bar{n}_1}\cdots\left(\frac{\partial^2}{\partial\alpha_{n-1}\partial\alpha_{n-1}^*}\right)^{\bar{n}_{n-1}} e^{\sum_j^{n-1}|\alpha_j|^2}|\boldsymbol{\alpha}\rangle\langle\boldsymbol{\alpha}|\rho\Bigg|_{\boldsymbol{\alpha}=0} \\
&= \left[\frac{1}{\bar{\mathbf{n}}!}\prod_{j=0}^{n-1}\left(\frac{\partial^2}{\partial\alpha_j\partial\alpha_j^*}\right)^{\bar{n}_j} e^{\sum_j^{n-1}|\alpha|_j^2}|\boldsymbol{\alpha}\rangle\langle\boldsymbol{\alpha}|\right]_{\boldsymbol{\alpha}=0},
\end{aligned} \tag{12.15}$$

where $\bar{\mathbf{n}}! = \bar{n_0}!\bar{n}_1!\ldots\bar{n}_{n-1}!$ [11]. By using Eq. (12.15) in (12.14), we have

$$\Pr(\bar{\mathbf{n}}) = \frac{\pi^n}{\bar{\mathbf{n}}!}\prod_{j=0}^{n-1}\left(\frac{\partial^2}{\partial\alpha_j\partial\alpha_j^*}\right)^{\bar{n}_j} e^{\sum_{j=0}^{n-1}|\alpha_j|^2} Q_\rho(\boldsymbol{\alpha},\boldsymbol{\alpha}^*)\Bigg|_{\boldsymbol{\alpha}=0}, \tag{12.16}$$

where the multimode Q-function is

$$Q_\rho(\boldsymbol{\alpha},\boldsymbol{\alpha}^*) = \frac{1}{\pi^n}\langle\boldsymbol{\alpha}|\rho|\boldsymbol{\alpha}\rangle. \tag{12.17}$$

To express Eq. (12.16) in terms of real variables, we define

$$\begin{aligned}
k_{2j} &= \frac{\alpha_j + \alpha_j^*}{\sqrt{2}} \\
k_{2j+1} &= \frac{\alpha_j^* - \alpha_j}{\sqrt{2}\mathrm{i}},
\end{aligned} \tag{12.18}$$

and

$$\begin{aligned}
x_{2j} &= \frac{z_j - z_j^*}{\sqrt{2}\mathrm{i}} \\
x_{2j+1} &= \frac{z_j + z_j^*}{\sqrt{2}},
\end{aligned} \tag{12.19}$$

with $j = 0, 1, \ldots, n-1$, and we have

$$\Pr(\bar{\mathbf{n}}) = \frac{\pi^n}{\bar{\mathbf{n}}!\, 2^{\bar{n}_T}} \prod_{j=0}^{n-1} \left(\tilde{\nabla}_j^2\right)^{\bar{n}_j} e^{\sum_{m=0}^{n-1} \frac{k_{2m}^2 + k_{2m+1}^2}{2}} Q_\rho(\boldsymbol{k}) \Bigg|_{\boldsymbol{\alpha}=0} \tag{12.20}$$

with $\bar{n}_T = \sum_{j=0}^{n-1} \bar{n}_j$ the total number of particles in the pattern, and

$$\tilde{\nabla}_j^2 = \frac{\partial^2}{\partial k_{2j}} + \frac{\partial^2}{\partial k_{2j+1}} . \tag{12.21}$$

12.4 Gaussian States

For Gaussian states, Q_ρ is explicitly computed as a multidimensional Gaussian integral. First, we consider the Q-representation

$$Q_\rho(\boldsymbol{\alpha}, \boldsymbol{\alpha}^*) = \frac{1}{\pi^{2n}} \int \chi(z, z^*) e^{-\frac{|z|^2}{2}} e^{\boldsymbol{\alpha}\cdot z^* - \boldsymbol{\alpha}^* \cdot z} \mathrm{d}^2 \boldsymbol{z} \tag{12.22}$$

being $|\boldsymbol{z}|^2 = z \cdot z^* = \sum_{j=0}^{n-1} |z_j|^2$, and $\mathrm{d}^2\boldsymbol{z} = \prod_{j=0}^{n-1} \mathrm{d}\Re(z_j)\mathrm{d}\Im(z_j)$.

In terms of $\boldsymbol{x}$ and $\boldsymbol{k}$, Eq. (12.22) reads

$$Q_\rho(\boldsymbol{k}) = \frac{1}{2^n \pi^{2n}} \int \chi(\boldsymbol{x}) e^{-\sum_{j=0}^{N-1} \frac{x_j^2}{4}} e^{-\mathrm{i} \sum_{j=0}^{N-1} k_j x_j} \mathrm{d}\boldsymbol{x} , \tag{12.23}$$

with $\mathrm{d}\boldsymbol{x} = \prod_{j=0}^{N-1} \mathrm{d}x_j$.

For a Gaussian state

$$\chi(\boldsymbol{x}) = e^{-\frac{1}{4} \sum_{pq} g_{pq} x_p x_q + \mathrm{i} \sum_m d_m x_m} , \tag{12.24}$$

with $p, q, m, n = 0 \ldots N-1$ and, using (12.23), gives

$$Q_\rho(\boldsymbol{k}) = \frac{1}{2^n \pi^{2n}} \int e^{-\frac{1}{2} \sum_{pq} Q_{pq} x_p x_q} e^{\sum_n B_n x_n} \mathrm{d}\boldsymbol{x} \tag{12.25}$$

with

$$\begin{aligned} Q_{pq} &= \frac{1}{2} \left(g_{pq} + \delta_{pq} \right) \\ B_n &= -\mathrm{i}(k_n - d_n) . \end{aligned} \tag{12.26}$$

As

$$Q = \frac{\mathbf{1}_N}{2} + \frac{\boldsymbol{g}}{2} = \frac{\mathbf{1}_N}{2} + \boldsymbol{\sigma}^R \tag{12.27}$$

with elements Q_{pq} is a positive definite matrix because $\boldsymbol{g} > 0$, Eq. (12.25) is a multidimensional Gaussian integral evaluated explicitly as follows:

$$Q_\rho(\boldsymbol{k}) = \frac{e^{\frac{1}{2}\sum_{pq} Q^{-1}_{pq} B_p B_q}}{\pi^n \sqrt{\det \mathbf{Q}}} = \frac{e^{-\frac{1}{2}\sum_{jk} Q^{-1}_{pq}(k_p - d_p)(k_q - d_q)}}{\pi^n \sqrt{\det \mathbf{Q}}} . \tag{12.28}$$

Finally, we have

$$\Pr(\bar{\mathbf{n}}) = \frac{1}{\bar{\mathbf{n}}!\, 2^{\bar{n}_T}} \left(\prod_{j=0}^{n-1} \tilde{\nabla}_j^{2\bar{n}_j} \right) \left. \frac{e^{\frac{k^2}{2}} e^{-\frac{1}{2}\sum_{pq} Q^{-1}_{pq}(k_p - d_p)(k_q - d_q)}}{\sqrt{\det \mathbf{Q}}} \right|_{\boldsymbol{k}=0} . \tag{12.29}$$

12.5 Independent Coherent States

We check Eq. (12.29) in some simple cases. If we have independent modes (a product state), the probability distribution factorizes as follows:

$$\Pr(\bar{\mathbf{n}}) = \prod_{j=0}^{n-1} P_j(\bar{n}_j) , \tag{12.30}$$

where $P_j(\bar{n}_j)$ is the probability of finding $\bar{n}_j$ particles in mode j.

The modes are independent when Q_{jk} is a block diagonal matrix, with each block of dimension 2, corresponding to the q_j and p_j variables of mode j. As following Eq. (12.27)

$$\mathbf{Q} = \frac{1}{2}\mathbf{1}_N + \boldsymbol{\sigma}^R , \tag{12.31}$$

each block of $\boldsymbol{Q}$ is related to the covariance matrix of each mode in $\boldsymbol{\sigma}^R$, which is block diagonal for independent modes. Also the inverse Q^{-1}_{jk} is block diagonal, and $\det \mathbf{Q}$ is the product of the determinants of the blocks.

In this simple case, we represent the block diagonal $\mathbf{Q}$ matrix as follows:

$$\mathbf{Q} = \begin{pmatrix} \mathbf{Q}_0 & & & \\ & \mathbf{Q}_1 & & \\ & & \ddots & \\ & & & \mathbf{Q}_{n-1} \end{pmatrix} \tag{12.32}$$

Denoting the blocks of Q^{-1}_{jk} as $[Q^{-1}_{jk}]_q$ with $q = 0, 1, \ldots, n-1$, Eq. (12.29) becomes Eq. (12.30) with

$$P_j(\bar{n}_j) = \frac{1}{\bar{n}_j! 2^{\bar{n}_j}} \left(\tilde{\nabla}_j^2\right)^{\bar{n}_j} \frac{e^{\frac{k_{2j}^2 + k_{2j+1}^2}{2}}}{\sqrt{\det[Q]_j}} e^{-\frac{1}{2}\sum_{pq}[Q^{-1}]_{j,pq}(k_p - d_p)(k_q - d_q)}, \Bigg|_{\mathbf{k}=0} \tag{12.33}$$

where p and q run in the components of the block $[Q^{-1}]_j$.

The simplest case is for n independent coherent states, given by $g_{jk} = \delta_{jk}$, which corresponds to $Q_{jk} = \delta_{jk}$.

Performing the derivatives in Eq. (12.33), one finds

$$P_j(\bar{n}_j) = \frac{1}{\bar{n}_j!} e^{\frac{d_{2j}^2 + d_{2j+1}^2}{2}} \left(\frac{d_{2j}^2 + d_{2j+1}^2}{2}\right)^{n_j} = \frac{1}{\bar{n}_j!} e^{-|\alpha_j|^2} |\alpha_j|^{2n_j} \tag{12.34}$$

which is the expected distribution function for a coherent states being

$$|\alpha_j|^2 = \frac{d_{2j}^2 + d_{2j+1}^2}{2}. \tag{12.35}$$

We test these particular cases after implementing the GBS neural network model.

12.6 Zero Displacement Case in Complex Variables and the Hafnian

We detail the significant case when one has only 0 or 1 particle per channel, and zero displacement, i.e.,

$$\bar{n}_j \in \{0, 1\} \tag{12.36}$$

$$\boldsymbol{d} = \mathbf{0}. \tag{12.37}$$

In this case, one links computing $\Pr(\bar{\mathbf{n}})$ to the problem of computing the so-called Hafnian of a matrix.

When $d_j = 0$, Eq. (12.29) reads

$$\Pr(\bar{\mathbf{n}}) = \frac{1}{\bar{\mathbf{n}}! 2^{\sum_{j=0}^{n-1}\bar{n}_j}} \left(\prod_{j=0}^{n-1} \tilde{\nabla}_j^{2\bar{n}_j}\right) \frac{e^{\frac{k^2}{2}} e^{-\frac{1}{2}\sum_{pq} Q_{pq}^{-1} k_p k_q}}{\sqrt{\det \mathbf{Q}}} \Bigg|_{\mathbf{k}=0}. \tag{12.38}$$

Equation (12.38) is cast in complex variables as follows [1, 11]:

$$\Pr(\bar{\mathbf{n}}) = \frac{1}{\bar{\mathbf{n}}!}\left(\prod_{j=0}^{n-1}\frac{\partial^2}{\partial\alpha_j^*\partial\alpha_j}\right)^{\bar{n}_j}\left.\frac{e^{\frac{1}{2}\boldsymbol{\alpha}_\nu^\top\cdot\boldsymbol{A}\cdot\boldsymbol{\alpha}_\nu}}{\sqrt{\det\boldsymbol{\sigma}_Q}}\right|_{\boldsymbol{\alpha}=0}, \tag{12.39}$$

where

$$\boldsymbol{A} = \begin{pmatrix}\mathbf{0}_n & \mathbf{1}_n\\ \mathbf{1}_n & \mathbf{0}_n\end{pmatrix}\left(\mathbf{1}_N - \boldsymbol{\sigma}_Q{}^{-1}\right), \tag{12.40}$$

with

$$\boldsymbol{\sigma}_Q = \frac{1}{2}\mathbf{1}_N + \boldsymbol{\sigma}^A \tag{12.41}$$

with $\boldsymbol{\sigma}^A$ the complex covariance matrix (see Sect. 7.6.2):

$$\boldsymbol{\alpha}_\nu = \left(\alpha_0,\,\alpha_1,\ldots\alpha_{n-1}\ldots\alpha_0^*,\,\alpha_1^*,\,\alpha_{n-1}^*\right)^\top,$$

and $\det\sigma_Q = \det\boldsymbol{Q}$. The proof of Eq. (12.39) is reported in Sect. 12.6.2.

In Eq. (12.39), we retain the dot symbol for the scalar product for consistency with the existence literature [1, 11], being $\boldsymbol{\alpha}_\nu$ a column vector. In our notation, as $\boldsymbol{\alpha}$ is a row vector, we can write in the exponent $\boldsymbol{\alpha}_\nu^\top\cdot\boldsymbol{A}\cdot\boldsymbol{\alpha}_\nu = \boldsymbol{\alpha}\boldsymbol{A}\boldsymbol{\alpha}^\top$, without the dot.

From Eq. (12.39) one shows that the boson sampling probability is expressed as the higher-order derivative of a Gaussian, which is written in terms of Hafnian $\mathrm{Haf}(A_S)$ [15] of a submatrix $\boldsymbol{A}_S$ of the matrix $\boldsymbol{A}$

$$\Pr(\bar{\mathbf{n}}) = \frac{1}{\bar{\mathbf{n}}!\det\boldsymbol{\sigma}_Q}\mathrm{Haf}(\boldsymbol{A}_s) \tag{12.42}$$

We recall that $\bar{\mathbf{n}}! = \bar{n}_0!\bar{n}_1!\ldots\bar{n}_{n-1}$.

In boson sampling, the matrix $\boldsymbol{A}$ is denoted the *kernel* matrix. The particle pattern $\bar{\mathbf{n}}$ fixes the submatrix $\boldsymbol{A}_S$, which is obtained from the matrix $\boldsymbol{A}$ by deleting the row and columns i and $i+n$ if $\bar{n}_i = 0$.

If mode i is not retained as an output channel, the corresponding row and column are removed from $\boldsymbol{A}$ to obtain $\boldsymbol{A}_S$. Otherwise, if $\bar{n}_i = 1$, we retain the column i and $i+n$. The reduced matrix $\boldsymbol{A}_S$ arises from the fact that derivatives in Eq. (12.42) are present only with respect to the displacement of the output channels with $\bar{n}_j \neq 0$.

For example, considering $n = 4$, the matrix $\mathbf{A}$ has dimension 8×8:

$$\mathbf{A} = \begin{pmatrix} A_{00} & A_{01} & A_{02} & A_{03} & A_{04} & A_{05} & A_{06} & A_{07} \\ A_{10} & A_{11} & A_{12} & A_{13} & A_{14} & A_{15} & A_{16} & A_{17} \\ A_{20} & A_{21} & A_{22} & A_{23} & A_{24} & A_{25} & A_{26} & A_{27} \\ A_{30} & A_{31} & A_{32} & A_{33} & A_{34} & A_{35} & A_{36} & A_{37} \\ A_{40} & A_{41} & A_{42} & A_{43} & A_{44} & A_{45} & A_{46} & A_{47} \\ A_{50} & A_{51} & A_{52} & A_{53} & A_{54} & A_{55} & A_{56} & A_{57} \\ A_{60} & A_{61} & A_{62} & A_{63} & A_{64} & A_{65} & A_{66} & A_{67} \\ A_{70} & A_{71} & A_{72} & A_{73} & A_{74} & A_{75} & A_{76} & A_{77} \end{pmatrix}$$

If we want $\Pr(\bar{\mathbf{n}})$, with $\bar{\mathbf{n}} = (1, 0, 1, 0)$, we are considering as output channels modes with indices $j = 0$ and $j = 2$. As a result, the reduced matrix $\mathbf{A}_S$ is obtained by removing the elements for modes $j = 1$ and $j = 3$ and has dimensions 4×4. These elements are the rows and the columns corresponding to mode $j = 1$, i.e., the column and the row 1 and the column and the row 5. Also, we delete the rows and the columns corresponding to mode $j = 3$, i.e., the column and the row 3 and the column and the row 7, such that

$$\mathbf{A}_S = \begin{pmatrix} A_{00} & A_{02} & A_{04} & A_{05} & A_{06} \\ A_{20} & A_{22} & A_{24} & A_{25} & A_{26} \\ A_{40} & A_{42} & A_{44} & A_{45} & A_{46} \\ A_{60} & A_{62} & A_{64} & A_{65} & A_{66} \end{pmatrix} .$$

12.6.1 The Resulting Algorithm

Summarizing, in the simplest case in which $\bar{n}_j \in 0, 1$, and $d_j = 0$, one reduces computing $\Pr(\bar{\mathbf{n}})$ to the computation of the Hafnian of a matrix $\mathbf{A}_S$.

The algorithm is as follows:

- Compute the complex covariance matrix $\boldsymbol{\sigma}^A$.
- Compute the matrix

$$\boldsymbol{\sigma}_Q = \frac{1}{2}\mathbf{1}_N + \boldsymbol{\sigma}^A . \tag{12.43}$$

- Compute the kernel matrix

$$\boldsymbol{A} = \mathbf{1_N} - \boldsymbol{\sigma}_{\mathbf{Q}}^{-1} . \tag{12.44}$$

- Compute the reduced matrix $\boldsymbol{A}_S$ by removing the channels with $\bar{n}_j = 0$.
- Use Eq. (12.42)

$$\Pr(\bar{\mathbf{n}}) = \frac{1}{\bar{\mathbf{n}}! \det \boldsymbol{\sigma}_Q} \mathrm{Haf}(\boldsymbol{A}_S) . \tag{12.45}$$

The Hafnian of a matrix $\mathbf{A}_S$ is detailed in [16], to which the interested reader may refer. Quesada, Gupt, and Izaac developed a specialized library for the Hafnian.[1]

The Hafnian is computationally hard to approximate, and the runtime of the best known algorithms for computing Hafnians scales exponentially with the dimension of the input matrix. The hardness of computing Hafnians for classical computers is still a matter of investigation, as discussed in [16].

Here, we will not focus on the computational resources needed in computing the Hafnian, but we will show how to use machine learning to obtain the boson sampling probabilities, without using the Hafnian.

12.6.2 Proof of Eq. (12.39)*

Proof We derive Eq. (12.39) from (12.38), which can be recast as

$$\Pr(\bar{\mathbf{n}}) = \frac{1}{\bar{\mathbf{n}}!2^{\sum_{j=0}^{n-1}\bar{n}_j}} \left(\prod_{j=0}^{n-1} \tilde{\nabla}_j^{2\bar{n}_j}\right) \frac{e^{\frac{1}{2}\sum_{pq}\mathcal{A}_{pq}k_p k_q}}{\sqrt{\det Q}}\Bigg|_{\boldsymbol{k}=0} \tag{12.46}$$

with

$$\mathcal{A}_{pq} = \delta_{pq} - Q^{-1}_{pq} \ . \tag{12.47}$$

Following Sect. 7.6.2, the $N \times 1$ column vectors $\boldsymbol{k}$ and $\boldsymbol{\alpha}_\mathrm{v}$ are related by

$$\boldsymbol{k} = P^\top \Omega^\dagger \boldsymbol{\alpha}_\mathrm{v} \ . \tag{12.48}$$

Hence, we have

$$\sum_{pq} \mathcal{A}_{pq}k_p k_q = \boldsymbol{k}^\top \cdot \mathcal{A} \cdot \boldsymbol{k} = \boldsymbol{\alpha}_\mathrm{v}^\dagger \cdot \Omega P \mathcal{A} P^\top \Omega^\dagger \cdot \boldsymbol{\alpha}_\mathrm{v}. \tag{12.49}$$

Being

$$\boldsymbol{\alpha}_\mathrm{v}^\top = \boldsymbol{\alpha}_\mathrm{v}^\dagger \begin{pmatrix} \mathbf{0}_n & \mathbf{1}_n \\ \mathbf{1}_n & \mathbf{0}_n \end{pmatrix} , \tag{12.50}$$

we write (12.49) as

$$\sum_{pq} \mathcal{A}_{pq}k_p k_q = \boldsymbol{\alpha}_\mathrm{v}^\top \cdot \boldsymbol{A} \cdot \boldsymbol{\alpha}_\mathrm{v}, \tag{12.51}$$

[1] https://hafnian.readthedocs.io/en/latest/.

with

$$\boldsymbol{A} = \begin{pmatrix} \mathbf{0}_n & \mathbf{1}_n \\ \mathbf{1}_n & \mathbf{0}_n \end{pmatrix} \Omega P \mathcal{A} P^\top \Omega^\dagger = \begin{pmatrix} \mathbf{0}_n & \mathbf{1}_n \\ \mathbf{1}_n & \mathbf{0}_n \end{pmatrix} \Omega P \left(\mathbf{1_N} - \boldsymbol{Q}^{-1} \right) P^\top \Omega^\dagger \,. \tag{12.52}$$

As the matrix $P^\top \Omega^\dagger$ is unitary, we have from Eq. (12.27)

$$\Omega P \boldsymbol{Q}^{-1} P^\top \Omega^\dagger = \Omega P \left(\frac{1}{2} \mathbf{1}_N + \boldsymbol{\sigma}^R \right)^{-1} P^\top \Omega^\dagger = \left(\frac{1}{2} \mathbf{1}_N + \Omega P \boldsymbol{\sigma}^R P^\top \Omega^\dagger \right)^{-1} . \tag{12.53}$$

Being

$$\Omega P \boldsymbol{\sigma}^R P^\top \Omega^\dagger = \boldsymbol{\sigma}^A, \tag{12.54}$$

we have Eq. (12.40) and

$$\boldsymbol{\sigma}_Q^{-1} = \Omega P \boldsymbol{Q}^{-1} P^\top \Omega^\dagger = \left(\frac{1}{2} \mathbf{1}_N + \boldsymbol{\sigma}^A \right)^{-1} . \tag{12.55}$$

Also following (12.53) we have $\det(\boldsymbol{\sigma}_Q) = \det(\boldsymbol{Q})$. □

12.7 The Q-Transform

Computing the boson pattern probability requires the Q_ρ representation, which is a Gaussian function determined by the covariance matrix $\boldsymbol{g}$ and the displacement vector $\boldsymbol{d}$. We introduce a layer specialized in determining Q_ρ and its derivatives. We adopt an approach based on functions returning functions, which are useful when dealing with derivatives.

First, we assume that we have a NN model and want the boson distribution. We recall that $\Pr(\bar{\mathbf{n}})$ is real-valued and depends on $\boldsymbol{g}$ and $\boldsymbol{d}$ obtained from the `CovarianceLayer`. We write Eq. (12.29) as

$$\Pr(\mathbf{n}) = \frac{1}{\bar{\mathbf{n}}!} \prod_{j=0}^{n-1} \left(\frac{1}{2} \tilde{\nabla}_j^2 \right)^{\bar{n}_j} Q[\boldsymbol{g}, \boldsymbol{d}](\boldsymbol{k}) \tag{12.56}$$

where we introduced the *Q-transform* of the characteristic function:

$$Q[\boldsymbol{g}, \boldsymbol{d}](\boldsymbol{k}) = \frac{e^{\frac{1}{2}\boldsymbol{k}^\top \boldsymbol{k}} e^{-\frac{1}{2}\boldsymbol{k}^\top \boldsymbol{Q} \boldsymbol{k}}}{\det \boldsymbol{Q}} \,. \tag{12.57}$$

The Q-transform has as inputs the covariance matrix $\boldsymbol{g}$ and the displacement vector $\boldsymbol{d}$ and returns a function of the wavevector $\boldsymbol{k}$. Hence, to implement the Q-transform, we define a `TensorFlow` function with inputs $\boldsymbol{g}$ and $\boldsymbol{d}$ and a callable function as output, which is allowed in `Python`.

We start from a generic model, among those we have defined in previous examples `model(x)`, and then we use the covariance layer.[2]

```
cov_layer = CovarianceLayer(N)
covariance_matrix, mean_R, hessian = cov_layer(chir,chii, model)
cov_out = tf.keras.Model(inputs = xin,
outputs=[covariance_matrix, mean_R, hessian])
xtrain = np.random.rand(1, N)-0.5
cov0,d0, _ =cov_out(xtrain); print(d0); tf.print(cov0)
```

Here `xtrain` is a dummy input to return the covariance matrix and the displacement in the tensors `cov0` and `d0`.

`d0` is the expected value of $\hat{\boldsymbol{R}}$ and has shape `[1,N]`.

`d0` and `cov0` are the inputs to the following `Qtransform` function.

```
def QTransform(g, d):
    # return the Q-transform of a model, starting
    # from its covariance matrix and displacement vector
    #
    # Parameters
    # ----------
    # g : covariance matrix [N,N]
    # d : displacement vector [1,N]
    #
    # Return a function of the k-vector with shape [1,N]

    N, _ =g.shape
    n = N//2
    EN = tf.eye(N, dtype=g.dtype)
    Q= tf.multiply(0.5,g+EN) #Q matrix
    QI = tf.linalg.inv(A)     #Inverse Q matrix
    # scale factor
    scale = tf.math.pow(tf.linalg.det(Q), tf.constant(-0.5))

    def functionQ(kin):
        k = kin-d
        # exp(k Q k/2)
        yR  = tf.matmul(QI, k, transpose_b=True)
        gDot = tf.matmul(-0.5*k, yR)
        f = tf.exp(gDot)
        # exp(-k^2/2)
        f = tf.multiply(f,tf.exp(tf.matmul(0.5*kin,kin,
```

[2] The code is in `jupyternotebooks/bosonsampling/BosonSamplingExample1.ipynb`.

```
            transpose_b=True)))
        f = tf.multiply(f, scale)
        return f
    return functionQ
```

In `QTransform`, we first compute `Q` corresponding to $\boldsymbol{Q}$ in Eq. (12.57), the inverse `QI`, and the scaling factor given by the square root of the inverse of the determinant $\det(\boldsymbol{Q})^{-1/2}$. Finally, we define a function as output to return a Gaussian with `QI` as a covariance matrix.

We use `QTransform` to create a function of a tensor `kin` corresponding to the vector $\boldsymbol{k}$ as follows.

```
Qrho = QTransform(cov0, d0) # create the function
kin = tf.zeros_like(d0) # input vector
Qrho(kin) # call the function
```

Note that in this formulation, `Qrho` is not a layer and cannot be embedded in a `keras.Model`.

12.8 The Q-Transform Layer

To embed the Q-transform in a trainable model, we create the transform layer after a `CovarianceLayer` and define the input vector `kin`. A further implementation is a function returning a `keras.Model` and the vector `kin`.

First, we define a layer, which returns a layer, as follows.

```
class QTransformLayer(tf.keras.layers.Layer):
    """ Return a layer for the Q-transform
    Usage
    QLayer = QTransformLayer(N, model)
    kin=tf.keras.layers.Input([N],dtype=model.dtype)
    Q = QLayer(kin)
    Qrho = tf.keras.Model(inputs=kin, outputs=Q)
    """

    def __init__(self, N, model):
        super(QTransformLayer, self).__init__()
        self.N = N
        self.EN = tf.eye(self.N, dtype=self.dtype)
        self.model=model

        x0 = tf.zeros([1,N],dtype=model.dtype)
        #covlayer = ps.CovarianceLayer2(N)
```

```
        covlayer = CovarianceLayer(N)
        cr, ci = model(x0)
        cov1, mean_R1, _ = covlayer(cr, ci,model)
        self.g = cov1
        self.d = mean_R1
        self.Q = tf.multiply(0.5, self.g + self.EN)
        self.QI = tf.linalg.inv(self.Q)
        self.scale = tf.math.pow(tf.linalg.det(self.Q),
        ↪  tf.constant(-0.5))

    def call(self, kin):

        k = kin - self.d
        yR = tf.matmul(self.QI, k, transpose_b=True)
        gDot = tf.matmul(-0.5 * k, yR)
        f = tf.exp(gDot)
        f = tf.multiply(f, tf.exp(tf.matmul(0.5 * kin, kin,
        ↪  transpose_b=True)))
        f = tf.multiply(f, self.scale)

        return f
```

Given a model as input, the `QTransformLayer` first computes the covariance, the displacement, the Q matrix, and its inverse. Then, in the `call`, the layer returns the Q-transform Gaussian.

By the `QTransformLayer`, we build a model by a function returning the model itself as follows.

```
def getQTransformModel(model):
    """ Return a model with the Qtransform

    Parameters
    ----------
    model: model to be transformed

    Returns
    -------
    The vector kin and the model Qrho corresponding
    to the modified Q-representation of the model
    """
    N=model.input_shape[1]
    QLayer = QTransformLayer(N, model)
    kin=tf.keras.layers.Input([N],dtype=model.dtype)
    Q = QLayer(kin)
    Qrho = tf.keras.Model(inputs=kin, outputs=Q)
    return kin, Qrho
```

We call the model-returning function `getQTransformModel` simply passing the model, as follows.

```
kin, Qrho =getQTransformModel(model)
```

This gives a model that has as input the wavevector `kin` and returns the Q-transform; we obtain its derivatives by automatic differentiation.

12.9 The Multiderivative Operator

Once we have a function with the Q-transform, we need the multiple higher-order derivatives as in Eq. (12.56); specifically, we need a product of powers of the Laplacian with different exponents n_j.

We start building functions that return derivatives w.r.t. k_j, where $j = 0, 1, \ldots, n-1$. Once again these are functions that return functions, i.e., the derivatives of $Q[\boldsymbol{g}, \boldsymbol{d}](\boldsymbol{k})$, as, e.g., $\partial Q/\partial k_0$, which is also a function of $\boldsymbol{k}$.

The first-order derivative is computed with the method `tf.gradient`, as in the following.

```
def partial(j, fun , k):
    # Return the partial derivative function wrt to k[j]
    # as a function of k, the shape of the output is [[1]]
    def L(k):
        with tf.GradientTape() as tape:
            tape.watch(k)
            y = fun(k)
        dy = tf.squeeze(tape.gradient(y,k))
        return tf.gather_nd(dy, [[j]])
    return L
```

The function `partial(j,fun,k)` returns one single derivative, extracted from the tensor `dy` output of `tf.gradient`. The target component with index `j` is obtained by `tf.gather_nd`. Note that we use `tf.squeeze` to remove dimensions with one element, e.g., to transform a tensor with shape `[1,N]` to shape `[N]`.

The expert reader will remark that this is not the most efficient computation, as we compute all the derivatives and then select the index `j` instead of just finding the target derivative. We keep this approach as it is more transparent and versatile and may be improved at a later stage by changing the function `partial`.

For example, to compute the derivative of the Q-transform above `Qrho`, we use the following code.

```
Qrho_j=partial(j, Qrho, kin) # compute the derivative function
# kin is needed as input to track the gradient tape
Qrho_j(kin) # evaluate the derivative at kin
```

We call `partial` multiple times to compute higher-order derivatives. We define the following function for the second derivative $\partial^2 Q/\partial k_j^2$.

```
def doublepartial(j, fun, k):
    # return the second derivative of fun wrt to x[j]
    dy1 = partial(j, fun, k)
    dy2 = partial(j, dy1, k)
    return dy2
```

Here, we first obtain the first-order partial derivative `dy1`, which is further derived for the second derivative of `fun` in the tensor `dy2`. We remark that `dy2` is also a function that has input `k` and returns a scalar as, e.g., in `dy2(kin)`.

We use the second partial derivatives for the operator $\tilde{\nabla}_j^2$ in (12.29) by summing the derivatives with respect k_{2j} and k_{2j+1}, as follows.

```
def laplacian(j, fun, k):
    # return the Laplacian with respect to mode index
    dj0 = doublepartial(2*j, fun, k)
    dj1 = doublepartial(2*j+1, fun, k)
    def L(k):
        return dj0(k)+dj1(k)
    return L
```

Note here that `j` is in the range $0, 1, \ldots, n-1$, corresponding to one of the modes; still `laplacian` returns a function that can be further derived.

Now we have all the components for building the operator in (12.29). We have to iterate the application of $\tilde{\nabla}_j^2$, n_j times and for all the modes for which $n_j \neq 0$. We employ a for loop as in the following code.[3]

```
def Pr(nbar, Qfun, mydtype=tf.float32):
# Return the boson sample operator
# nbar is list of integers
# Qfun is a function
    n=len(nbar) # number of modes al the length of the list
```

[3] The code is in `phasespace.py`.

```
    N=2*n
    x = tf.Variable(tf.zeros([1, N],dtype=mydtype))
    # compute the derivatives
    dm = Qfun
    for j in tf.range(n):
        for _ in tf.range(nbar[j]):
            dm = laplacian(j, dm, x)

    # normalization factor
    nbar_sum=tf.reduce_sum(nbar).numpy()
    scale=1
    for j in range(n):
        nj = nbar[j]
        scale = scale*np.math.factorial(nj)
    scale = scale*np.power(2.0,nbar_sum)
    scale = 1.0/scale

    out=tf.multiply(tf.constant(scale, dtype=mydtype),dm(x))

    return out
```

`nbar` is a list of N integers corresponding to the sampling pattern

$$(\bar{n}_0, \bar{n}_1, \ldots, \bar{n}_{n-1}) \ , \tag{12.58}$$

as, for example, `nbar=[2,3]` for $n = 2$. `Qfun` is a Q-transform, as `Qrho`.

We call the `Pr` operator as, e.g., `Pr([2,3], Qrho)`.

In `Pr`, we make a double loop, one ranging over the modes with index `j` and the other for each `j` calling $\bar{n}_j$ times the Laplacian `dm`. `dm` denotes the multiple derivatives of `Qfun`. Finally, we scale with the normalization factor `scale`.

We remark the `Pr` operator acts on callable quantities, either functions or models. Correspondingly, `Pr` works both on the output of `QTransform` and on the `QTranformLayer`. We detail in the following some examples and tests.

12.10 Single-Mode Coherent State

As a first example, we test the single-mode case with a coherent state with nonzero displacement.[4] In this case we have that $\Pr(\bar{\mathbf{n}}!) = P_c(\bar{n}!)$ after Eq. (12.34).

We build a single-mode model with a given displacement vector.

```
vacuum = VacuumLayer(N)
alpha =1.5 # complex displacement constant
```

[4] The example is in `BosonSamplingExample1.iynp`.

```
dinput = tf.constant(
          [[np.sqrt(2)*np.real(alpha)],
          ↪    [np.sqrt(2)*np.imag(alpha)]],
        dtype=vacuum.dtype)
displacer = ps.DisplacementLayerConstant(dinput) # Glauber layer
# build the model
xin = tf.keras.layers.Input(N)
x1, a1 = displacer(xin)
chir, chii = vacuum(x1, a1)
model = tf.keras.Model(inputs = xin, outputs=[chir, chii])
```

We use a `CovarianceLayer` to retrieve $\boldsymbol{g}$ and $\boldsymbol{d}$ in the tensors `cov0` and `d0`.

```
cov_layer = covariance(N) # covariance layer
covariance_matrix, mean_R, _ = cov_layer(chir,chii, model)
# model that returns g and d
model_cov = tf.keras.Model(inputs = xin,
             outputs=[covariance_matrix, mean_R, _])
xtrain = np.random.rand(1, N)-0.5; # dummy input
cov0,d0, _ =model_cov(xtrain); # compute g and d
# model_cov has 3 outputs, but we retain only cov0 and d0
```

After obtaining `cov0` and `d0`, we perform the Q-transform by the `Qtransform` function.

```
Qrho = Qtransform(cov0, d0)
```

We use the boson sampling operator `Pr` on the Q-transform for the particle distribution as follows.

```
# probability of zero photons P(0)
Pr([0], Qrho)
# output is
# tf.Tensor: shape=(1, 1), dtype=float32,
↪   numpy=array([[0.10539922]], dtype=float32)>
# probability of zero photons P(0)
Pr([2], Qrho)
# output is
# tf.Tensor: shape=(1,), dtype=float32, numpy=array([0.2667918],
↪   dtype=float32)>
```

To compare with theory, we define a function `Pcoherent(n, alpha)` returning the expected particle number of a coherent state, namely, [14]

$$P_\alpha(\bar{n}) = \frac{1}{\bar{n}!} e^{-|\alpha|^2} |\alpha|^{2\bar{n}} \,. \tag{12.59}$$

The Python function reads

```
def Pcoherent(nj, alpha):
    # return the theory value for photon distribution in a
    ↪  coherent state
    scale = 1
    aj2 = np.abs(alpha)**2
    scale = scale/np.math.factorial(nj)
    scale = scale*np.exp(-aj2)*np.power(aj2,nj)
    return scale
```

We make a loop to find out the particle number distribution with the neural network model.

```
nmax=6
Pn = np.zeros([nmax,], dtype=np.float32)
Pth = np.zeros_like(Pn)
xaxis=np.zeros_like(Pn)
for nbar in range(nmax):
    print('Computing Pn at n '+repr(nbar))
    Pn[nbar]=ps.Pr([nbar],Qrho).numpy()
    Pth[nbar]=Pcoherent(nbar,alpha)
    xaxis[nbar]=nbar
print('Done')
```

Figure 12.1 shows the agreement between the output of the GBS operator and $P_\alpha(\bar{n})$

12.11 Single-Mode Squeezed Vacuum State

In the second example, we consider a single-mode model with a squeezed state. At variance with Sect. 12.10, we use a SingleModeSqueezerLayer instead of a displacement layer.[5]

[5] The code is in jupyternotebooks/bosonsampling/BosonSamplingExample2.ipynb.

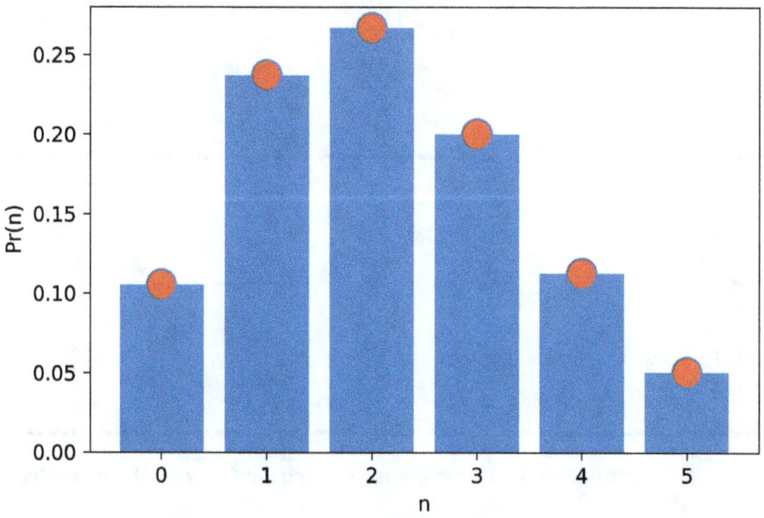

Fig. 12.1 Particle number distribution for a single-mode coherent state ($\alpha = 3$) computed with the `Pr` function on the neural network model (bars) and expected from theory (dots)

```
vacuum = VacuumLayer(N)
r_np=0.8;
theta_np=np.pi/4;
squeezer=ps.SingleModeSqueezerLayer(N,
          r_np=r_np, theta_np=theta_np, n_squeezed=0)
# build the model
xin = tf.keras.layers.Input(N)
x1, a1 = squeezer(xin)
chir, chii = vacuum(x1, a1)
model = tf.keras.Model(inputs = xin, outputs=[chir, chii])
```

In this case, the expected theoretical distribution of the particles is [14]

$$\begin{aligned} P_\zeta(2n) &= \text{sech}(r)\frac{(2n!)}{(n!)^2 2^{2n}}\tanh(r)^{2n} \\ P_\zeta(2n+1) &= 0 \end{aligned} \tag{12.60}$$

with $n = 0, 1, 2, \ldots$ up to the chosen maximal particle number.

Figure 12.2 shows the comparison between theory [Eq. (12.60)] and the output of the `Pr` function.

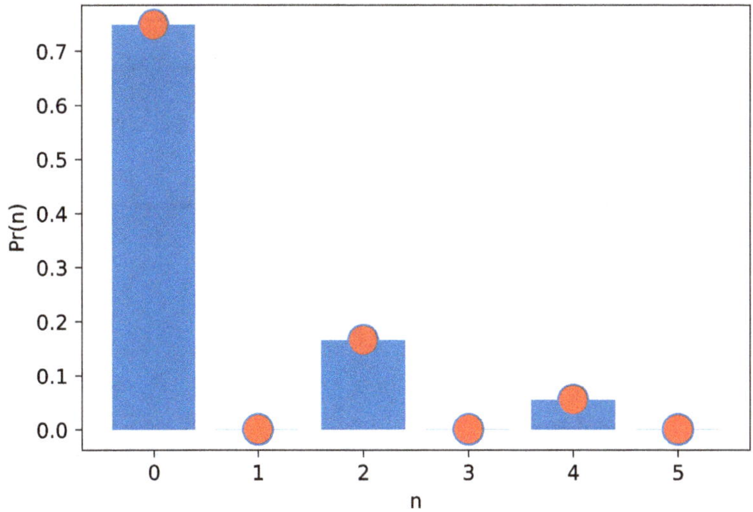

Fig. 12.2 Particle number distribution for a single-mode squeezed vacuum ($r = 0.8$, $\theta = \pi/4$) computed with the `Pr` function on the neural network model (bars) and expected from theory (dots). For odd n, the model returns a zero probability within numerical precision

12.12 Multimode Coherent Case

Here we consider a two-mode coherent state to verify Eq. (12.34).[6] We build the model as follows.

```
# define the complex vector of displacement
alpha=np.zeros([2],dtype=np_complex)
alpha[0]=np.exp(1j*np.pi/3)
alpha[1]=2.0
# corresponding real displacement vector
dinput_np=np.zeros([N,1], dtype=np_real)
for j in range(n):
    dinput_np[2*j]=np.sqrt(2)*np.real(alpha[j])
    dinput_np[2*j+1]=np.sqrt(2)*np.imag(alpha[j])
# displacement layer
dinput = tf.constant( dinput_np, dtype=vacuum.dtype)
displacer = ps.DisplacementLayerConstant(dinput)
# build the model
xin = tf.keras.layers.Input(N)
x1, a1 = displacer(xin)
chir, chii = vacuum(x1, a1)
model = tf.keras.Model(inputs = xin, outputs=[chir, chii])
```

[6] The code is in `jupyternotebooks/bosonsampling/BosonSamplingExample3.ipynb`.

alpha is a vector of complex variables corresponding to the complex displacement coefficients of the two modes.
We use the CovarianceLayer to compute $\boldsymbol{g}$ and $\boldsymbol{d}$. The QTransform is applied as above.

```
# compute the Q-transform
Qrho = QTransform(cov0, d0)
# call the Pr function for various patterns
Pr([0,0], Qrho)
# returns 0.038
Pr([0,2], Qrho)
# returns 0.098
```

The only remark here is each sampling pattern is a list of $n = 2$ integers, for example, nbar=[1,2].
For each sampling pattern, we compute the theoretically expected value by Eq. (12.34) and the following function.

```
def theoryPr(nbar, alpha):
    # returh the theory value for the Boson sampling for coherent
     ↪ states
    n=len(nbar)

    scale = 1
    for j in range(n):
        nj = nbar[j]
        aj2 = np.abs(alpha[j])**2
        scale = scale/np.math.factorial(nj)
        scale = scale*np.exp(-aj2)*np.power(aj2,nj)

    return scale
```

To retrieve all the possible sampling patterns $\bar{\mathbf{n}}$, we use the itertools.product function as follows.

```
nmax=3 # max number of particles
# generate the combinations
import itertools as it
nlist=it.product(range(nmax+1),repeat=n)
ln=list(nlist)
# print(ln) gives
#[(0, 0), (0, 1), (0, 3), (1, 0), (1, 1), (1, 2),...
```

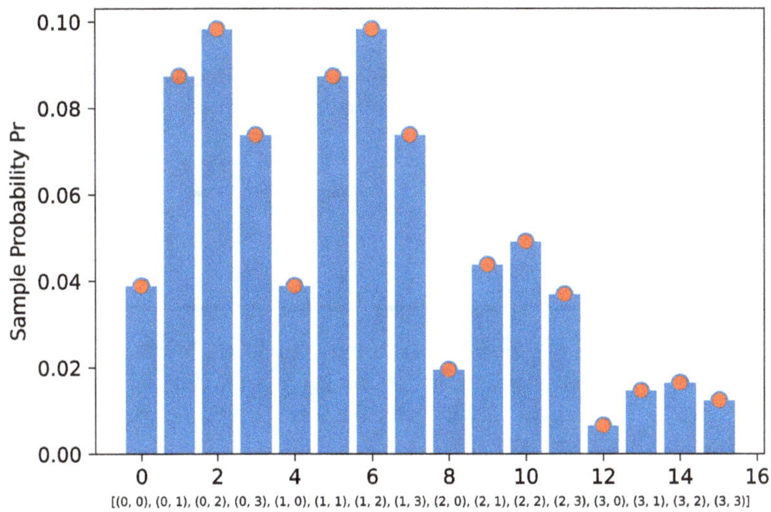

Fig. 12.3 Particle number distribution for two-mode sampling patterns and two uncorrelated coherent modes. Circles correspond to theoretical values after Eq. (12.34). Patterns are indicated in the labels

Here `nmax` is the maximum number of particles in the patterns.
We compute the `Pr` function in the generated patterns and compare with theory as in Fig. 12.3.

12.13 A Coherent Mode and a Squeezed Vacuum

In the next example, we consider two independent modes, one coherent state and one squeezed state to test Eq. (12.30), i.e., the fact that the pattern probability factorizes.

The model is built as follows.

```
# define the complex vector of displacement
alpha=np.zeros([2],dtype=np_complex)
alpha[0]=0 # mode 0 is vacuum
alpha[1]=2.0
# corresponding real displacement vector
dinput_np=np.zeros([N,1], dtype=np_real)
for j in range(n):
    dinput_np[2*j]=np.sqrt(2)*np.real(alpha[j])
    dinput_np[2*j+1]=np.sqrt(2)*np.imag(alpha[j])
# displacement layer
dinput = tf.constant( dinput_np, dtype=vacuum.dtype)
displacer = DisplacementLayerConstant(dinput)
# squeezer for mode 0
```

```
r_np=0.8;
theta_np=np.pi/4;
squeezer=SingleModeSqueezer(N, r_np=r_np, theta_np=theta_np,
↪   n_squeezed=0)
# build the model
xin = tf.keras.layers.Input(N)
x1, a1 = displacer(xin)
x2, a2 = squeezer(x1, a1)
chir, chii = vacuum(x2, a2)
model = tf.keras.Model(inputs = xin, outputs=[chir, chii])
```

All the computation follows the previous example in Sect. 12.12.[7]

In this case, the particle distribution of the coherent state (Eq .(12.59)) is different from the squeezed state (Eq. (12.60)). We have the theoretically expected value:

$$\Pr(\bar{n}_0, \bar{n}_1) = P_\zeta(\bar{n}_0) P_\alpha(\bar{n}_1) \tag{12.61}$$

Figure 12.4 compares the output of the model with Eq. (12.61).

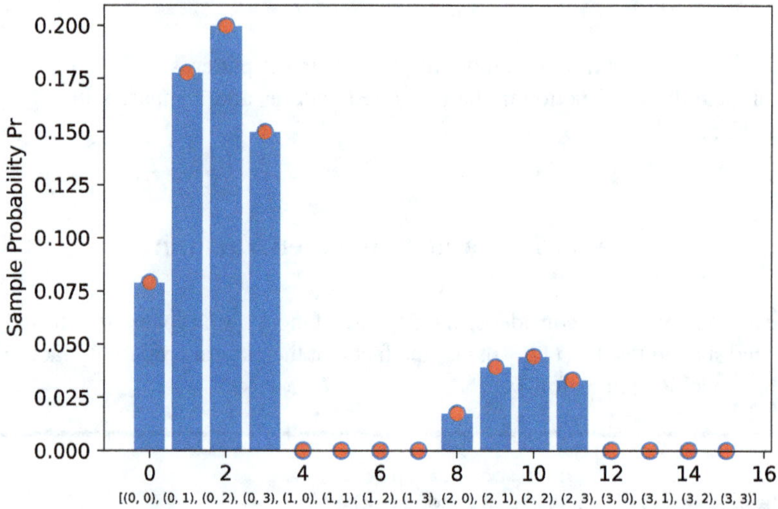

Fig. 12.4 Particle number distribution for 16 two-mode sampling patterns and two uncorrelated modes. The two modes are in a squeezed state ($r = 0.8$,$\theta = \pi/4$) and a coherent state ($\alpha = 1.5$). Circles correspond to theoretical values after Eq. (12.61). Note that odd values for $\bar{n}_0$ give $\Pr(\bar{\mathbf{n}}) = 0$, as expected because $P_\zeta(\bar{n}_0) = 0$. Patterns are indicated in the labels

[7] The code in `BosonSamplingExample4.iynb` notebook, `BosonSampling Example4b.iynb` shows an implementation with the `QTransformLayer`.

12.14 A Squeezed Mode and a Coherent Mode in a Random Interferometer

The two modes are not independent, and the probabilities do not factorize as in Eq. (12.30). This happens when the modes are mixed by beam splitters or interferometers and entanglement occurs.

We obtain the boson sampling patterns for the two modes as in Sect. 12.13 with an additional layer for the random interferometer.

The model is built as follows.

```
# define the complex vector of displacement
alpha=np.zeros([2],dtype=np_complex)
alpha[0]=0 # mode 0 is vacuum
alpha[1]=2.0
# corresponding real displacement vector
dinput_np=np.zeros([N,1], dtype=np_real)
for j in range(n):
    dinput_np[2*j]=np.sqrt(2)*np.real(alpha[j])
    dinput_np[2*j+1]=np.sqrt(2)*np.imag(alpha[j])
# displacement layer
dinput = tf.constant( dinput_np, dtype=vacuum.dtype)
displacer = DisplacementLayerConstant(dinput)
# squeezer for mode 0
r_np=0.8;
theta_np=np.pi/4;
squeezer=SingleModeSqueezerLayer(N, r_np=r_np, theta_np=theta_np,
↪   n_squeezed=0)
# non trainable random layer
R=RandomLayerConstant(N)
# build the model
xin = tf.keras.layers.Input(N)
x1, a1 = R(xin)
x2, a2 = displacer(x1, a1)
x3, a3 = squeezer(x2, a2)
chir, chii = vacuum(x3, a3)
model = tf.keras.Model(inputs = xin, outputs=[chir, chii])
```

The computation is as in Sect. 12.13.[8]

Figure 12.5 shows a bug train representation including the `QTransformLayer` and the `Pr` function. Figure 12.6 shows the computed Pr for 16 sampling patterns. At variance with the example in Sect. 12.13, $\Pr(\bar{\mathbf{n}})$ does not vanish when $P_\zeta(\bar{n}_0)$ is zero because the modes are correlated by the random interferometer and the pattern probability does not factorize [Eq. (12.61) does not hold].

[8] The code in `BosonSamplingExample5.ipynb` notebook.

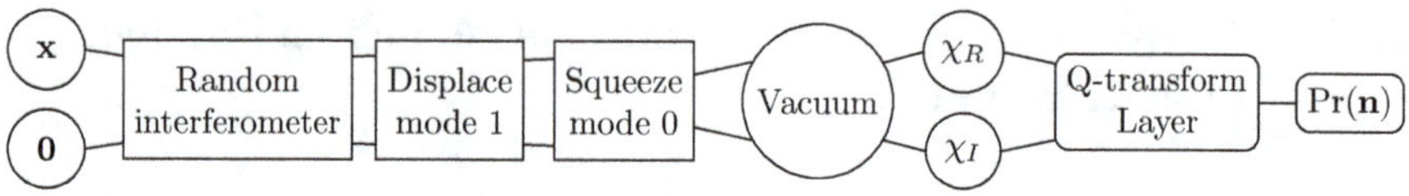

Fig. 12.5 Bug train model with a squeezed vacuum, a coherent state, and a random interferometer with the boson sampling pattern in Fig. 12.6

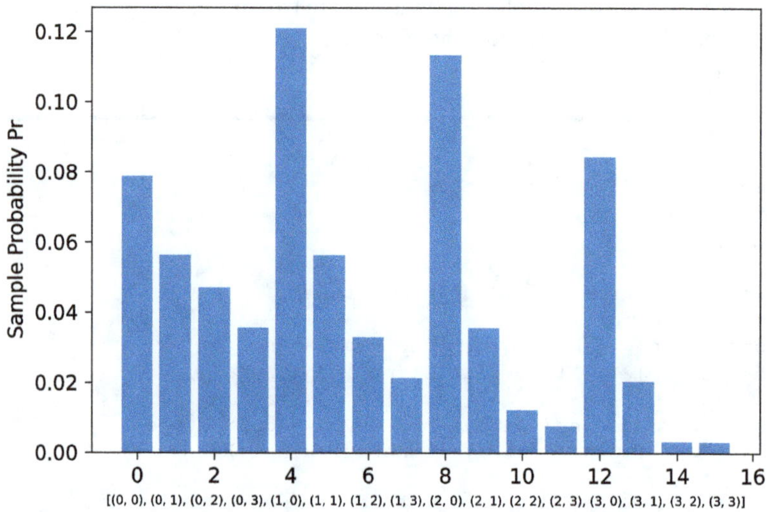

Fig. 12.6 Boson number distribution for 16 two-mode sampling patterns. One coherent state (α = 1.5) and one squeezed state (r = 0.8, $\theta = \pi/4$) are mixed by a random medium (model in Fig. 12.5). Boson patterns are indicated in the labels

12.15 Gaussian Boson Sampling with Haar Random Unitary Matrices

In the simplest formulation, the Gaussian boson sampling protocol [1] determines the probabilities of particle pairs ("double-clicks") from n squeezed vacuum states entering a Haar interferometer, which is modeled by a $n \times n$ random Haar unitary matrix U_H.

U_H is a random matrix belonging to the Haar class, which has the advantage of having individual elements approximately Gaussianly distributed, and this enables the theoretical analysis of Gaussian boson sampling and assessing its complexity class [10]. Haar matrices can be generated by using specific matrix decompositions [17]; we will use a specific routine from the `Python Quantum Toolbox in Python (QuTiP)` package [18].

For implementing this protocol as a neural network, we need

- A layer class corresponding to the Haar matrix
- A model with an arbitrary number of squeezed states
- The `Pr` function on all the patterns with *two* particles in n modes

12.15.1 The Haar Random Layer

We use the `QuTiP` to generate a Haar unitary matrix [18].[9] The following `HaarLayerConstant` returns the corresponding layer.[10]

```
class HaarLayerConstant(layers.Layer):
    """
        Define a random layer corresponding to the
        \tilde a = U a
        with U a Haar unitary complex matrix
        In the R space, corresponds to
        \tilde x = R x
        In input has input n, the size of the R vector (must be
↪   even)

        The Haar unitary matrix is generated
            by the QuTiP package (qutip.org)
    """

    def __init__(self, N=10,
                 d_np=None,
                trainable_d=False,
                 **kwargs):
        super(HaarLayerConstant, self).__init__(**kwargs)
        assert N % 2 == 0, " Dimension must be even "
        self.trainable_d = trainable_d
        self.N = N
        n = np.floor_divide(self.N, 2)
        Rx, Rp, J = RQRP(N)
        self.Rx = tf.constant(Rx)
        self.Rp = tf.constant(Rp)
        self.J = tf.constant(J)
        # generate the Haar unitary
        U=random_objects.rand_unitary_haar(n)
        # convert to a tensor
        self.U = tf.constant(U.full())
        # return the real and imaginary part
        UR = tf.constant(np.real(U.full()),dtype=self.dtype)
```

[9] `qutip.org` licensed under the terms of the 3-clause BSD license.

[10] The extended version of the function is in `phasespace.py`.

```
        UI = tf.constant(np.imag(U.full()),dtype=self.dtype)
        # Build the symplectic matrix M
        M = \
            tf.matmul(self.Rx, tf.matmul(UR, self.Rx,
            ↪ transpose_b=True)) +\
            tf.matmul(self.Rp, tf.matmul(UR, self.Rp,
            ↪ transpose_b=True)) -\
            tf.matmul(self.Rx, tf.matmul(UI, self.Rp,
            ↪ transpose_b=True)) +\
            tf.matmul(self.Rp, tf.matmul(UI, self.Rx,
            ↪ transpose_b=True))
        # Inverse of M
        MI = tf.matmul(tf.matmul(M, self.J), self.J,
        ↪ transpose_a=True)
        self.M=M
        self.MI=MI
        if d_np is None:
            # the layer has a constant zero displacement
            self.d = tf.constant(np.zeros((N, 1), dtype=np_real),
                                 dtype=tf_real)
            trainable_d = False
        else:
            # the layer has a trainable displacement
            # if trainable_d is set to true (default is false)
            self.d = tf.Variable(d_np, dtype=tf_real,
                                       trainable=trainable_d)

    def get_M(self):
        # this function returns the symplectic matrix and its
        ↪ inverse
        # usefult for inspection
        return self.M, self.MI

    def call(self, x, di=None):
        # no transpose here for M, as M is already transpose
        if di is None:
            d2 = tf.constant(np.zeros((self.N, 1),
            ↪ dtype=np_real),
                                 dtype=tf_real)
        else:
            d2 = di
        return [tf.matmul(x, self.M), tf.matmul(self.MI,
        ↪ d2+self.d)]
```

We use the method `random_objects.rand_unitary_haar` to create numpy matrix, and then we create a constant tensor `U`. Then we proceed as in Sect. 7.8 to create the corresponding symplectic matrix `M`.

12.15.2 A Model with a Varying Number of Layers

We want a model such that the number of layers varies with the n modes, for which we need a `for loop`.

We consider n single-mode squeezer layers and the Haar interferometer as follows.

```
# define an HAAR layer
HAAR=HaarLayerConstant(N)
# parameters of the squeezing layer
# all the layer have the same r and theta)
r_np=0.88
theta_np=np.pi/4
# build the model
xin = tf.keras.layers.Input(N)
x1, a1 = HAAR(xin)
for j in range(nmodes):
    #define the layer and connect to the graph
    x1, a1 = SingleModeSqueezer(N, r_np=r_np,
                    theta_np=theta_np, n_squeezed=j)(x1,a1)
chir, chii = vacuum(x1, a1)
model = tf.keras.Model(inputs = xin, outputs=[chir, chii])
```

Note that in the `for` loop, we use the same input and output tensors `x1` and `a1` recursively. Also we define the layer and add to the graph in the same line.

The parameters of the squeezer `r_np` and `theta_np` are the same for all layers, while the parameter `n_squeezed` changes with mode index, as each layer acts on a different mode.

To display the resulting model, we use the `model.summary()` method that returns the following output. We remark that the model has 12 trainable parameters corresponding to the *six* squeezers.[11]

```
Model: "model"
__________________________________________________________________________________________________
 Layer (type)                   Output Shape         Param #     Connected to
==================================================================================================
 input_1 (InputLayer)           [(None, 12)]         0           []

 haar_layer_constant (HaarLayer  [(None, 12),        0           ['input_1[0][0]']
 Constant)                       (12, 1)]

 single_mode_squeezer_layer (Si  [(None, 12),        2           ['haar_layer_constant[0][0]',
 ngleModeSqueezerLayer)          (12, 1)]                         'haar_layer_constant[0][1]']

 single_mode_squeezer_layer_1 (  [(None, 12),        2           ['single_mode_squeezer_layer[0][0
 SingleModeSqueezerLayer)        (12, 1)]                        ]',
                                                                  'single_mode_squeezer_layer[0][1
                                                                 ]']

 single_mode_squeezer_layer_2 (  [(None, 12),        2           ['single_mode_squeezer_layer_1[0]
 SingleModeSqueezerLayer)        (12, 1)]                        [0]',
```

[11] The code is in boson_sampling/BosonSamplingExample6.ipynb.

```
                                                                   'single_mode_squeezer_layer_1[0]
                                                                   [1]']

single_mode_squeezer_layer_3 (  [(None, 12),       2               ['single_mode_squeezer_layer_2[0]
SingleModeSqueezerLayer)        (12, 1)]                           [0]',
                                                                    'single_mode_squeezer_layer_2[0]
                                                                   [1]']

single_mode_squeezer_layer_4 (  [(None, 12),       2               ['single_mode_squeezer_layer_3[0]
SingleModeSqueezerLayer)        (12, 1)]                           [0]',
                                                                    'single_mode_squeezer_layer_3[0]
                                                                   [1]']

single_mode_squeezer_layer_5 (  [(None, 12),       2               ['single_mode_squeezer_layer_4[0]
SingleModeSqueezerLayer)        (12, 1)]                           [0]',
                                                                    'single_mode_squeezer_layer_4[0]
                                                                   [1]']

gaussian_layer (GaussianLayer)  [(None, 1),        156             ['single_mode_squeezer_layer_5[0]
                                (None, 1)]                         [0]',
                                                                    'single_mode_squeezer_layer_5[0]
                                                                   [1]']

==================================================================================================
Total params: 168
Trainable params: 12
Non-trainable params: 156
__________________________________________________________________________________________________
```

We use the `tf.keras.utils.plot_model` to have a graphical representation of the model as in Fig. 12.7.

12.15.3 Generating the Sampling Patterns

In order to obtain the statistics of boson sampling, we need all the possible particle patterns. For example, for $n = 2$, we have $\bar{\mathbf{n}} = (0, 0)$ for zero particles, $\bar{\mathbf{n}} = (1, 0)$ for one particle in mode 0 and zero in mode 1, $\bar{\mathbf{n}} = (0, 1)$, $\bar{\mathbf{n}} = (1, 1)$, $\bar{\mathbf{n}} = (2, 0)$, $\bar{\mathbf{n}} = (2, 1)$, and so forth.

As we deal with Gaussian states, the number of potential particles in any mode at the output is not limited.

However, depending on the mean particle number, the probability of high particle states is low.

In GBS experiments, the target patterns are limiting the maximum of *one* particle per mode [3].

Thus, we are interested in the probabilities with $n_{max} = 1$. For example, for two modes, we have $four$ targets $\bar{\mathbf{n}} = (0, 0)$, $\bar{\mathbf{n}} = (0, 1)$, $\bar{\mathbf{n}} = (1, 0)$, and $\bar{\mathbf{n}} = (1, 1)$. For an arbitrary number of modes n, we need to identify all of these patterns and then compute their probability.

A simple way is using the numpy package `itertools` as follows (we refer to $n = 6$) :

- Generate a seed string like 1, 1, 0, 0, 0, 0.
- Generate all combinations by the method `permutations`, which includes repetitions of all the elements, which are retained as distinguishable.
- Remove the duplicates.

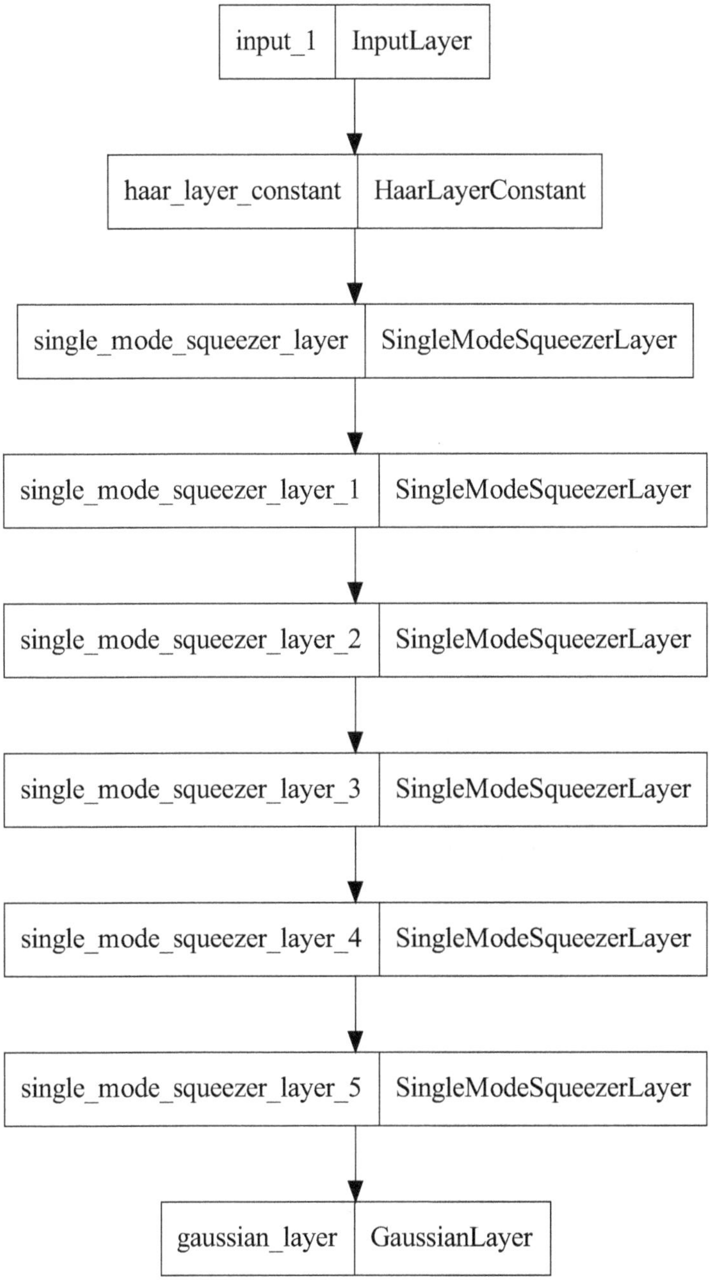

Fig. 12.7 Representation of the model with six squeezing layers and the Haar interferometer generated by `tf.keras.utils.plot_model`

We use the following python functions

```
def patterns(nphotons, nmodes):
    # Return the patterns of nmodes with maximal nphotons

    # generate a list of zero
    l1=[0]*nmodes
    # set nphotons 1
    for j in range(nphotons):
        l1[j]=1
    # compute all the permutations (zeros and one are
    ↪  distinguishable)
    nlist=it.permutations(l1,nmodes) # return iterators
    # convert to list the iterators
    ln=list(nlist)
    # sort the list in reverse order to have the patterns with
    ↪  more ones first
    ln.sort(reverse=True)
    # remove adjacent identical elements after sorting
    ln=list(ln for ln,_ in it.groupby(ln))
    return ln
```

In `patterns`, `ln` is the list of patterns produced by the output of `itertools.permutations`. We sort the list in reverse order, and we remove duplicates by the `itertools.groupby`.

12.15.4 Computing the Pattern Probability

Finally, we loop over the generated patterns and compute the probabilities. The loop may be computationally demanding mainly for large n and n_{max}. Figure 12.8 shows an example for $n = 6$ and $n_{max} = 1$ [12] One can also be interested in summing up over all the particle pairs to determine the cumulative probability of a double-click, as follows.

```
%%time
nmax = 2 # number of couples
probn=np.zeros([nmax+1,], dtype=np_real)
for j in range(nmax+1):
    photon_number=2*j
    # generate the patterns
    ln=patterns(photon_number,nmodes)
    npattern=len(ln)
```

[12] The code is available at `bosonsampling/BosonSamplingExample6.ipynb`.

```
    print('Generated '+repr(npattern)+\
          ' patterns with photon number '+repr(photon_number))
    # compute probability for each pattern
    Pn=np.zeros([npattern,], dtype=np_real)
    for nbar in range(npattern):
        Pn[nbar]=ps.Pr(ln[nbar],Qrho).numpy()
        print('Photons '+repr(photon_number)+' Sample
        ↪  '+repr(nbar+1)+\
              ' of '+repr(npatterns)+' Pr=' +repr(Pn[nbar])+\
              ' pattern '+repr(ln[nbar]))
    # sum the probability of each pattern
    probn[j]=sum(Pn)
print('Done')
```

Figure 12.9 shows the result.

12.16 Training Boson Sampling

The implementation described in Sect. 12.7 enables computing the probability distribution but does not furnish a trainable model.

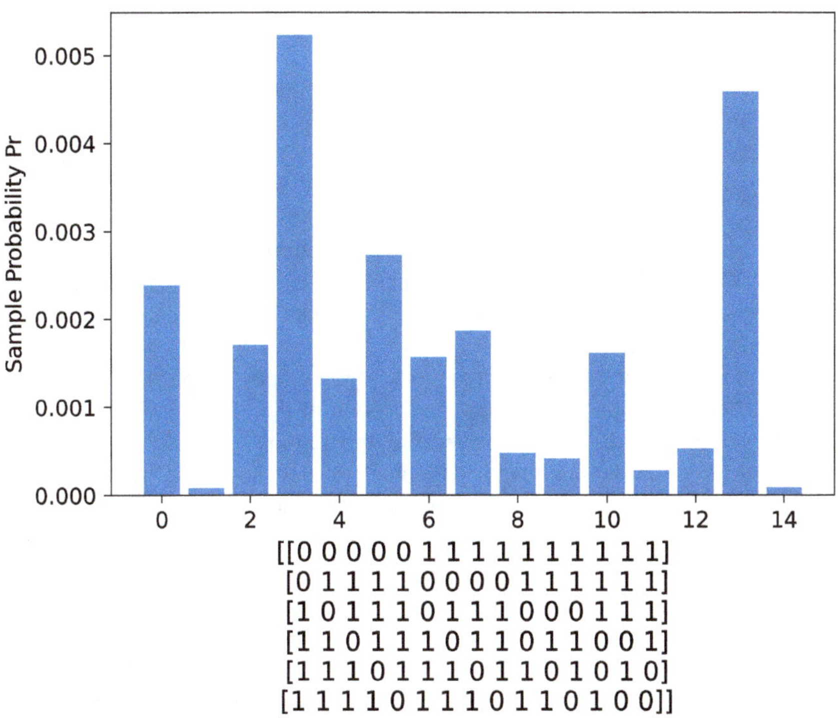

Fig. 12.8 Probability distribution of patterns with two photons and maximum one photon per mode $n = 6$ with squeezed vacuum modes and Haar interferometer. The bottom inset reports the patterns. Parameters: `r_np=0.88, theta_np=np.pi/4`

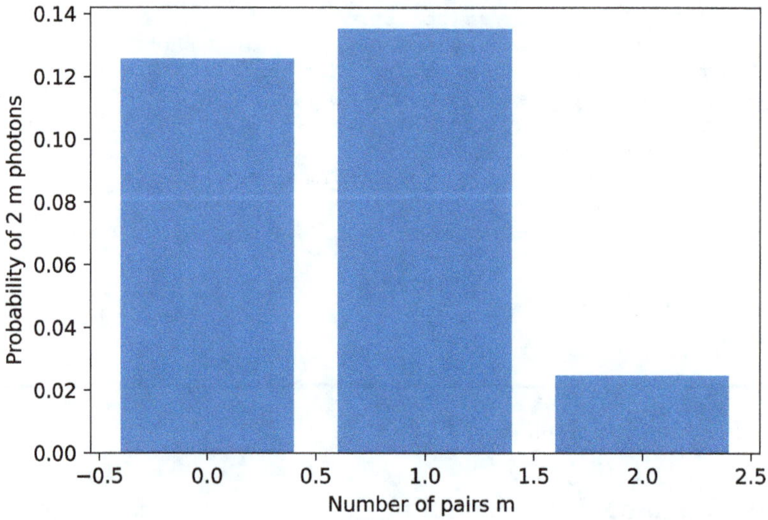

Fig. 12.9 Probability of m pairs with maximum one particle per mode ($m = 6$) with squeezed vacuum and Haar interferometer. Parameters: `r_np=0.88, theta_np=np.pi/4`

Our interest is understanding if we can train the model to maximize the generation of specific patterns, e.g., those containing two or more particles in target modes.

Using complex media to train the response of linear systems is a renowned technique in many different fields, e.g., to engineer specific gates [19, 20]. Here, we use the phase-space approach to optimize multiparticle events by using Gaussian boson sampling and a trainable complex medium [21, 22].

A first strategy is using the `Pr` operator to build a loss function that is minimized during training to maximize the probability of a pattern $\bar{\mathbf{n}}$. For example, we may use $\exp(-\,\mathrm{Pr}(\bar{\mathbf{n}}))$. Unfortunately, computing $\mathrm{Pr}(\bar{\mathbf{n}})$ is hard on a digital computer, especially at large particle number. As training requires many calls to $\mathrm{Pr}(\bar{\mathbf{n}})$, it is not efficient in conventional hardware.

Also, the automatic buildup of the computational graph that occurs in `TensorFlow` is demanding for the computation of the derivatives of $\mathrm{Pr}(\bar{\mathbf{n}})$. The derivatives of the loss functions are needed in the training.

We resort to a simpler but effective approach to show how to tailor the statistical distribution of the boson patterns: given n modes, our goal is to maximize the probability of patterns that contains a particle pair in modes 0 or 1; this may be useful, e.g., for sources of entangled photons in optics.

For example, for $n = 6$, we maximize the probability of $\mathbf{n} = (1, 1, 0, 0, 0, 0)$ with respect to the other patterns such as $\mathbf{n} = (1, 0, 0, 1, 0, 0)$. We assume that the modes are distinguishable at detection, which is the case in recent optical GBS experiments [23–25]. As a side effect, this approach also maximizes higher-order events for modes 0 and 1, as in the pattern $\mathbf{n} = (1, 1, 1, 1, 0, 0)$ with $four$ particles.

12.17 The Loss Function

Computing the particle number in each mode is attained by the `PhotonCounterLayer` described in Sect. 10.10. Here, we use the layer to return all the expected number of particles in each mode. Then we consider modes 0 and 1, with mean particle numbers n_0 and n_1, respectively. We define the loss function:

$$\mathcal{L} = e^{-n_0 - n_1} \tag{12.62}$$

and we train the system to minimize $\mathcal{L}$. We show below that this radically affects the particle statistics.

However, we need to define a model that has trainable layers. We adopt as weights the values of the squeezing parameters and the displacements.

12.18 Trainable GBS Model

A common configuration in experiments to train linear gates is using a tunable interferometer, typically realized by amplitude and phase spatial light modulators.

Here we retain a similar approach. We consider n coherent squeezed modes impinging on a trainable interferometer and then traveling through the Haar mode-mixer.

At variance with the conventional GBS, we have a displacement operator, as we use coherent squeezed states instead of squeezed vacuum. We take the same displacement for each channel because we take identical sources (the modes have the same expected particle number at input). The states are first mixed in the trainable interferometer and then travel through the Haar gate. One could use a single layer but we prefer to distinguish the trainable part from the mode-mixing part.

As nonclassical states are mixed by linear interferometers, we have entangled light at the output, as one verifies by looking at the eigenvalues of the covariance matrix.

Figure 12.10 shows the model with the summary we obtain from the `keras.utils.plot_model` method. We have a multiple pullback with `SingleModeSqueezeLayer` in Fig. 12.10a. When including the Haar interferometer, we get the GBS setup without a training interferometer (Fig. 12.10b).

Figure 12.10c shows the complete model with the trainable interferometer to control the output patterns.

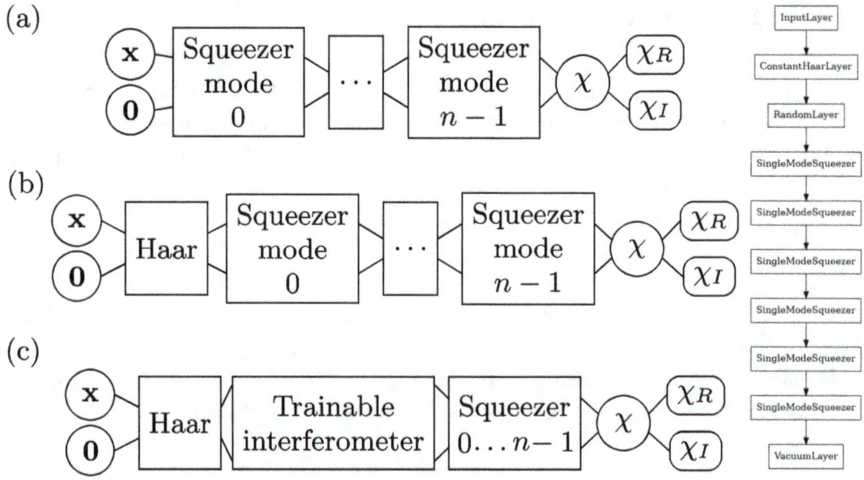

Fig. 12.10 (**a**) A multiple pullback that represents a many-body squeezed vacuum, obtained by a vacuum state χ by cascading n identical single-mode squeezers. The order of the squeezers is not relevant as they act on different modes. χ_R and χ_I are the real and imaginary parts of the resulting characteristic function. (**b**) GBS setup, a n-body squeezed vacuum, enters a Haar interferometer. Note that the order of the operators, from the vacuum to the interferometer, goes from right to left. (**c**) GBS setup including a trainable random interferometer before entering the Haar interferometer. The multiple squeezers are represented as a single block. The trainable interferometer can optimize the probability of pair generation as detailed in Sect. 12.20. The right panel shows the architecture of the `TensorFlow` model for $n = 6$

12.19 Boson Sampling the Model

The model definition is as follows.[13]

```
HAAR=HaarLayerConstant(N)
R=RandomLayer(N) # trainable random layer
D=DisplacementLayerConstant(np.ones((N,1))
xin = tf.keras.layers.Input(N)
x1, a1 = HAAR(xin)
x1, a1 = R(x1,a1)
for j in range(nmodes):
    x1, a1 = ps.SingleModeSqueezer(N,
                r_np=r_np, theta_np=theta_np,
                n_squeezed=j, trainable=False)(x1,a1)
x1, a1=D(x1,a1)
chir, chii = vacuum(x1, a1)
model = tf.keras.Model(inputs = xin, outputs=[chir, chii])
```

[13] The code is in `jupyternotebooks/bosonsampling/BosonSamplingExample7.ipynb`.

We introduce a `PhotonCountingLayer` for the particle number in each mode.

```
# define the layer
photon_counter=ps.PhotonCountingLayer(N)
# define the output tensor
n_out = photon_counter(chir,chir, model);
 # define the model with inputs and ouputs
Nphoton = tf.keras.Model(inputs = xin, outputs=n_out)
tf.print(Nphoton(xtrain));
%%% OUTPUT IS
# [[1.72245288 2.77835131 2.77175 1.38288355 1.38208294
↪   1.92756987]]
```

`xtrain` is a dummy input, as the output does not depend on `xtrain`. Note that –after the model is created– the average particle number is between 1 and 3 for all modes.

Figure 12.11 shows the model including the `PhotonCountingLayer`.

Before training, we compute the probability distribution of the boson patterns. We compute the Q−transform needed for the Pr function and generate all the possible patterns with maximum one particle per mode. (see Sect. 12.7)

```
# transform the model
kin, Qrho =getQTransformModel(model)
# function to generate the patterns
def patterns(nphotons, nmodes):
    # Return the patterns of nmodes with maximal nphotons

    # generate a list of zero
    l1=[0]*nmodes
    # set nphotons 1
    for j in range(nphotons):
        l1[j]=1
    # compute all the permutations (zeros and one are
    ↪   distinguishable)
    nlist=it.permutations(l1,nmodes) # return iterators
    # convert to list the iterators
    ln=list(nlist)
    # sort the list in reverse order to have the patterns with
    ↪   more ones first
    ln.sort(reverse=True)
    # remove duplicates (remove adjacent identical elements after
    ↪   sorting)
    ln=list(ln for ln,_ in it.groupby(ln))
    return ln
# patterns
ln =patterns(2,nmodes)
# we obtain 15 patterns as follows
[(1, 1, 0, 0, 0, 0), (1, 0, 1, 0, 0, 0), (1, 0, 0, 1, 0, 0),
(1, 0, 0, 0, 1, 0), (1, 0, 0, 0, 0, 1), (0, 1, 1, 0, 0, 0),
```

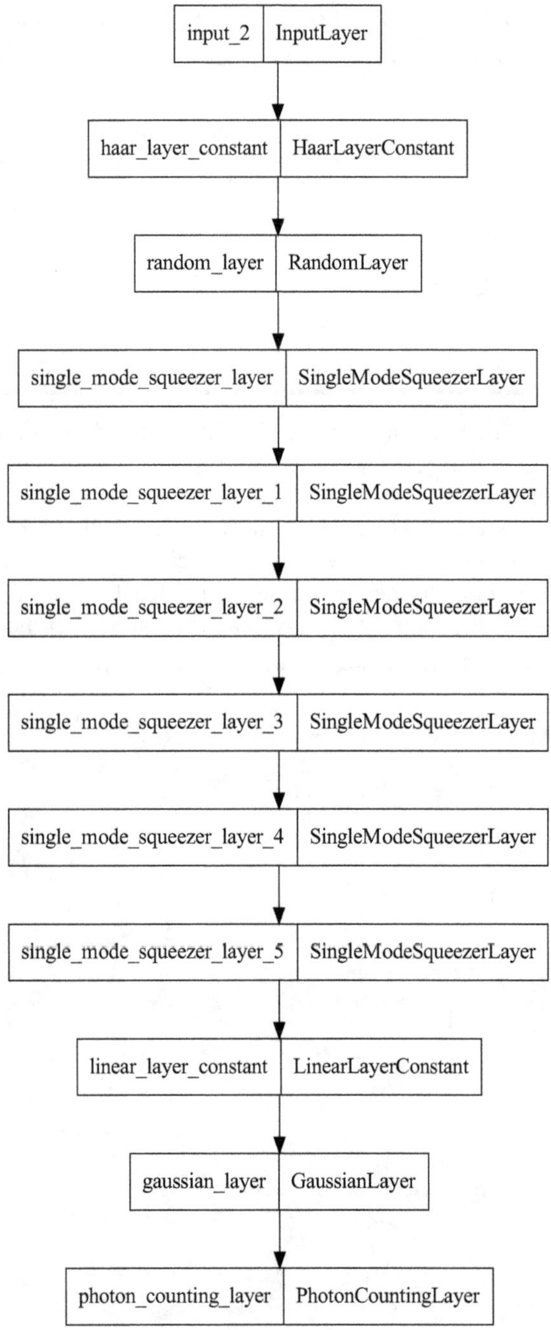

Fig. 12.11 Plot of the model obtained by the `keras.utils.plot_model` method

```
(0, 1, 0, 1, 0, 0), (0, 1, 0, 0, 1, 0), (0, 1, 0, 0, 0, 1),
(0, 0, 1, 1, 0, 0), (0, 0, 1, 0, 1, 0), (0, 0, 1, 0, 0, 1),
(0, 0, 0, 1, 1, 0), (0, 0, 0, 1, 0, 1), (0, 0, 0, 0, 1, 1)]
```

Then we compute the probability for each pattern by the `Pr` function as follows.

```
npatterns=len(ln)
Pn = np.zeros([npatterns,], dtype=np.float32)
Pth = np.zeros_like(Pn)
xaxis=np.zeros_like(Pn)
for nbar in range(npatterns):
    print('Sample '+repr(nbar)+' of '+repr(npatterns)+'
    ↪  '+repr(ln[nbar]))
    Pn[nbar]=ps.Pr(ln[nbar],Qrho).numpy()
    tf.print(Pn[nbar])
    xaxis[nbar]=nbar
print('Done')
# the output is
Sample 0 of 15 (1, 1, 0, 0, 0, 0)
0.0004652245
Sample 1 of 15 (1, 0, 1, 0, 0, 0)
0.00073040975
Sample 2 of 15 (1, 0, 0, 1, 0, 0)
0.0013922576
Sample 3 of 15 (1, 0, 0, 0, 1, 0)
2.280654e-05
Sample 4 of 15 (1, 0, 0, 0, 0, 1)
0.0005218925
Sample 5 of 15 (0, 1, 1, 0, 0, 0)
0.0013013579
Sample 6 of 15 (0, 1, 0, 1, 0, 0)
0.0033604752
Sample 7 of 15 (0, 1, 0, 0, 1, 0)
0.0022158423
Sample 8 of 15 (0, 1, 0, 0, 0, 1)
0.0021270977
Sample 9 of 15 (0, 0, 1, 1, 0, 0)
0.00021208788
Sample 10 of 15 (0, 0, 1, 0, 1, 0)
0.001363168
Sample 11 of 15 (0, 0, 1, 0, 0, 1)
0.0020175062
Sample 12 of 15 (0, 0, 0, 1, 1, 0)
0.00030628894
Sample 13 of 15 (0, 0, 0, 1, 0, 1)
0.0053390833
Sample 14 of 15 (0, 0, 0, 0, 1, 1)
0.00083708746
Done
```

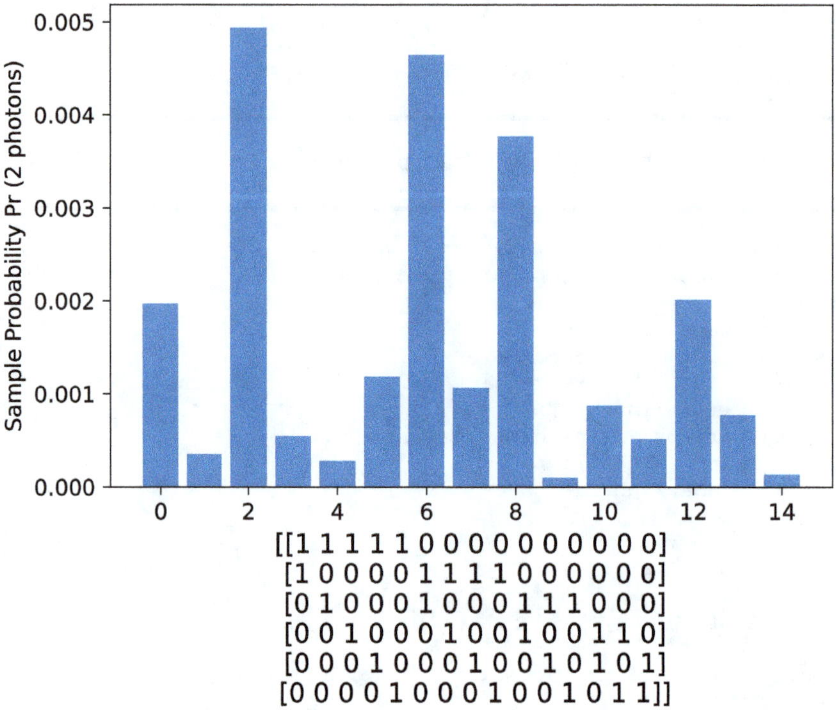

Fig. 12.12 Probability distribution of patterns with two particles for $n = 6$ modes in the model in Fig. 12.11. The inset report the patterns. Parameters: `r_np=0.88, theta_np=np.pi/4`

Figure 12.12 shows the results for the probability of pairs before training.

We then consider the probability of *four* particle events (Fig. 12.13).

```
# generate patterns with 4 particles
print(len(ln))
npatterns=len(ln)
Pn = np.zeros([npatterns,], dtype=np.float32)
Pth = np.zeros_like(Pn)
xaxis=np.zeros_like(Pn)
for nbar in range(npatterns):
    print('Sample '+repr(nbar+1)+' of '+repr(npatterns)+'
     ↪ '+repr(ln[nbar]))
    Pn[nbar]=ps.Pr(ln[nbar],Qrho).numpy()
    tf.print(Pn[nbar])
    xaxis[nbar]=nbar
print('Done')
# the output is
Sample 1 of 15 (1, 1, 1, 1, 0, 0)
0.000948885
Sample 2 of 15 (1, 1, 1, 0, 1, 0)
```

```
0.00017854349
Sample 3 of 15 (1, 1, 1, 0, 0, 1)
7.2374794e-05
Sample 4 of 15 (1, 1, 0, 1, 1, 0)
0.000712906
Sample 5 of 15 (1, 1, 0, 1, 0, 1)
0.0016828682
Sample 6 of 15 (1, 1, 0, 0, 1, 1)
0.00053918053
Sample 7 of 15 (1, 0, 1, 1, 1, 0)
0.00040825928
Sample 8 of 15 (1, 0, 1, 1, 0, 1)
0.0004931734
Sample 9 of 15 (1, 0, 1, 0, 1, 1)
0.0003072312
Sample 10 of 15 (1, 0, 0, 1, 1, 1)
0.0004550742
Sample 11 of 15 (0, 1, 1, 1, 1, 0)
0.0003555224
Sample 12 of 15 (0, 1, 1, 1, 0, 1)
0.00040174997
Sample 13 of 15 (0, 1, 1, 0, 1, 1)
0.0008511516
Sample 14 of 15 (0, 1, 0, 1, 1, 1)
0.00060582825
Sample 15 of 15 (0, 0, 1, 1, 1, 1)
0.00071170204
Done
```

12.20 Training the Model

To train the GBS, we define a further model in which the output corresponds to $\mathcal{L}$, as follows.

```
n_out0=tf.gather(tf.squeeze(n_out),0) #check this
n_out1=tf.gather(tf.squeeze(n_out),1)
min_out=tf.exp(-n_out0-n_out1)
# define the model with inputs and ouputs
Train_model = tf.keras.Model(inputs = xin, outputs=min_out)
```

We first gather from the particle number output tensor `n_out` the bosons in mode 0 and 1, stored in `n_out0` and `n_0out1`, respectively. Then we build the new tensor `min_out` that returns the quantity to be minimized and define the corresponding model `Train_model`.

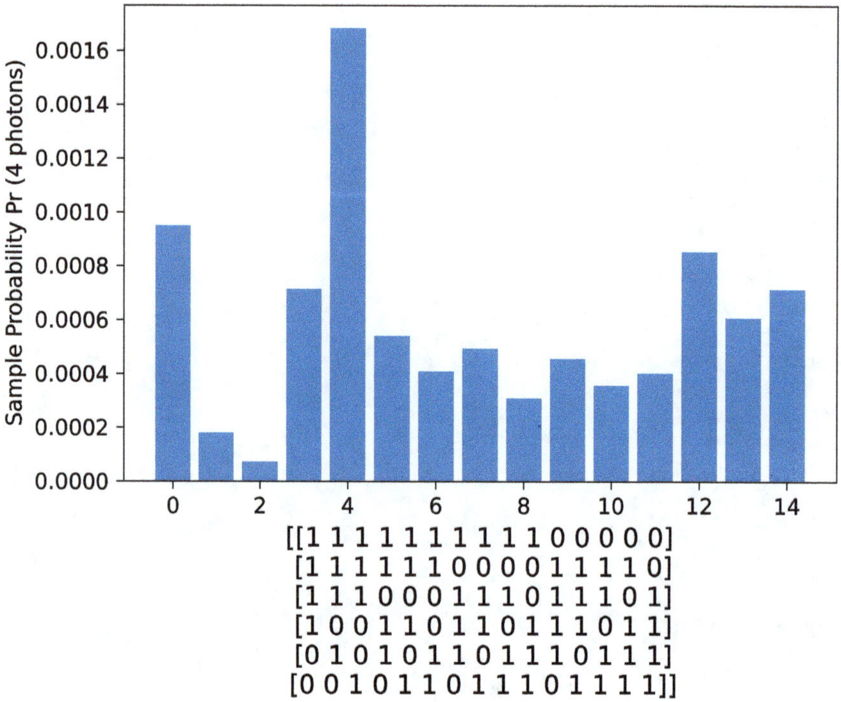

Fig. 12.13 Probability distribution of patterns with four particles for $n = 6$ modes in the model in Fig. 12.11. The inset reports the patterns. Parameters: `r_np=0.88, theta_np=np.pi/4`

To define a custom loss, we define a new python function, which takes as input the target value `yt` and the output value `yp`, which is the value returned by the `Train_model`.

```
def custom_loss(yt,yp):
    # the loss only return the output of the network yp
    # that need to be minimized
    return yp
# compile the model with the custom loss function
Train_model.compile(loss=custom_loss, optimizer='Adam')
# we choose Adam for no specific reason
```

Note that the `custom_loss` simply gives the positive quantity `yp`, as the target `yt` is zero. In this way, both the input and the output are dummy quantities, and we are using `TensorFlow` to make a regression to optimize the states.

A different, and more direct, way is to use the `tf.add_loss` method, which lets us use a specific tensor as a quantity to be minimized, as follows.

```
Train_model.add_loss(min_out)
# compile the model with the custom loss function
Train_model.compile(optimizer='Adam')
# we choose Adam for no specific reason
```

The model is trained to obtain the minimum of the `min_out` tensor [Eq. (12.62)], as follows.

```
history=Train_model.fit(xtrain,np.zeros_like(xtrain),epochs=10000)
```

Figure 12.14 shows the training history of the loss. After the training, we check the boson distribution and realize that the particle number is increased in modes 0 and 1. Then we can compute the pair probability and higher-order events. As shown in Figs. 12.15 and 12.16, we find an order of magnitude increase in the probability of events involving one pair in modes 0 and 1.

We have shown that one can use a variational quantum circuit to train boson sampling. Different circuit configurations, training parameters (as degrees of squeezing and displacements), and loss function can be used.[14] The trained circuit may optimize quantum circuits to be used as sources with tailored output states for various applications.

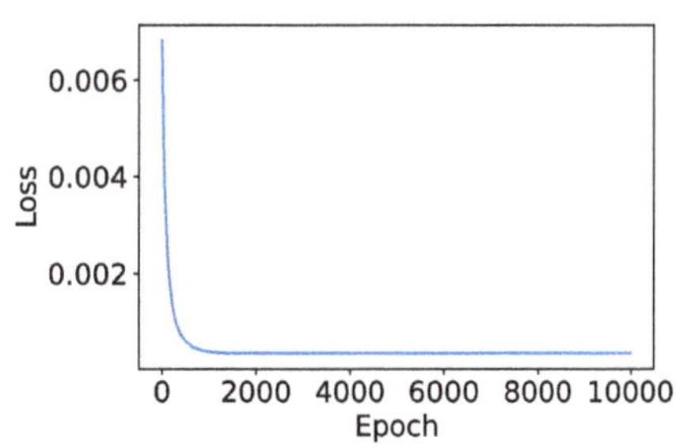

Fig. 12.14 Training history of the model in Fig. 12.11

[14] See the notebooks `jupyternotebooks/bosonsampling/BosonSamplingExample8.ipynb` and `jupyternotebooks/bosonsampling/BosonSamplingExample9.ipynb`.

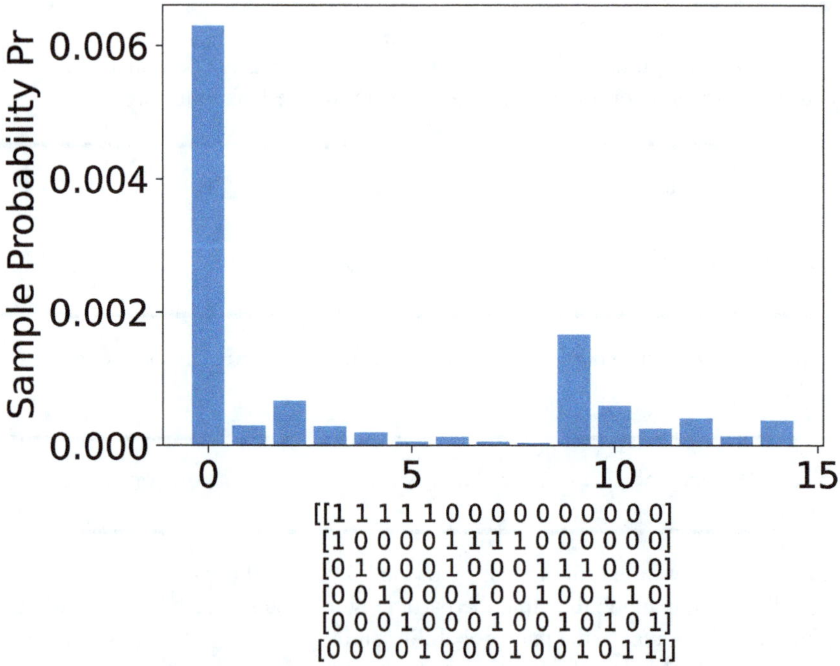

Fig. 12.15 Probability distribution of patterns with two particles ($n = 6$) in the model in Fig. 12.11 after training with loss as in Eq. (12.62). Compare with Fig. 12.12. The inset reports the patterns. Parameters: `r_np=0.88, theta_np=np.pi/4`

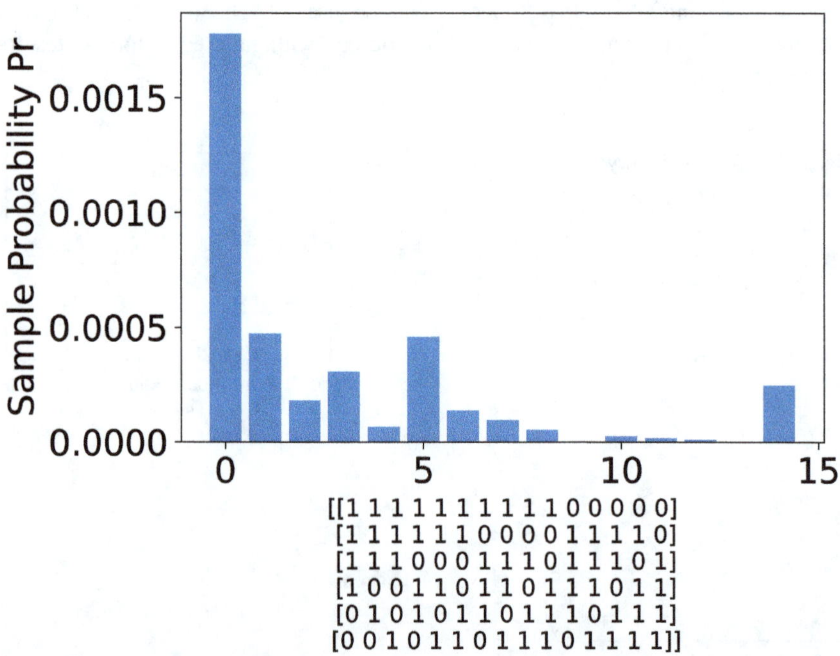

Fig. 12.16 Probability distribution of patterns with four particles ($n = 6$) in the model in Fig. 12.11 after training with loss as in Eq. (12.62). Compare with Fig. 12.13. The inset reports the patterns. Parameters: `r_np=0.88, theta_np=np.pi/4`

12.21 Further Reading

- The trainable Gaussian boson sampling with automatic differentiation has been originally reported in [26].

References

1. C.S. Hamilton, R. Kruse, L. Sansoni, S. Barkhofen, C. Silberhorn, I. Jex, Phys. Rev. Lett. **119**, 170501 (2017). https://doi.org/10.1103/PhysRevLett.119.170501
2. N. Quesada, J.M. Arrazola, N. Killoran, Phys. Rev. A **98**, 062322 (2018). https://doi.org/10.1103/PhysRevA.98.062322. https://link.aps.org/doi/10.1103/PhysRevA.98.062322
3. H.S. Zhong, H. Wang, Y.H. Deng, M.C. Chen, L.C. Peng, Y.H. Luo, J. Qin, D. Wu, X. Ding, Y. Hu, P. Hu, X.Y. Yang, W.J. Zhang, H. Li, Y. Li, X. Jiang, L. Gan, G. Yang, L. You, Z. Wang, L. Li, N.L. Liu, C.Y. Lu, J.W. Pan, Science **370**, 1460 (2020)
4. M. Tillmann, B. Dakić, R. Heilmann, S. Nolte, A. Szameit, P. Walther, Nat. Photonics **7**, 540 (2012)
5. M.A. Broome, A. Fedrizzi, S. Rahimi-Keshari, J. Dove, S. Aaronson, T.C. Ralph, A.G. White, Science **339**(6121), 794 (2013)
6. J.B. Spring, B.J. Metcalf, P.C. Humphreys, W.S. Kolthammer, X.M. Jin, M. Barbieri, A. Datta, N. Thomas-Peter, N.K. Langford, D. Kundys, J.C. Gates, B.J. Smith, P.G.R. Smith, I.A. Walmsley, Science **339**(6121), 798 (2013)
7. N. Spagnolo, C. Vitelli, M. Bentivegna, D.J. Brod, A. Crespi, F. Flamini, S. Giacomini, G. Milani, R. Ramponi, P. Mataloni, R. Osellame, E.F. Galvão, F. Sciarrino, Nat. Photonics **8**(8), 615 (2014). https://doi.org/10.1038/nphoton.2014.135
8. J. Carolan, J.D.A. Meinecke, P.J. Shadbolt, N.J. Russell, N. Ismail, K. Wörhoff, T. Rudolph, M.G. Thompson, J.L. O'Brien, J.C.F. Matthews, A. Laing, Nat. Photonics **8**(8), 621 (2014). https://doi.org/10.1038/nphoton.2014.152
9. H. Wang, J. Qin, X. Ding, M.C. Chen, S. Chen, X. You, Y.M. He, X. Jiang, L. You, Z. Wang, C. Schneider, J.J. Renema, S. Höfling, C.Y. Lu, J.W. Pan, Phys. Rev. Lett. **123**, 250503 (2019). https://doi.org/10.1103/PhysRevLett.123.250503. https://link.aps.org/doi/10.1103/PhysRevLett.123.250503
10. S. Aaronson, A. Arkhipov, Theory Comput. **9**(4), 143 (2013). https://doi.org/10.4086/toc.2013.v009a004. http://www.theoryofcomputing.org/articles/v009a004
11. R. Kruse, C.S. Hamilton, L. Sansoni, S. Barkhofen, C. Silberhorn, I. Jex, Phys. Rev. A **100**, 032326 (2019)
12. Y.H. Deng, S.Q. Gong, Y.C. Gu, Z.J. Zhang, H.L. Liu, H. Su, H.Y. Tang, J.M. Xu, M.H. Jia, M.C. Chen, H.S. Zhong, H. Wang, J. Yan, Y. Hu, J. Huang, W.J. Zhang, H. Li, X. Jiang, L. You, Z. Wang, L. Li, N.L. Liu, C.Y. Lu, J.W. Pan, Phys. Rev. Lett. **130**, 190601 (2023). https://doi.org/10.1103/PhysRevLett.130.190601. https://link.aps.org/doi/10.1103/PhysRevLett.130.190601
13. C.W. Gardiner, P. Zoller, *Quantum Noise*, 3rd edn. (Springer, Berlin, 2004)
14. S.M. Barnett, P.M. Radmore, *Methods in Theoretical Quantum Optics* (Oxford University Press, New York, 1997)
15. M. Rudelson, A. Samorodnitsky, O. Zeitouni, Ann. Probab. **44**(4), 2858 (2016). https://doi.org/10.1214/15-AOP1036
16. N. Quesada, J.M. Arrazola, Phys. Rev. Res. **2**, 023005 (2020). https://doi.org/10.1103/PhysRevResearch.2.023005. https://link.aps.org/doi/10.1103/PhysRevResearch.2.023005
17. K. Zyczkowski, M. Kus, J. Phys. A: Math. Gen. **27**, 4235 (1994)

18. J. Johansson, P. Nation, F. Nori, Comput. Phys. Commun. **184**(4), 1234 (2013). https://doi.org/https://doi.org/10.1016/j.cpc.2012.11.019. https://www.sciencedirect.com/science/article/pii/S0010465512003955
19. S. Leedumrongwatthanakun, L. Innocenti, H. Defienne, T. Juffmann, A. Ferraro, M. Paternostro, S. Gigan, Nat. Photonics **14**, 139 (2020)
20. G. Marcucci, D. Pierangeli, P.W.H. Pinkse, M. Malik, C. Conti, Opt. Express **28**(9), 14018 (2020). https://doi.org/10.1364/OE.389432. http://www.opticsexpress.org/abstract.cfm?URI=oe-28-9-14018
21. L. Banchi, N. Quesada, J.M. Arrazola, Phys. Rev. A **102**, 012417 (2020). https://doi.org/10.1103/PhysRevA.102.012417. https://link.aps.org/doi/10.1103/PhysRevA.102.012417
22. C. Conti, Quantum Mach. Intell. **3**(2), 26 (2021). https://doi.org/10.1007/s42484-021-00052-y
23. Y. Li, M. Chen, Y. Chen, H. Lu, L. Gan, C. Lu, J. Pan, H. Fu, G. Yang, arXiv:2009.01177 (2020)
24. J.M. Arrazola, V. Bergholm, K. Brádler, T.R. Bromley, M.J. Collins, I. Dhand, A. Fumagalli, T. Gerrits, A. Goussev, L.G. Helt, J. Hundal, T. Isacsson, R.B. Israel, J. Izaac, S. Jahangiri, R. Janik, N. Killoran, S.P. Kumar, J. Lavoie, A.E. Lita, D.H. Mahler, M. Menotti, B. Morrison, S.W. Nam, L. Neuhaus, H.Y. Qi, N. Quesada, A. Repingon, K.K. Sabapathy, M. Schuld, D. Su, J. Swinarton, A. Száva, K. Tan, P. Tan, V.D. Vaidya, Z. Vernon, Z. Zabaneh, Y. Zhang, Nature **591**(7848), 54 (2021). https://doi.org/10.1038/s41586-021-03202-1
25. F. Hoch, S. Piacentini, T. Giordani, Z.N. Tian, M. Iuliano, C. Esposito, A. Camillini, G. Carvacho, F. Ceccarelli, N. Spagnolo, A. Crespi, F. Sciarrino, R. Osellame, arXiv:2106.08260 (2021)
26. C. Conti, arXiv:2102.12142 (2021)

Chapter 13
Variational Circuits for Quantum Solitons

No need to succeed in training

Abstract We use quantum neural networks in the phase space as a Gaussian variational ansatz for the ground state of many-body Hamiltonians. Specifically, we consider a Bose-Hubbard model or quantum discrete nonlinear Schrodinger equation. In specific regimes, the model supports localized states and collections of localized states determined by the particle number and interaction strength. We study entanglement in the trained quantum network and perform boson sampling to outline the generation of particle pairs.

13.1 Introduction

Neural network variational ansatzes have been introduced to study quantum many-body models and quantum circuits [1]. The main goal is finding high-dimensional ground states [2–6]. We report on a detailed example that uses the machinery we discussed in the previous chapters. Specifically, we show that QML in the phase space allows a comprehensive quantum description of discrete nonlinear localization phenomena, namely, quantum solitons.

By a trainable boson sampling quantum processor [7, 8], we identify self-localized states solutions in a many-body nonlinear Hamiltonian. We also study their bound states that exhibit entanglement. We show that QML can be used for the boson sampling pattern probability that reveals the generation of correlated photons.

In many-body physics, one starts from a Hamiltonian with many modes, which correspond to particle fields in different locations or to their momenta. The first goal is finding the ground state.

Typically, the ground state depends on some parameters, as the interaction strength or the amount of disorder. When varying the parameters, one observes transitions, e.g., from localized to delocalized wave functions. Correspondingly, one builds up phase diagrams or investigates how entanglement occurs in the system.

C. Conti, *Quantum Machine Learning*, Quantum Science and Technology,
https://doi.org/10.1007/978-3-031-44226-1_13

NNs provide a versatile and general ansatz for the many-body ground states. Here we show how to realize this methodology by using gates in the phase space.

We use the boson sampling circuit studied in the previous chapter to generate a guess for the ground state. The guess is a characteristic function $\chi(\boldsymbol{x}, \boldsymbol{\theta})$ and depends on all of the parameters $\boldsymbol{\theta}$ of the model.

By using regression, we minimize the Hamiltonian as a function of $\boldsymbol{\theta}$. In these terms, the Hamiltonian is a functional that has as input $\chi(\boldsymbol{x}, \boldsymbol{\theta})$ and returns a real number.

13.2 The Bose-Hubbard Model

We refer to the state generated by the quantum circuit as the boson sampling variational ansatz (BSVA) used in the quantum discrete nonlinear Schrödinger equation (QDNLS):

$$\mathrm{i}\partial_t \hat{\psi}_j = (\hat{\psi}_{j+1} + \hat{\psi}_{j-1}) + \gamma \hat{\psi}_j^\dagger \hat{\psi}_j \hat{\psi}_j \tag{13.1}$$

on a lattice with size n (see Fig. 13.1a). Here we use $|\mathbf{vac}\rangle$ to identify the many-body vacuum state.

The physical meaning of fields $\hat{\psi}_j$ depends on the applications. They may be the amplitudes at specific physical sites or the samples at discretized spatial positions.

Within a scaling factor, the field $\hat{\psi}_j$ is identified with the onsite ladder operators $\hat{a}_j$. We will neglect the scaling factors and assume $\hat{\psi}_j = \hat{a}_j$ hereafter.

Quantum discrete solitons are localized solutions of the time-independent QDNLS. The Hamiltonian is

$$\hat{\mathrm{H}} = -\sum_j \left(\hat{\psi}_j^\dagger \hat{\psi}_{j+1} + \hat{\psi}_j^\dagger \hat{\psi}_{j-1}\right) + \frac{\gamma}{2} \hat{\psi}_j^\dagger \hat{\psi}_j^\dagger \hat{\psi}_j \hat{\psi}_j = \hat{\mathrm{K}} + \hat{\mathrm{V}} \tag{13.2}$$

with the potential energy

$$\hat{\mathrm{V}} = \frac{\gamma}{2} \sum_j \left(\hat{n}_j^2 - \hat{n}_j\right) \tag{13.3}$$

being $\hat{n}_i = \hat{\psi}_i^\dagger \hat{\psi}_i$, and the kinetic energy

$$\hat{\mathrm{K}} = \sum_{jk} \omega_{jk} \hat{\psi}_j^\dagger \hat{\psi}_k \tag{13.4}$$

being $\omega_{jk} = -\delta_{j,k-1} - \delta_{j,k+1}$ with j, k in $[0, n-1]$ letting $\omega_{0,-1} = \omega_{n-1,n} = 0$, as homogeneous boundary conditions for a finite lattice. The considered ω_{jk} represent a short-range interaction, such that particles jump from one site to the nearest sites.

The total particle number operator is

$$\hat{\mathrm{N}} = \sum_j \hat{n}_j \,. \tag{13.5}$$

13.3 Ansatz and Quantum Circuit

Figure 13.1 shows the boson sampling quantum processor in a discrete array with n sites. The circuit is composed of n squeezers with complex parameters ζ_j, and displacement operators with complex displacements δ_j, being $j = 0, 1, \ldots, n-1$ hereafter.

A unitary interferometer $\hat{U}$ mixes the modes before entering the waveguide array.

We map the quantum processor to a neural network in the phase space. All the parameters of the squeezers, displacements, and interferometers are trainable and initialized as random variables.

To build the ansatz as a neural network, we use the n-body complex characteristic function $\chi(\boldsymbol{x}) = \chi_R(\boldsymbol{x}) + \mathrm{i}\chi_I(\boldsymbol{x})$, with $\boldsymbol{x}$ real-valued vector with dimension $1 \times N$ with $N = 2n$.

For Gaussian states, we have

$$\chi(\boldsymbol{x}) = e^{-\frac{1}{4}\boldsymbol{x}\boldsymbol{g}\boldsymbol{x}^\top + \mathrm{i}\boldsymbol{x}\boldsymbol{d}}. \tag{13.6}$$

with $\boldsymbol{g}$ the real covariance $N \times N$ matrix and $\boldsymbol{d}$ the real displacement $N \times 1$ vector.

Given the vacuum state $|\mathbf{vac}\rangle$ with characteristic function χ, one builds the NN of an arbitrary state by multiple pullbacks. Figure 13.1e shows a n-mode squeezed vacuum, followed by the displacement operator and interferometer as multiple pullbacks. Figure 13.1e is the NN model for the BSVA in Fig. 13.1a.

We obtain the mean value of observable quantities as $\hat{\mathrm{H}}$ and $\hat{\mathrm{N}}$ as derivatives of the characteristic function at $\boldsymbol{x} = 0$.

We have in symmetric ordering (see Chap. 7)

$$\langle\hat{\mathrm{K}}\rangle = \sum_{jk} \omega_{jk} \left[\frac{\partial^2}{\partial\alpha_j \partial(-\alpha_k^*)} \chi(\boldsymbol{\alpha}) - \frac{\delta_{jk}}{2}\chi \right]\Bigg|_{\boldsymbol{\alpha}=0} \tag{13.7}$$

where $\boldsymbol{\alpha} = \left(\alpha_0 \ldots \alpha_{n-1}, \alpha_0^* \ldots \alpha_{n-1}^*\right)$, and $\sqrt{2}\alpha_j = x_{2j} + \mathrm{i}x_{2j+1}$. The proof of (13.7) is in Sect. 13.3.1.

For the interaction term (see Sect. 13.3.2), we have

$$\langle\hat{\mathrm{V}}\rangle = \frac{\gamma}{2} \sum_j \frac{\partial^4 \chi}{\partial\alpha_j^2 \partial(-\alpha_j^*)^2} - 2\frac{\partial^2 \chi}{\partial\alpha_j \partial(-\alpha_j^*)} + \frac{\chi}{2}\Bigg|_{\boldsymbol{\alpha}=0} . \tag{13.8}$$

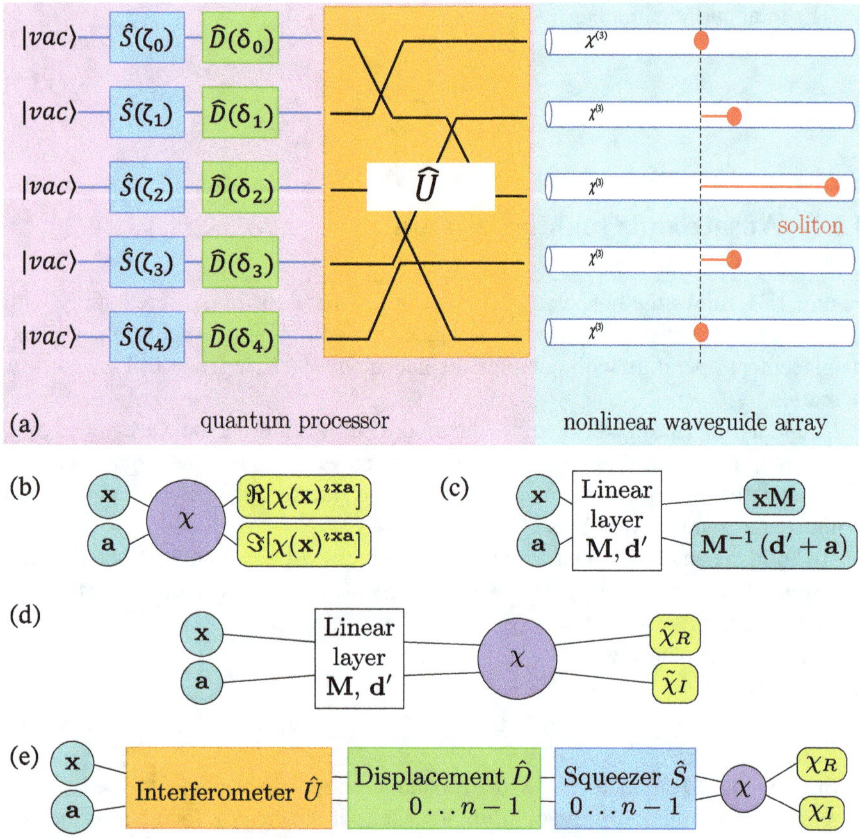

Fig. 13.1 (**a**) A boson sampling processor prepares the state to launch the quantum discrete soliton in an array of coupled optical waveguides with cubic $\chi^{(3)}$ nonlinearity. n spatial modes initially in the vacuum state $|\text{vac}\rangle$ are first squeezed and displaced and then mixed by a trainable interferometer $\hat{U}$ ($n = 5$ in the figure). The state is represented in the phase space as a quantum neural network that returns the characteristic function χ. (**b**) The simplest neural network model. Two inputs, a data vector $\boldsymbol{x}_{1\times N=2n}$, and a bias vector $\boldsymbol{a}_{N\times 1}$ seed the model that computes χ and returns the real and imaginary parts of $\chi(\boldsymbol{x})e^{\text{i}\boldsymbol{x}\boldsymbol{a}}$. (**c**) A layer for a linear transformation of the state by a unitary operator represented by a symplectic matrix $\mathbf{M}_{N\times N}$ and displacement vector $\boldsymbol{d}'_{N\times 1}$. (**d**) A model representing a state with characteristic function χ, subject to a unitary transformation. This is a pullback of a linear transform from the original state, which produces a new state with characteristic function $\tilde{\chi}$ [see Eq. (8.10)]. $\tilde{\chi}_R$ and $\tilde{\chi}_I$ are the real and imaginary parts of $\tilde{\chi}$. (**e**) Pullback for the many-body state in panel (**a**). The multiple squeezers and displacements are in a single block. Note that the order of the operators, from the vacuum to the interferometer, goes from right to left

We evaluate $\langle\hat{\text{K}}\rangle$ and $\langle\hat{\text{V}}\rangle$ by automatic differentiation. However, as we are dealing with a Gaussian state, we find derivatives algebraically by the tensors $\boldsymbol{g}$ and $\boldsymbol{d}$, which accelerates computing and training.

13.3.1 Proof of Eq. (13.7)*

Proof Letting $\hat{a}_j = \hat{\psi}_j$, we have

$$\hat{\mathrm{K}} = \sum_{j,k=0}^{n-1} \omega_{jk} \hat{a}_j^\dagger \hat{a}_k \ , \tag{13.9}$$

such that

$$\langle \hat{\mathrm{K}} \rangle = \sum_{j,k=0}^{n-1} \omega_{jk} \langle \hat{a}_j^\dagger \hat{a}_k \rangle \ , \tag{13.10}$$

We follow Sect. 11.5 and we have for the characteristic function

$$\chi(z, z^*) = \mathrm{Tr}\left[\hat{\rho} \exp\left(\sum_k z_k \hat{a}_k^\dagger - z_k^* a_k\right)\right] = \mathrm{Tr}\left[\hat{\rho} e^{\sum_k z_k \hat{a}_k^\dagger} e^{-\sum_k z_k^* a_j} e^{-\frac{1}{2}\sum_k z_k z_k^*}\right], \tag{13.11}$$

and its derivatives with respect to z_j and z_j^*:

$$\frac{\partial \chi}{\partial z_j}, \quad \frac{\partial \chi}{\partial(-z_j^*)}, \quad \frac{\partial^2 \chi}{\partial z_j^2}, \quad \ldots \frac{\partial^2 \chi}{\partial(-z_j^*)^2}, \quad \frac{\partial \chi}{\partial z_j^2 \partial(-z_j^*)^2} \ . \tag{13.12}$$

We simplify the notation by considering the following operator:

$$\hat{\chi}_j = e^{z_j \hat{a}_j^\dagger} e^{-z_j^* \hat{a}_j} e^{-\frac{1}{2} z_j z_j^*}, \tag{13.13}$$

such that

$$\chi = \mathrm{Tr}\left(\hat{\rho} \prod_i \hat{\chi}_i\right) \tag{13.14}$$

and

$$\langle \hat{a}_j^\dagger \hat{a}_k \rangle = \mathrm{Tr}\left(\hat{\rho} \hat{a}_j^\dagger \hat{a}_k \prod_i \hat{\chi}_i\right)\bigg|_{z=0} \ . \tag{13.15}$$

Note that $\hat{\chi}_j$ and $\hat{\chi}_k$ commute as $j \neq k$.

Also, we define the symbols

$$\begin{aligned} \partial_j &\equiv \frac{\partial}{\partial z_j} \\ \bar{\partial}_j &\equiv \frac{\partial}{\partial(-z_j)^*} \end{aligned} \tag{13.16}$$

and

$$\begin{aligned} \hat{E}_j &= e^{z_j \hat{a}_j^\dagger} \\ \hat{\hat{E}}_j &= e^{-z_j^* \hat{a}_j} \\ e_j &= e^{-|z_j|^2/2} \end{aligned} \tag{13.17}$$

such that $\hat{\chi}_j = \hat{E}_j \hat{\hat{E}}_j e_j$.

We have

$$\begin{aligned} \partial_j \hat{E}_j &= \hat{a}_j^\dagger \hat{E}_j \\ \bar{\partial}_j \hat{\hat{E}}_j &= \hat{a}_j \hat{\hat{E}}_j \\ \partial_j e_j &= -\frac{z_j^*}{2} e_j \\ \bar{\partial}_j e_j &= \frac{z_j}{2} e_j \, , \end{aligned} \tag{13.18}$$

and

$$\begin{aligned} \bar{\partial}_k \prod_i \hat{\chi}_i &= \prod_i \hat{E}_i (\hat{a}_k)^{\delta_{ki}} \hat{\hat{E}}_i e_i + \frac{z_k}{2} \prod_i \hat{\chi}_i \\ \partial_j \bar{\partial}_k \prod_i \hat{\chi}_i &= \prod_i \hat{E}_i (\hat{a}_j^\dagger)^{\delta_{ij}} (\hat{a}_k)^{\delta_{ki}} \hat{\hat{E}}_i e_i \\ &\quad - \frac{z_j^*}{2} \prod_i \hat{E}_i (\hat{a}_k)^{\delta_{ik}} \hat{\hat{E}}_i e_i \\ &\quad + \frac{z_k}{2} \prod_i (\hat{a}^\dagger)^{\delta_{ij}} \hat{E}_i \hat{\hat{E}}_i e_i - \\ &\quad + \frac{z_j z_k^*}{4} \prod_i \hat{\chi}_i + \frac{\delta_{jk}}{2} \prod_i \hat{\chi}_i \end{aligned} \tag{13.19}$$

As $\hat{E}_j \to 1$, $\hat{\hat{E}}_j \to 1$, and $\hat{\chi} \to 1$ when $z \to 0$, we have

$$\partial_j \bar{\partial}_k \prod_i \hat{\chi}_i = \hat{a}_j^\dagger \hat{a}_k + \frac{\delta_{jk}}{2} \prod_i \hat{\chi}_i + o(z) \tag{13.20}$$

where $o(z)$ is after the Bachmann-Landau notation, i.e., $o(z) \to 0$ as $z \to 0$. Finally,

$$\langle \hat{a}_j^\dagger \hat{a}_k \rangle = \mathrm{Tr}\left(\rho \hat{a}_j^\dagger \hat{a}_k\right) = \partial_j \bar{\partial}_k \chi - \frac{\delta_{jk}}{2} \chi \bigg|_{z=0} \tag{13.21}$$

which is the proof.

13.3.2 *Proof of Eq. (13.8)**

This proof follows Sect. 11.5.

13.4 Total Number of Particles in Real Variables

To code with real variables, we decomplexity the operators $\hat{\mathrm{N}}$, $\hat{\mathrm{K}}$, and $\hat{\mathrm{V}}$. We start considering $\hat{\mathrm{N}}$, which is (see Sect. 11.1)

$$\hat{\mathrm{N}} = \sum_j \hat{n}_j = \sum_{j=0}^{n-1} \frac{\partial^2}{\partial \alpha_j \partial(-\alpha_j^*)} \chi(\boldsymbol{\alpha}, \boldsymbol{\alpha}^*) \bigg|_{\boldsymbol{\alpha}=0} . \tag{13.22}$$

In terms of real variables, we have (Eq. 11.3)

$$\langle \hat{a}_j^\dagger \hat{a}_j \rangle = \langle \hat{n}_j \rangle = -\frac{\partial^2 \chi}{\partial z_j \, \partial z_j^*} \bigg|_{z=0} = -\frac{1}{2} \left(\partial_{q_j}^2 + \partial_{p_j}^2\right) \chi_R \bigg|_{x=0} - \frac{1}{2} . \tag{13.23}$$

Correspondingly,

$$\hat{\mathrm{N}} = \sum_{j=0}^{n-1} \langle \hat{a}_j^\dagger \hat{a}_j \rangle = \sum_{j=0}^{n-1} -\frac{1}{2} \left(\partial_{q_j}^2 + \partial_{p_j}^2\right) \chi_R \bigg|_{x=0} - \frac{1}{2} . \tag{13.24}$$

13.5 Kinetic Energy in Real Variables

To write the kinetic energy in terms of quadratures, we have to take into account the general case in which ω_{jk} in (13.4) is a complex function. Letting

$$\omega_{jk} = \omega^R_{jk} + \mathrm{i}\omega^I_{jk} \tag{13.25}$$

we get, following Chap. 11,

$$\langle \hat{\mathrm{K}} \rangle = -\frac{1}{2}\sum_{mn}\left[\frac{\partial^2\chi}{\partial q_m \partial q_n} + \frac{\partial^2\chi}{\partial p_m \partial p_n} + \chi\right]\omega^R_{mn} + \left(\frac{\partial^2\chi}{\partial q_m \partial p_n} - \frac{\partial^2\chi}{\partial p_n \partial p_q}\right)\omega^I_{mn}\Bigg|_{x=0} . \tag{13.26}$$

We remark that for a Gaussian state, the second derivatives of χ_I at $\boldsymbol{x} = 0$ are vanishing (see Sect. 9.7.1); hence, in Eq. (13.26), we can use χ_R instead of χ.

13.6 Potential Energy in Real Variables

For the potential energy, we have

$$\langle \hat{\mathrm{V}} \rangle = \sum_{nm} V_{nm} \langle \hat{a}^\dagger_n \hat{a}^\dagger_m \hat{a}_n \hat{a}_m \rangle , \tag{13.27}$$

and for a local interaction,

$$V_{nm} = \frac{1}{2}\delta_{nm} . \tag{13.28}$$

We also have

$$\langle \hat{a}^\dagger_j \hat{a}^\dagger_j \hat{a}_j \hat{a}_j \rangle = \frac{1}{4}\left(\partial^2_{q_j} + \partial^2_{p_j}\right)^2 \chi + \left(\partial^2_{q_j} + \partial^2_{p_j}\right)\chi + \frac{1}{2}\chi\Bigg|_{x=0} \tag{13.29}$$

13.7 Layer for the Particle Number

We build a model that returns the expected value of the particle number, the kinetic energy, and the potential energy. We first define `Python` functions and then convert them to `TensorFlow` layers.[1]

We consider the particle number functions, which follows the methods for Gaussian functions defined in Chap. 11.

```
# decorator to optimize
@tf.function
```

[1] The methods for quantum solitons are defined in the module `thqml.quantumsolitons`.

```
# function definition
def gaussian_boson_numbers(tensor, **kwargs):
    """Function that returns n,n2, and Dn2 given g,d,hessian
    (used in variational ansatz)

    In the call

    Parameters
    ----------
    tensor[0]: covariance matrix (N,N)
    tensor[1]: avg displacement vector (1,N)
    tensor[2]: hessian matrix (N,N)

    g=tensor[0]
    d=tensor[1]
    hessian=tensor[2]

    Returns
    -------
    param: nboson: number of bosons per mode (1,n)
    param: nboson2: squared number of bosons per mode (1,n)

    Remark
    ------
    N, Rq, Rp are global here


    """
    g = tensor[0]
    d = tensor[1]
    hessian = tensor[2]

    # evaluate the Laplacian by gathering the diagonal elements
    ↪  of gaussian
    cr_2x = tf.reshape(tf.linalg.diag_part(hessian), [1, N])

    # sum the derivatives to have the squared Laplacian
    dqq = tf.matmul(cr_2x, Rq)  # [1,N]
    dpp = tf.matmul(cr_2x, Rp)  # [1,N]
    laplacian = dqq + dpp  # [1,N]

    # evaluate the biharmonic by diag g and d
    d = tf.squeeze(d)  # [N,] dj
    gd = tf.linalg.diag_part(g)  # [N,] gjj
    d2 = tf.square(d)  # [N,] dj^2
    d4 = tf.tensordot(d2, d2, axes=0)  # [N,N] dj^2 dk^2
    dd = tf.tensordot(d, d, axes=0)  # [N,N] dj dk

    djjkk = (
        0.25 * tf.tensordot(gd, gd, axes=0)
        + 0.5 * tf.square(g)
        + 0.5 * tf.tensordot(gd, d2, axes=0)
        + 0.5 * tf.tensordot(d2, gd, axes=0)
        + 2.0 * tf.multiply(g, dd)
```

```
        + d4
    )

    biharmonic = tf.zeros_like(laplacian)  # [1,N]
    for j in tf.range(n):
        qj = 2 * j
        pj = 2 * j + 1
        dqqpp = tf.gather_nd(djjkk, [[qj, pj]])
        dqqqq = tf.gather_nd(djjkk, [[qj, qj]])
        dpppp = tf.gather_nd(djjkk, [[pj, pj]])
        biharmonic = tf.tensor_scatter_nd_add(
            biharmonic, [[0, j]], dqqqq + dpppp + 2 * dqqpp
        )
    # evaluate the number of bosons
    nboson = -0.5 * laplacian - 0.5

    # evaluate the n2
    nboson2 = 0.25 * biharmonic + 0.5 * laplacian

    return nboson, nboson2
```

Given the `Python` function `gaussian_boson_number`, we convert it to a `TensorFlow` model by using a `Lambda` layer as follows.

```
#import the module with the definitions
# of gaussian_boson_number
import thqml.quantumsolitons as qs
# build the layer as Lambda
# use a Lambda layer that converts
# a python function in a tensor
BL=tf.keras.layers.Lambda(qs.gaussian_boson_numbers,
        name="Bosons", dtype=tf.float32)
```

Note that according to the definition of `gaussian_boson_number`, the layer returns two tensors that we call `nboson` and `nboson2`.

`nboson` is an array with the expected value of the boson number $\langle \hat{n}_j \rangle$ per site. `nboson2` is an array with the expected value of the squared boson number $\langle \hat{n}_j^2 \rangle$ per site.

To obtain the expected value of the total particle number, we define a layer that takes as input `nboson` and return the sum of its elements (`reduce_sum`).

```
# define layer the sum the elements of an input tensor
NL=tf.keras.layers.Lambda(tf.reduce_sum, dtype=tf_real,name="N")
```

The `NL` furnishes $\langle \hat{\mathrm{N}} \rangle$ in the variable `Ntotal` with the call `Ntotal=NL(nboson)`.

13.8 Layer for the Kinetic Energy

Once we have the expected value of the kinetic energy as a function of the components of the covariance matrix `cov`, we code a function that returns $\langle \hat{\mathrm{K}} \rangle$ with input `conv`.

The first step is to define the constant matrix ω_{ij} for the interaction.

```
def evaluate_omega_nnt(**kwargs):
    """returns tf tensor (n,n) k matrix nearest neighbor
     ↪  aperiodical

    Returns
    -------
    Omegar constant tensor
    Omegai constant tensor

    """

    # compute the complex omega matrix
    omega = np.zeros([n, n], dtype=np.complex64)
    for p in range(n):
        if p + 1 < n:
            pp1 = p + 1
            omega[p, pp1] = -1.0
        if p - 1 >= 0:
            pm1 = p - 1
            omega[p, pm1] = -1.0

    # return two constant tf.tensors
    # with real and imaginary part
    # **kwargs is used to pass some arguments from
    # input as the precision to be adopted
    tfomegar = tf.constant(np.real(omega), **kwargs)
    tfomegai = tf.constant(np.imag(omega), **kwargs)
    return tfomegar, tfomegai
```

The two matrices are then defined at the initialization and used as global variables.

```
# define the matrices with single precision
OMEGAR, OMEGAI = evaluate_omega_nnt(dtype=tf.float32)
```

By using `OMEGAR` and `OMEGAI`, we find the kinetic energy by the following function.

```
# decorator to speed up
@tf.function
```

```
# function definition
def kinetic_energy(tensor, **kwargs):
    """Compute the kinetic energy by covariance, d and Omega
↪   matrix

    Parameters
    ----------
    tensor[0]: covariance matrix (N, N)
    tensor[1]: displacement vector <R> (1, N)

    Returns
    -------
    Ktotal  total kinetic energy (scalar)

    Remark
    ------
    N, n, OMEGAR and OMEGAI are global variables

    """
    cov = tensor[0]
    displ = tensor[1]

    # compute the second derivatives
    djk = tf.squeeze(tf.tensordot(displ, displ, axes=0))
    chimn = tf.add(tf.math.scalar_mul(-0.5, cov),
    ↪   tf.math.scalar_mul(-1.0, djk))
    # build the tensor of mixed derivatives
    chi2R = tf.math.scalar_mul(-0.5, tf.eye(n, dtype=cov.dtype,
    ↪   **kwargs))
    chi2I = tf.zeros([n, n], dtype=cov.dtype, **kwargs)
    for j in tf.range(n):
        qj = 2 * j
        pj = qj + 1
        tmpR = tf.gather_nd(chimn, [[qj, qj]]) +
        ↪   tf.gather_nd(chimn, [[pj, pj]])
        tmpR = -0.5 * tmpR
        chi2R = tf.tensor_scatter_nd_add(chi2R, [[j, j]], tmpR)
        for k in tf.range(j + 1, n):
            qk = 2 * k
            pk = qk + 1
            tmpR = tf.gather_nd(chimn, [[qj, qk]]) +
            ↪   tf.gather_nd(chimn, [[pj, pk]])
            tmpR = -0.5 * tmpR
            tmpI = tf.gather_nd(chimn, [[qj, pk]]) -
            ↪   tf.gather_nd(chimn, [[pj, qk]])
            tmpI = -0.5 * tmpI
            # build symmetric part
            chi2R = tf.tensor_scatter_nd_add(chi2R, [[j, k]],
            ↪   tmpR)
            chi2R = tf.tensor_scatter_nd_add(chi2R, [[k, j]],
            ↪   tmpR)
            # build antisymmetric
            chi2I = tf.tensor_scatter_nd_add(chi2I, [[j, k]],
            ↪   tmpI)
```

```
            chi2I = tf.tensor_scatter_nd_add(chi2I, [[k, j]],
            ↪   -tmpI)

    # compute the total kinetic energy
    ktotal = tf.math.reduce_sum(
        tf.add(tf.multiply(OMEGAR, chi2R), tf.multiply(OMEGAI,
        ↪   chi2I))
    )
    return ktotal
```

Given the Python function kinetic_energy, we convert it to a TensorFlow model by using a Lambda layer as follows.

```
#import the module with the definitions
# of kinetic_energy
import thqml.quantumsolitons as qs
# build the layer as Lambda
# use a Lambda layer that converts
# a python function in a tensor
KinL=tf.keras.layers.Lambda(qs.kinetic_energy,
↪   name="K",dtype=tf_real)
```

13.9 Layer for the Potential Energy

We define a function returning the expected value of potential energy.

```
@tf.function
def potential_energy(tensor, **kwargs):
    """
    Return the potential energy


    Parameters
    ----------
    tensor[0] number of bosons (1,n)
    tensor[1] number of bosons squared (1,n)

    Returns
    -------
    Vtotal: total potential energy

    Remark, chi is a global variable here
    """
    nb = tensor[0]
    nb2 = tensor[1]
```

```
    # potential energy (chi/2) sum (nj^2 - nj)
    pe = tf.math.scalar_mul(0.5 * chi, tf.math.reduce_sum(nb2 -
    ↪  nb))
    return pe
```

The function `potential_energy` takes as input a list of two tensors corresponding to $\langle \hat{n}_j \rangle$ and $\langle \hat{n}_j^2 \rangle$, obtained as described in Chap. 11. The function then returns their sums as appearing in $\langle \hat{V} \rangle$. Note that the value of `chi` is used globally, i.e., it must be defined at the initialization. `chi` corresponds to the strength of the nonlinear coupling γ in Eq. (13.2).

By using `potential_energy`, we define a layer returning the potential energy as a `Lambda` function.

```
# import module thml.quantumsolitons
# that contains the definition of potential_energy
import thqml.quantumsolitons as qs
# define a layer returning the potential energy
# use a Lambda layer that converts
# a python function in a tensor
VL=tf.keras.layers.Lambda(qs.potential_energy,
↪   dtype=tf.float32,name="V")
```

13.10 Layer for the Total Energy

We need a layer to combine the expected values of kinetic and potential energy to return the expected value of the Hamiltonian.

First, we define a function for the total energy. The function `total_energy` has as input a list of two tensors and returns their sum.

```
@tf.function
def total_energy(tensor, **kwargs):
    """Return the potential energy

    Parameters
    ----------
    tensor[0] kinetic energy (scalar)
    tensor[1] potential energy (scalar)

    Returns
    -------
    Vtotal: total potential energy
```

```
    """
    kt = tensor[0]
    vt = tensor[1]
    return kt + vt
```

Second, we use a Lambda layer to convert the function in a tensor in the model.

```
#import the module with the definitions
# of potential_energy
import thqml.quantumsolitons as qs
# build the layer as Lambda
# use a Lambda layer that converts
# a python function in a tensor
HL=tf.keras.layers.Lambda(qs.total_energy,
↪   dtype=tf_real,name="H")
```

If we have the expected value of the kinetic energy in the tensor Ktotal and of the potential energy in Vtotal, we obtain the expected value of the Hamiltonian by

```
Htotal=HL([Ktotal, Vtotal])
```

13.11 The Trainable Boson Sampling Ansatz

We have the layer for computing the observable quantities on a Gaussian state. We now build a model that returns the characteristic function of the state emerging from the cascade of the trainable interferometer and the squeezer.

The BSVA is a pullback from the vacuum of the different gates. We start creating a vacuum layer.[2]

```
# import the modules with the layer definition
from thqml import phasespace as qx
# create a vacuum layer
vacuum = qx.VacuumLayer(N,dtype=tf_real, name="Vacuum")
```

We then build a model representing vacuum, squeezers, a trainable displacer, and a trainable interferometer, as in Fig. 13.1e. We start from an input layer and we proceed in the reverse order.

[2] The definition of the layers is in thqml.phasespace.

```
# create the input layer
xin = tf.keras.layers.Input(N, dtype=tf_real, name="Input")
# define the internal variables
x1 = xin
a1 = None
# define and connect a trainable
# random interferometer
x1, a1=qx.RandomLayer(N,dtype=tf.float32,name="R")(x1,a1)
# define and connect a trainable
# displacer with initial displacement dtarget
dtarget=0.1*np.ones([N,1],dtype=np_real)
D=qx.TrainableDisplacementLayer(dtarget,dtype=tf.float32,
  name="Displacement")
x1, a1 = D(x1,a1)
# create and connect n trainable squeezers
# with random initialization
# for the squeezing parameters
for ij in range(n): # loop over the modes
        # random squeezing parameter
        r_np=0.1*np.random.rand(1)
        # mode index for the squeezer
        n_squeeze=ij
        # random squeezing angle
        theta_np=2.0*np.pi*np.random.rand(1)
        # create and connect the layer
        x1, a1 = qx.SingleModeSqueezerLayer(N, r_np=r_np,
                 theta_np=theta_np, n_squeezed=n_squeeze,
                 ↪  trainable=True,
                 dtype=tf.float32, name="S"+repr(ij))(x1,a1)
chir, chii = vacuum(x1, a1)
```

Now we have the graph that returns the characteristic function for the BSVA in the two tensors `chir` and `chii`.

We create a model to train and connect to the measurement layers by

```
PSImodel = tf.keras.Model(inputs = xin, outputs=[chir, chii])
```

13.12 Connecting the BSVA to the Measurement Layers

We want a model that returns the expected value of the Hamiltonian on a Gaussian state. First, we seed the covariance matrix to the measurement layers. Thus, we define a layer to measure the covariance matrix `cov`.

```
# create a covariance layer
CovL=qx.CovarianceLayer(N, name="covariance",dtype=tf_real)
# connect the model
# to obtain the covariance matrix,
# and also the displacement d
# and the hessian
cov, d, hessian = CovL(chir,chii, PSImodel)
```

In addition to the covariance matrix, we also need the mean particle number per site $\langle \hat{n}_j \rangle$ and the mean particle number squared $\langle \hat{n}_j^2 \rangle$, which we obtain by `BL` layer in Sect. 13.7.

```
# the BL has input [cov,d,hessian]
# returns nboson and nboson2
[nboson,nboson2]=BL([cov,d,hessian])
```

Finally, we have all the ingredients for the mean observable quantities as[3]

```
Ntotal=NL(nboson)
Vtotal=VL([nboson,nboson2])
Htotal=HL([Ktotal, Vtotal])
```

13.13 Training for the Quantum Solitons

We train the model by conventional regression using steepest descent. We use as cost function $\exp(\langle \hat{\mathrm{H}} \rangle / n)$, which is minimized as $\langle \hat{\mathrm{H}} \rangle < 0$ is at its lowest value.

The cost function is coded as

```
expL=tf.exp(Htotal/n)
# add the loss
model.add_loss(expL)
```

As an additional cost function, we use $(\langle \hat{\mathrm{N}} \rangle - N_T)^2$, to constraint the target boson number $\langle \hat{\mathrm{N}} \rangle = N_T$. We define a variable `Ntarget` corresponding to N_T and an additional cost function, corresponding to the mean $(\langle \hat{\mathrm{N}} \rangle - N_T)^2$.

[3] The code is in `jupyternotebooks/soliton/BoseHubbardNNT.ipynb`.

```
# define and add loss for the
# target particle number
model.add_loss(tf.reduce_mean(tf.square(Ntotal-Ntarget_tf)))
# the reduce_mean returns a scalar loss
```

The classical DNLS admits single solitons with the field localized in one or two sites as $\gamma < 0$. We want to understand if the trained BSVA furnishes the quantum version of the single soliton (Fig. 13.2).

The soliton is found after training the NN model which minimizes the cost function and the expected Hamiltonian $\langle \hat{\mathrm{H}} \rangle$ at fixed $\langle \hat{\mathrm{N}} \rangle = N_T$.

We first consider small interaction strength $|\gamma|$, at which we expect delocalized solutions. We show in Fig. 13.3a the profile of the expected values for the displacements $|\langle \psi_j \rangle|^2 = (d_{2j}^2 + d_{2j+1}^2)/2$ after the training and in Fig. 13.3b the profile of the expected number of bosons per site $\langle \hat{n}_j \rangle$. These profiles are obtained for $N_T = 10$ and $\gamma = -0.01$ after training the model for thousands of epochs.

Strong localization is obtained for $N_T = 10$ and $\gamma = -1$ as in Fig. 13.3b, c.

In Fig. 13.4 we report $\langle \hat{\mathrm{H}} \rangle$ and $\langle \hat{\mathrm{N}} \rangle$ when varying the number of training epochs for $N_T = 10$. The system converges after some thousands of epochs and during the training explores different solutions.

Figure 13.4a shows the training of the model when $\gamma = -0.01$: the system converges steadily to the delocalized state in Fig. 13.3a, b. This can be approximated in the continuum limit, as a non-squeezed coherent state with a sinusoidal profile, and the value of the Hamiltonian $\langle \hat{\mathrm{H}} \rangle \simeq -2N_T$, which is approximately the plateau in Fig. 13.4a for $N_T = 10$.

Figure 13.4b shows the strong interaction case $\gamma = -1$. During the training, the systems settle in two plateaus: $\langle \hat{\mathrm{H}} \rangle \cong -50$ and $\langle \hat{\mathrm{H}} \rangle \cong -155$. These values are understood in terms of approximate solutions, in which $\langle \hat{\psi}_{j \neq A} \rangle = 0$, with the exception of central site $A = \lfloor \frac{n}{2} \rfloor + 1$, where the soliton is localized.

If we consider the many-body state $|\boldsymbol{\alpha}\rangle = \hat{D}_A(\boldsymbol{\alpha})|\mathbf{vac}\rangle$ with displacement vector $\alpha_i = \delta_{iA}\alpha$, we have

$$\langle \hat{\mathrm{H}} \rangle \simeq \frac{\gamma}{2} \left(\sum_i \langle |\hat{n}_i{}^2| \rangle - \langle |\hat{n}_i| \rangle \right) = \frac{\gamma}{2} \sum_i |\alpha_i|^4 = \frac{\gamma}{2} |\alpha|^4 \tag{13.30}$$

One has $\langle \hat{\mathrm{N}} \rangle = |\alpha|^2 = N_T$ and $\langle \hat{\mathrm{H}} \rangle = \frac{\gamma}{2} N_T^2$ which gives $\langle \hat{\mathrm{H}} \rangle \simeq -50$, approximately the first plateau in Fig. 13.4.

The difference due to the neglected $\langle \hat{\mathrm{K}} \rangle$. $|\boldsymbol{\alpha}\rangle$ is a coherent state with negligible entanglement, as detailed below.

The lowest energy asymptotic solution found by the training is obtained at a larger number of epochs as shown in Fig. 13.4b. The lowest energy plateau is estimated by considering a squeezed coherent state obtained as

Fig. 13.2 Model for the BSVA with various squeezing layers and the measurement layers

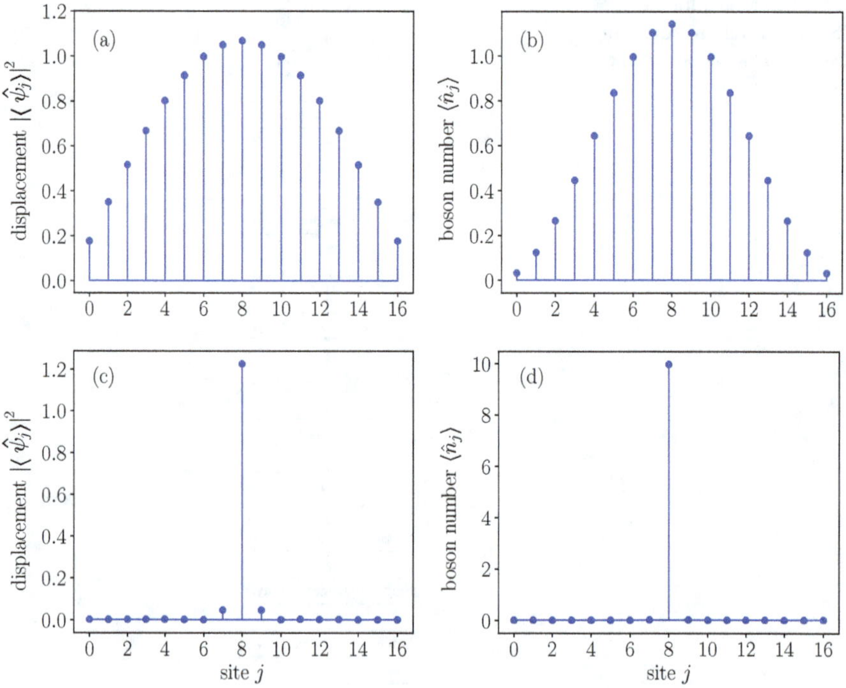

Fig. 13.3 (**a, b**) Profile of the ground state for $N_T = 10$ and $\gamma = -0.01$: (**a**) Square modulus of the expected displacement; (**b**) expected value of the boson number at different sites; (**c, d**) as in (**a, b**) with $N_T = 10$ and $\gamma = -1$ ($n = 17$)

$$|\boldsymbol{\alpha}, \zeta\rangle = \hat{D}_A(\boldsymbol{\alpha})\hat{S}_A(\zeta)|\mathbf{vac}\rangle \tag{13.31}$$

where $\hat{S}_A$ is the squeezing operator for the site $j = A$ with squeezing parameter $\zeta = re^{i\theta}$. For $|\boldsymbol{\alpha}, \zeta\rangle$, one has

$$\begin{aligned}
\langle\hat{\mathrm{N}}\rangle &= \sinh(r)^2 + |\alpha|^2 = N_T \\
\langle\hat{\mathrm{H}}\rangle &= \frac{5}{8} - 2|\alpha^2| + |\alpha|^4 + (-1 + 2|\alpha|^2)\cosh(2r) + \\
&\frac{3}{8}\cosh(4r) - |\alpha|^2\cos(2\phi - \theta)\sinh(2r),
\end{aligned} \tag{13.32}$$

with minimum $\langle\hat{\mathrm{H}}\rangle \simeq -155$ for $N_T = 10$ as in Fig. 13.4b.

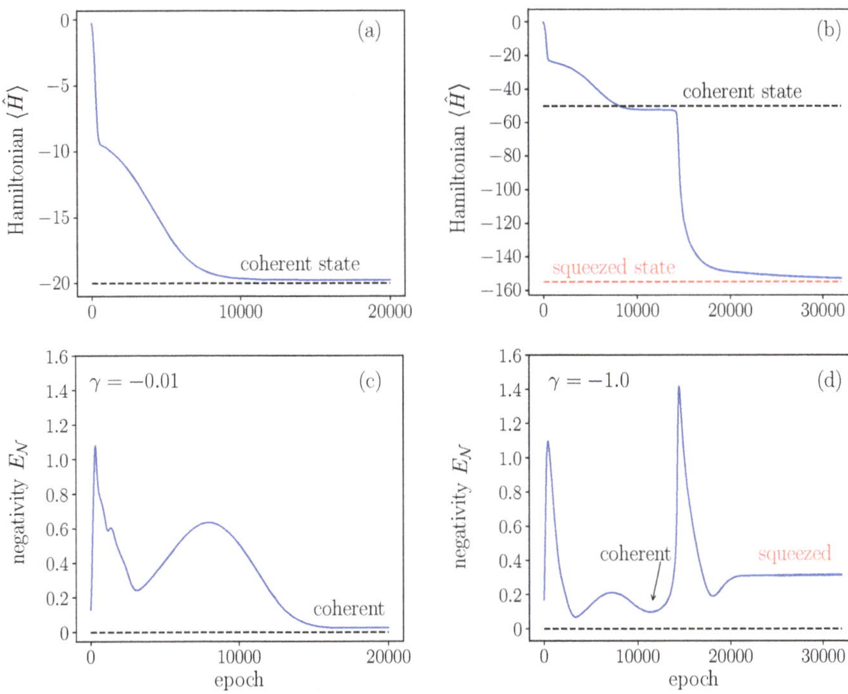

Fig. 13.4 Training history for the solutions in Fig. 13.3: (**a**) $\langle H\rangle$ versus the number of epochs, which rapidly reaches the target values for $\gamma = -0.01$; (**b**) as in (**a**) for $\gamma = -1$. The NN model explores different plateaus of the Hilbert space, corresponding to discrete solitons with different degrees of entanglement; (**c**) logarithmic negativity during training for $\gamma = -0.01$; no entanglement occurs for negligible interaction; (**d**) as in (**c**) for $\gamma = -1$, the lowest energy solution is entangled

13.14 Entanglement of the Single Soliton

Entanglement in Gaussian states (see Chap. 11) is quantified by the logarithmic negativity [9–11]. Specifically, we partition the system in a single site "Alice" A at $A = \lfloor \frac{n}{2} \rfloor + 1$ where the soliton is localized. All the remaining sites with $j \neq A$ correspond to "Bob." We retrieve the covariance matrix $\boldsymbol{g}$ from the NN model after training, and compute the partial transpose $\tilde{\boldsymbol{g}}$ [9], obtained by multiplying by -1 all the matrix elements connected to Alice's momenta $\hat{p}_A$. We then compute the symplectic eigenvalues $\tilde{c}_0, \tilde{c}_1, \ldots, \tilde{c}_{n-1}$ as the moduli of the eigenvalues of $\mathbf{J}^\top \tilde{\boldsymbol{g}}/2$.

The logarithmic negativity is

$$E_{\mathcal{N}} = -\sum_{j=0}^{n-1} \log_2 \min\{1, 2c_j\}. \tag{13.33}$$

Once we have the covariance matrix in cov, we use the following function, which returns the logarithmic negativity for a Gaussian state.[4]

```
def logarithmic_negativity(cov_np, mask=np.zeros(n)):
    """Return the entropy of entanglement as determined from the
↪   eigs of cov

    Reference: vidal, PRA 65, 032314 (2002)

    cov: is the transposed cov matrix according to the mask

    the logarithmic negativity is computed by the symplectic
↪   eigenvalues cj as
    EN=sum_j=0^(n-1) F(cj)
    with F(cj)=0 if c>=0.5 and F(cj)=-log2(2.0*cj) for c<0.5

    Input:
    ------
    cov, covariance matrix tensor [N,N]
    mask, np array mask transpose [n,] (one for the modes to
↪   transpose)


    Output:
    -------
    EN, logarithmic negativity
    negativity, negativity as EN=log2(2 N+1), or N=0.5*(2^EN-1)
    symplectic_eigs, symplectic eigenvalues
    """

    # build a diagonal with -1 corresponding to the momenta to be
    ↪  transposed
    diag1 = np.ones(N, dtype=np_real)
    for j in range(n):
        if mask[j] == 1:
            diag1[2 * j] = 1
            diag1[2 * j + 1] = -1
            # build the transformation matrix
    Ftrans = np.diag(diag1)
    # transform the covariance matrix
    cov_np_Ta = np.matmul(Ftrans, np.matmul(cov_np, Ftrans))

    Jnp = J.numpy()
    HG = np.matmul(Jnp.transpose(), (0.5 * cov_np_Ta))
    e1, _ = np.linalg.eig(HG)
    e2, v2 = np.linalg.eig(-np.matmul(HG, HG))

    ae1 = np.abs(e1)
    e1red = np.sort(np.sort(ae1)[0::2])
```

[4] The code is in the module thqml.quantumsolitons.

```
    ENeg = 0.0
    for ij in tf.range(n):
        etmp = e1red[ij]
        if etmp < 0.5:
            ENeg = ENeg - np.log2(2.0 * etmp)

    negativity = 0.5 * (np.power(2, ENeg) - 1.0)

    return ENeg, negativity, e1red, e2, v2
```

Figure 13.4c shows the trend of $E_\mathcal{N}$ for $\gamma = -0.01$ during the training. Initially, even though the system explores randomly generated entangled solution, the asymptotic state is a coherent non-squeezed state, with vanishing entanglement as expected for a negligible interaction.

Figure 13.4d shows $E_\mathcal{N}$ during the epochs for $\gamma = -1$, measuring the degree of entanglement of the soliton with the other sites. The NN explores the landscape while visiting different states, with varying $E_\mathcal{N}$. The local minimum of $\langle \hat{\mathrm{H}} \rangle$ is the coherent state with entanglement smaller than the value of the squeezed ground state at epoch 3×10^4. Nevertheless, the asymptotic value of $E_\mathcal{N}$ is negligible if compared with the case of bounded solitons considered in the following.

13.15 Entangled Bound States of Solitons

We study bound states of two solitons [12–14] by training the network with proper loss functions.

We fix the location of two sites, denoted Alice at A and Bob at $B = A + \Delta$, and we introduce as cost function $\exp(-\langle \hat{n}_A \rangle)$ to maximize the bosons at A and $\exp(\langle \hat{n}_A - \hat{n}_B \rangle)$; this is expected to generate a solution with the same boson numbers at A and B.

Figure 13.5 shows the model including the final `Lambda` layers for the losses adopted in training.[5] Figure 13.6 shows the coupled soliton profiles for different peak positions when $N_T = 40$, obtained after training. At variance with the single soliton in Fig. 13.3, we find that the expected boson number is not vanishing in all the sites of the array. Nevertheless, the expected value of the intensities $|\langle \psi_j \rangle|^2$ is not vanishing only at the soliton sites, outlining the onset of squeezed nonlocal states in the system.

Indeed, the found bound soliton solutions are entangled. We quantify entanglement by considering as a bipartite system composed of the site in A (Alice) and the rest of the sites. As indicated in Fig. 13.6, the computed $E_\mathcal{N}$ depends on the distance Δ between the two solitons.

---

[5] The code with details is in `jupyternotebooks/soliton/BoseHubbardTwinNNT.ipynb`.

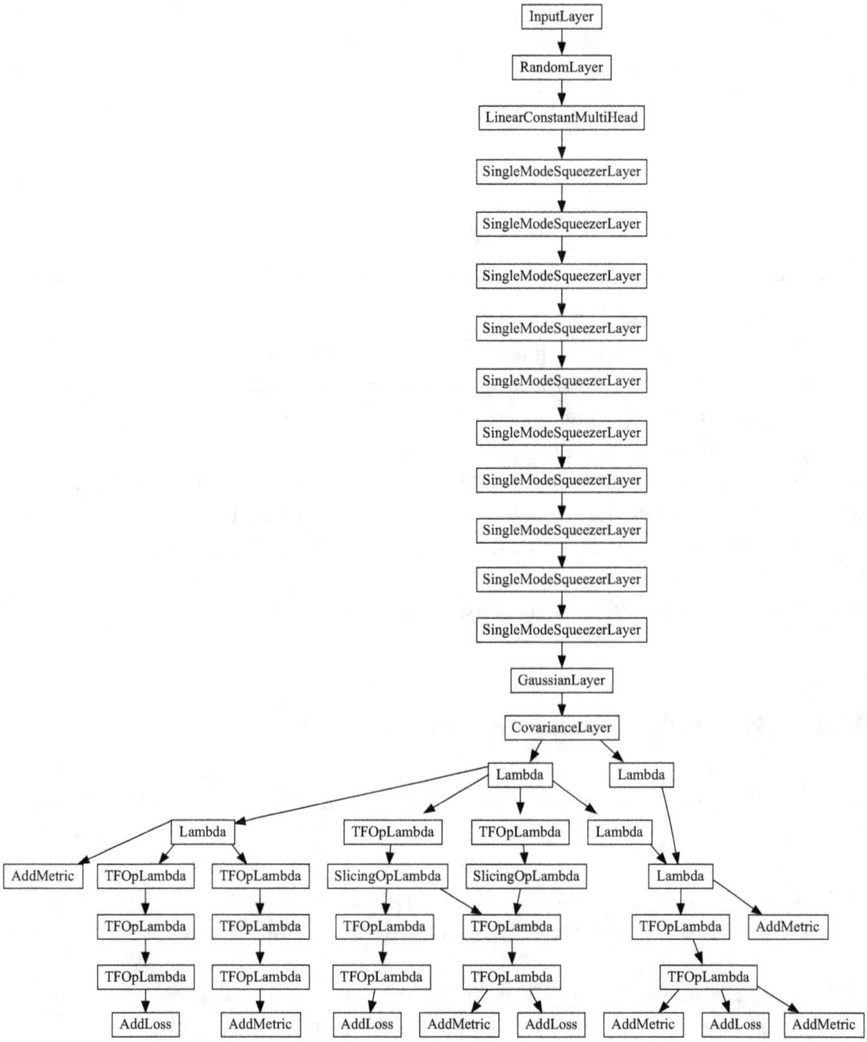

Fig. 13.5 Model for the BSVA for the study to soliton bound states

We also find that the logarithmic negativity depends on the number of total bosons N_T: Fig. 13.6c shows $E_{\mathcal{N}}$ versus N_T for various Δ.[6]

We observe that the degree of entanglement is small and comparable with the single soliton for case $N_T < 10$, and growing when $N_T > 10$.

[6] The code with details is in `jupyternotebooks/soliton/BoseHubbardTwinNNTVersusN.ipynb`.

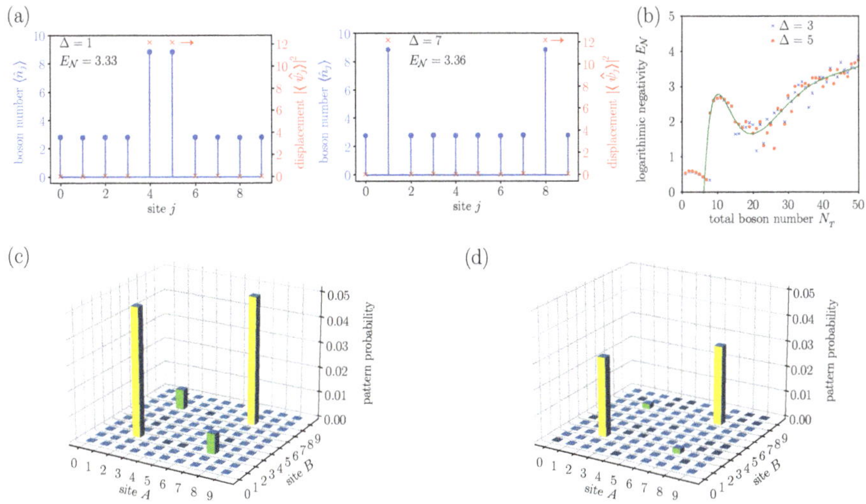

Fig. 13.6 (**a**) Boson distribution and mean displacement (right axis) for two representative bounded soliton solutions with $N_T = 40$ and $\gamma = -1.0$; the values of Δ and $E_{\mathcal{N}}$ are reported; (**b**) logarithmic negativity versus the boson number N_T for different Δ; (**c**) boson sampling probabilities for patterns containing $\bar{n}_T = 2$ photons in two varying sites A and B and $N_T = 10$; (**d**) as in (**c**) for patterns with two photons per site ($\bar{n}_T = 4$). The pattern probability is not vanishing only when the photons in the pattern are at the soliton sites, denoting the generation of entangled pairs ($n = 10$)

13.16 Boson Sampling

As we show in the following, the degree of entanglement is connected to the observation of correlated particles in different output channels.

Boson sampling gives the probability $\Pr(\bar{\mathbf{n}})$ of finding $\bar{n}_0$ photons in mode 0, $\bar{n}_1$ photons in mode 1, and so forth, where $\bar{\mathbf{n}} = (\bar{n}_0, \bar{n}_1, \ldots, \bar{n}_{n-1})$ is a given photon pattern. Letting $\hat{\rho}$ be the density matrix, one has $\Pr(\bar{\mathbf{n}}) = \mathrm{Tr}[\hat{\rho}|\bar{\mathbf{n}}\rangle\langle\bar{\mathbf{n}}|]$, with $|\bar{\mathbf{n}}\rangle\langle\bar{\mathbf{n}}| = \otimes_j |\bar{n}_j\rangle\langle\bar{n}_j|$. Correspondingly [15],

$$\Pr(\bar{\mathbf{n}}) = \frac{1}{\bar{\mathbf{n}}!} \prod_j \left(\frac{\partial^2}{\partial\alpha_j \partial\alpha_j^*} \right)^{\bar{n}_j} e^{\sum_j |\alpha|_j^2} Q_\rho(\boldsymbol{\alpha}) \Bigg|_{\boldsymbol{\alpha}=0} \tag{13.34}$$

where $\bar{\mathbf{n}}! = \bar{n_0}!\bar{n}_1!\ldots\bar{n}_{n-1}!$ and $Q_\rho = \pi^n \langle\boldsymbol{\alpha}|\rho|\boldsymbol{\alpha}\rangle$ is the Q-representation of the density matrix [16, 17]. We compute $\Pr(\bar{\mathbf{n}})$ as detailed in Chap.12 [18].

For a given bound soliton solution, we consider the probability of observing $\bar{n}_T$ particles in two sites A and B. Specifically, we consider the patterns such that (i) $\bar{n}_A = \bar{n}_B = n_T/2$ for $A \neq B$, or (ii) $\bar{n}_A = \bar{n}_B = \bar{n}_T$ when $A = B$, and $\bar{n}_j = 0$ elsewhere.

For $n_T = 2$, this corresponds to observing a pair of particles in a single site, or two particles in two distinct sites.

Figure 13.6c shows the pattern probability for $\bar{n}_T = 2$ varying A and B when $N_T = 10$. Such a probability is different from zero only when A and B are the positions of one soliton (located in $A = 3$ and $B = 8$ in the figure). When $A = B$ (yellow bar), we have the probability of observing the two photons in one single soliton. When A and B are the sites of the two solitons (green bar), we have the probability of observing a couple of photons in the two soliton sites. For all other patterns with $\bar{n}_T = 2$, there is a negligible probability of observing a particle pair. Hence, pairs of particles appear either in a single soliton or as entangled bosons in the two solitons.

Seemingly, when $\bar{n}_T = 4$, we have only nonvanishing probability.

Particles are hence observed simultaneously only when they are monitored at the positions of the two localized solitons, demonstrating their quantum correlation and the nonlocality in the system.

13.17 Conclusion

Quantum machine learning in many-body systems furnishes new tools for studying the physics and the applications of quantum processors. Also, one can use QML to design experiments to maximize entanglement or the probability of observing specific states, as bound states of quantum solitons.

Training of a neural network variational ansatz enables computing states for single solitons and their bound states. In this chapter, we considered Gaussian states as they are expected to conform to experimental observations at high intensities and also allow us to rigorously compute the degree of entanglement by the logarithmic negativity.

Furthermore, the neural network formulation is compatible with the Gaussian boson sampling protocol and is used to obtain the probabilities of photon patterns toward experimental investigations on the quantum advantage in the presence of nonlinearity.

We found that entangled pairs of photons emerge from self-localized bounded solitons. In addition, the results unveil that properly trained boson sampling circuits can be used to synthesize nonlinear quantum waves. The methodology can be extended beyond Gaussian states and generalized to continuous systems and multidimensional arrays.

13.18 Further Reading

- For an introduction and further references to quantum solitons, see [19].
- The boson sampling variational ansatz for Gaussian solitons was originally reported in [20].

References

1. G. Carleo, I. Cirac, K. Cranmer, L. Daudet, M. Schuld, N. Tishby, L. Vogt-Maranto, L. Zdeborová, Rev. Mod. Phys. **91**, 045002 (2019). https://doi.org/10.1103/RevModPhys.91.045002
2. M. Broughton, G. Verdon, T. McCourt, A.J. Martinez, J.H. Yoo, S.V. Isakov, P. Massey, M.Y. Niu, R. Halavati, E. Peters, M. Leib, A. Skolik, M. Streif, D.V. Dollen, J.R. McClean, S. Boixo, D. Bacon, A.K. Ho, H. Neven, M. Mohseni (2020). arXiv:2003.02989
3. G. Marcucci, D. Pierangeli, P.W.H. Pinkse, M. Malik, C. Conti, Opt. Express **28**(9), 14018 (2020). https://doi.org/10.1364/OE.389432
4. F. Marquardt (2021). arXiv:2101.01759
5. F. Tacchino, S. Mangini, P.K. Barkoutsos, C. Macchiavello, D. Gerace, I. Tavernelli, D. Bajoni, arXiv:2103.02498 (2021)
6. A. Kardashin, A. Pervishko, J. Biamonte, D. Yudin, Phys. Rev. A **104**, L020402 (2021). https://doi.org/10.1103/PhysRevA.104.L020402
7. J.M. Arrazola, V. Bergholm, K. Brádler, T.R. Bromley, M.J. Collins, I. Dhand, A. Fumagalli, T. Gerrits, A. Goussev, L.G. Helt, J. Hundal, T. Isacsson, R.B. Israel, J. Izaac, S. Jahangiri, R. Janik, N. Killoran, S.P. Kumar, J. Lavoie, A.E. Lita, D.H. Mahler, M. Menotti, B. Morrison, S.W. Nam, L. Neuhaus, H.Y. Qi, N. Quesada, A. Repingon, K.K. Sabapathy, M. Schuld, D. Su, J. Swinarton, A. Száva, K. Tan, P. Tan, V.D. Vaidya, Z. Vernon, Z. Zabaneh, Y. Zhang, Nature **591**(7848), 54 (2021). https://doi.org/10.1038/s41586-021-03202-1
8. F. Hoch, S. Piacentini, T. Giordani, Z.N. Tian, M. Iuliano, C. Esposito, A. Camillini, G. Carvacho, F. Ceccarelli, N. Spagnolo, A. Crespi, F. Sciarrino, R. Osellame (2021). arXiv:2106.08260
9. G. Vidal, R.F. Werner, Phys. Rev. A **65**, 032314 (2002). https://doi.org/10.1103/PhysRevA.65.032314
10. G. Adesso, A. Serafini, F. Illuminati, Phys. Rev. A **70**, 022318 (2004). https://doi.org/10.1103/PhysRevA.70.022318
11. V.O. Martynov, V.O. Munyaev, L.A. Smirnov (2020). arXiv:2011.07662
12. F. Lederer, G.I. Stegeman, D.N. Christodoulides, G. Assanto, M. Segev, Y. Silberberg, Phys. Rep. **463**(1), 1 (2008). https://doi.org/10.1016/j.physrep.2008.04.004
13. Y.V. Kartashov, B.A. Malomed, L. Torner, Rev. Mod. Phys. **83**, 247–305 (2011). https://doi.org/10.1103/RevModPhys.83.247
14. Y. Kivshar, G.P. Agrawal, *Optical Solitons* (Academic Press, New York, 2003)
15. R. Kruse, C.S. Hamilton, L. Sansoni, S. Barkhofen, C. Silberhorn, I. Jex, Phys. Rev. A **100**, 032326 (2019)
16. C.W. Gardiner, P. Zoller, *Quantum Noise*, 3rd edn. (Springer, Berlin, 2004)
17. S.M. Barnett, P.M. Radmore, *Methods in Theoretical Quantum Optics* (Oxford University Press, New York, 1997)
18. C. Conti (2021). arXiv:2102.12142
19. L.D.M. Villari, D. Faccio, F. Biancalana, C. Conti, Phys. Rev. A **98**, 043859 (2018). https://doi.org/10.1103/PhysRevA.98.043859
20. C. Conti, Phys. Rev. A **106**, 013518 (2022). https://doi.org/10.1103/PhysRevA.106.013518

Index

C. Conti, *Quantum Machine Learning*, Quantum Science and Technology,
https://doi.org/10.1007/978-3-031-44226-1

GPSR Compliance
The European Union's (EU) General Product Safety Regulation (GPSR) is a set of rules that requires consumer products to be safe and our obligations to ensure this.

If you have any concerns about our products, you can contact us on

ProductSafety@springernature.com

In case Publisher is established outside the EU, the EU authorized representative is:

Springer Nature Customer Service Center GmbH
Europaplatz 3
69115 Heidelberg, Germany

www.ingramcontent.com/pod-product-compliance
Lightning Source LLC
Chambersburg PA
CBHW072312070526
44736CB00003B/73
* 9 7 8 3 0 3 1 4 4 2 2 5 4 *